Advances in Soil Science

INTERACTING PROCESSES IN SOIL SCIENCE

Edited by

**R.J. Wagenet
P. Baveye
B.A. Stewart**

CRC Press
Taylor & Francis Group
Boca Raton London New York

CRC Press is an imprint of the
Taylor & Francis Group, an **informa** business

CRC Press
Taylor & Francis Group
6000 Broken Sound Parkway NW, Suite 300
Boca Raton, FL 33487-2742

First issued in paperback 2020

ISBN-13: 978-0-367-45019-9 (pbk)
ISBN-13: 978-0-87371-889-9 (hbk)

Visit the Taylor & Francis Web site at
http://www.taylorandfrancis.com

and the CRC Press Web site at
http://www.crcpress.com

Library of Congress Cataloging-in-Publication Data

Interacting processes in soil science / edited by R.J. Wagenet, P.
  Baveye, B.A. Stewart.
      p. cm -- (Advances in soil science)
    Includes bibliographical references and index.
    ISBN 0-87371-889-5
    1. Soil science. I. Wagenet, R. J.   II. Baveye, P. (Philippe)
  III. Stewart, B. A. (Bobby Alton), 1932–   . IV. Series: Advances
  in soil science (Boca Raton, Fla.)
  S591.I43   1992
  631.4--dc20
                                                            92-27776
                                                                CIP

Preface

Science has long recognized the complexity of natural systems. This complexity is manifested all around us, from the dynamics of global and local weather patterns to the activities of microorganisms as they transform, degrade and consume a vast variety of materials. Arguments have been very logically constructed recognizing that, in fact, all such processes in natural systems are interdependent and interrelated. Yet, we soon discover that study of such complexity must be simplified, focusing on a reduced number of processes and properties, if we are to comprehend essential elements of the system. In this way, it has been presumed that knowledge comes from integrating the results of individual, discrete and disparate studies.

Given this philosophy, scientists have pursued their investigations from a reductionist viewpoint, rather than from a more realistic holistic perspective. Until very recently, this has led to focusing the scientific approach of hypothesis, experimentation, and analysis upon very narrowly defined components of the complex network of interacting natural processes. Given the relative lack of ability to measure even one process in detail, let alone three or four or more simultaneously, this was perceived as the only possible starting point for scientific effort. While it has led us to some understanding of singular processes, there are many interactions that now must be considered before further understanding can be achieved. The next scientific steps must therefore treat the issues that arise from interactions between physical, biological, and chemical processes. This will take some effort and philosophical reorientation, but it is important that it occur.

The discipline of soil science is an excellent example. Ever since it acquired a legitimacy of its own, soil science has been oscillating between the two opposing tendencies of reductionism and holism (see, e.g., Chatelin, Y. 1979. *Une épistémologie des sciences du sol*. Mémoires de l'ORSTOM Volume 88, ORSTOM, Paris). An example of this duality is clear from casual inspection of the table of contents of two major soil science journals. Founded in 1916, the journal *Soil Science* has over the years consistently avoided dividing its contents into subdisciplines. This apparently reflected Jacob G. Lipman's philosophy about the unity and undivisiveness of soil science (see, e.g., Boulaine, J. 1989. *Histoire des pédologues et de la science des sols*. INRA Publications, Paris). Contrastingly, the first volume of the *Soil Science Society of America Proceedings*, in 1936, was organized along the same lines as the society itself, i.e., with six divisions. The first five were identical to the current divisions 1 to 5, while the sixth division was that of Soil Technology. Over the years, the structure of the journal has changed, and more divisions have been added, but the implicit reductionist view of soil science has been maintained. A little closer to us, the series *Advances in Soil Science*, started in 1985, does not explicitly acknowledge subdisciplines within soil science, while the journal *Soil Technology*, started in 1988, is entirely devoted to a particular area within soil science. Further, the annual meetings of the Soil Science Society of America have always been

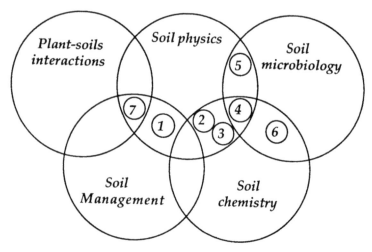

Figure 1. Schematic illustration of the contents of this volume. (Numbers correspond to the various chapters.)

separated into subdisciplines of soil physics, soil chemistry, soil microbiology, soil fertility, and others, further promoting reductionism, and in the process professionally separating soil scientists in fact as well as in title. This has done little to promote the interactions needed between these various subdisciplines of the "science of soils", although lip service is paid throughout the professions to the need for such interaction.

This separation has produced some substantial rewards regarding the basic nature of soils, but the fact remains that relatively few soil scientists cross the subdisciplinary boundaries discussed above to reap the benefits of increased understanding in another area. For example, few soil physicists understand the consequences of soil solution composition on soil hydraulic properties; the impact of soil water movement and the dynamics of heat flow in soil are rarely incorporated by soil microbiologists who are more content with laboratory studies under equilibrium conditions; and soil chemists seldom consider the transient nature of soil water or the thickness of water films as they consider the distribution of charged ions near the soil surface. Certainly much is to be gained in such areas from increased interaction between the subdisciplines of soil science.

One way to imagine the mutual concerns of these subdisciplines is to use a Venn diagram as presented in Figure 1. The subdisciplines included are only a subset of those possible, but present the concept of interacting processes as covered in this book. Other such mutuality of professional overlap could be developed. The key message of such a diagram, and the intent of this book, is to present a first attempt to develop a presentation of several areas of soil science in which we must consider interaction of processes in order to further understand the complexity of the soil system. Other overlaps can be identified by simple exercising of the imagination combined with the curiosity of a true scientist. It is our intent to stimulate such studies, and in the process to erode some of the barriers that currently exist between the subdisciplines of our profession.

As a final word, we wish to thank both the authors and reviewers for their efforts in producing the papers contained herein. Without their professionalism and commitment, this volume would not have been possible.

R.J. Wagenet
P. Baveye
B.A. Stewart

Contributors

C.A. Bellin, Department of Soil Science, University of Florida, Gainesville, FL 32611, USA

J. Bouma, Department of Soil Science and Geology, Agricultural University, P.O.B. 37, 6700 AA Wageningen, The Netherlands

M.L. Brusseau, Department of Soil and Water Science, 429 Shantz Hall, University of Arizona, Tucson, AZ 85721, USA

G. Destouni, Department of Hydraulics Engineering, Royal Institute of Technology, Brinellvägen 32,S-100 44 Stockholm, Sweden

H. Gan, Department of Agronomy, University of Illinois, 1102 South Goodwin Avenue, Urbana, IL 61801, USA

R.W. Harvey, U.S. Geological Survey, 325 Broadway, US 4, Boulder, CO 80303-3328, USA

D. Hettiaratchi, Department of Agricultural and Environmental Science, The University of Newcastle upon Tyne, Newcastle upon Tyne NE1 7RU, United Kingdom

G.J. Levy, Institute of Soils and Water, ARO, The Volcani Center, P.O. Box 6, Bet Dagan 50250, Israel

P.S.C. Rao, Department of Soil Science, University of Florida, Gainesville, FL 32611, USA

I. Shainberg, Institute of Soils and Water, ARO, The Volcani Center, P.O. Box 6, Bet Dagan 50250, Israel

J.W. Stucki, Department of Agronomy, University of Illinois, 1102 South Goodwin Avenue, Urbana, IL 61801, USA

S.E.A.T.M. van der Zee, Department of Soil Science and Plant Nutrition, Agricultural University, P.O.B. 8005, 6700 EC Wageningen, The Netherlands

M.A. Widdowson, Department of Civil Engineering, University of South Carolina, Columbia, SC 29208, USA

H.T. Wilkinson, Department of Agronomy, University of Illinois, 1102 South Goodwin Avenue, Urbana, IL 61801, USA

Contents

Effect of Soil Structure, Tillage, and Aggregation upon
Soil Hydraulic Properties 1
J. Bouma

Physico-Chemical Effects of Salts upon Infiltration and Water
Movement in Soils 37
I. Shainberg and G.J. Levy

Transport of Inorganic Solutes in Soil 95
S.E.A.T.M. van der Zee and G. Destouni

Modeling Coupled Processes in Porous Media: Sorption,
Transformation, and Transport of Organic Solutes 147
M.L. Brusseau, P.S.C. Rao, and C.A. Bellin

Microbial Distributions, Activities, and Movement
in the Terrestrial Subsurface: Experimental and
Theoretical Studies 185
R.W. Harvey and M.A. Widdowson

Effects of Microorganisms on Phyllosilicate Properties
and Behavior 227
J.W. Stucki, H. Gan, and H.T. Wilkinson

The Mechanics of Soil-Root Interactions 255
D.R.P. Hettiaratchi

Index 289

Effect of Soil Structure, Tillage, and Aggregation upon Soil Hydraulic Properties

J. Bouma

I.   Introduction 1
II.   Characterization of Soil Structure 3
     A. Introduction 3
     B. Soil Physical Techniques 3
     C. Soil Morphological Techniques 4
     D. Using Soil Structure Data to Predict Soil Hydraulic Properties . 6
     E. Using Soil Morphological Analyses to Improve Physical
        Measurements 7
III.   Effects of Soil Management on Soil Structure and the Associated
        Hydraulic Properties 9
     A. Methods for Measuring Hydraulic Properties 9
     B. Examples 12
IV.   Use of Simulation Models to Predict Soil Hydraulic Properties ... 28
     A. Introduction 28
     B. Modeling Soil Water Regimes and Crop Growth in a Typic
        Haplaquent 29
V.   Conclusions and Recommendations 32
References 34

I. Introduction

This chapter addresses the relationship between soil structure and soil hydraulic properties and how they are affected by soil management. The study of soil structure is as old as soil science itself. Many techniques for characterizing soil structure have been proposed by different schools of thought, of which the morphological and physical schools are perhaps the most important. Rather than discuss and compare definitions, this chapter relates soil structure to hydraulic properties and presents some case studies to demonstrate the effects of soil tillage and aggregation on soil structure. Relating soil structure to soil hydraulic properties represents only one type of interpretation of soil structure, which was

ISBN 0-87371-889-5

1

arbitrarily selected for this chapter to restrict its focus. Rather than discuss physical and morphological characterizations of soil structure separately, this chapter reviews the synergism provided when physical measurements are enhanced or interpreted by morphological observations. A case will be developed for direct measurement of hydraulic properties. Predictions of these properties using morphological and physical soil structure data will be critically reviewed.

Morphological techniques can be helpful to explain what may appear to be erratic physical measurements at first sight when the latter are based on the assumption that soil materials are homogeneous and isotropic; for example, samples for physical analyses including different soil horizons or having a smaller volume than the representative elementary volume to be estimated by morphological analysis. Also, some soils are so heterogeneous that it is not possible to make random measurements that are truly representative for the entire soil. A stratification on the basis of a morphological analysis may help to define subsamples that are relatively homogeneous. By knowing their relative volume in the entire soil, an estimate of the properties of the entire soil can be obtained. An example, to be discussed later, relates to the measurement of the hydraulic conductivity in a prismatic soil with sandy areas around the prisms. By making separate measurements in the sandy areas and in the prisms and by determining their relative areas, a correct picture may be obtained for the hydraulic properties of the entire soil. A useful guideline for soil physical measurements could be to:"Look first, then measure!".

Soil structure is strongly affected by soil management. Examples will be provided for different soils in the tropics and in temperate areas which show different soil structures as a result of tillage and soil traffic. Effects of biological activity on soil structure are considered by examining some soils under grassland. The effect of grassland on soil structure may be favorable in a temperate climate where modern tillage practices may lead to structure degradation in arable land. However, when tropical rain forest is cut and changed into grassland, soil structure may be degraded. Although a negative statement is used here, we generally will try to avoid quality judgments for soil structure in this chapter. The agricultural quality of a certain type of soil structure depends strongly on climatic conditions and on land use, including the type of crops that are grown. It is therefore preferable to characterize soil structure in quantitative terms and to associate its quality with specific types of land uses and associated interpretations. Characterization of soil structure is based on use of morphological and physical techniques. The physical techniques follow two procedures in this chapter.

First, soil structure is characterized with physical measurements of static and dynamic parameters, such as moisture retention and hydraulic conductivity. Many methods have been published, some methods are described in this chapter that have been of particular value for our research work.

Second, we explored the use of simulation models for characterizing soil hydraulic conditions as a function of space and time. These models use the hydraulic parameters discussed earlier, and provide the type of data that users are generally interested in.

The choice to emphasize the combined use of physical and morphological techniques to characterize soil structure, as discussed, restricted the selection of case studies to be reviewed.

II. Characterization of Soil Structure

A. Introduction

Soil structure is defined as: "The physical constitution of a soil material as expressed by the size, shape and arrangement of the solid particles and voids, including both the primary particles to form compound particles and the compound particles themselves" (Brewer, 1964). Compound particles are often referred to as: "peds". Both morphological and physical techniques are being used to characterize soil structure.

Soil profile descriptions typically include a field evaluation of soil structure that includes sizes, shapes, and degrees of development of natural aggregates (peds) (e.g., Soil Survey Staff, 1951). Also, the occurrence of macropores formed by roots or soil animals is observed. This information is rather qualitative in nature, but major differences among different structures in the same soil can be established quite reproducibly by different observers. A mistake is to try to establish many types of soil structure that cannot reproducibly be observed. However, soil morphological studies can also contribute more specific, quantitative information by analyzing soil peels, soil thin sections, or polished sections of plastic-impregnated, undisturbed soil samples. Recently, submicroscopic techniques have been developed that allow very detailed observations and chemical and mineralogical characterizations of soil compounds in situ. Such techniques could be valuable for investigating structure stability. Use of CT and NMI scanning techniques can elucidate three-dimensional pore patterns (e.g., Warner et al., 1989). Of particular significance is the application of morphological techniques to provide information that could not have been obtained by physical methods. For example, the study of different types of pores, shapes of peds, and large-pore continuity can only be made by morphological techniques. The efficiency of such studies is increased when they are made together with soil physical studies that characterize the occurrence and flow of water and air in soils.

B. Soil Physical Techniques

Standard methods for measuring bulk density and porosity in stony and non-stony soil as a measure for soil structure have been reviewed elsewhere (e.g., Klute, 1986; Burke et al., 1986). Equivalent pore-size distributions can be derived from moisture retention data by assuming that soil porosity is composed of bundles of capillary pores of a characteristic range of diameters. This can be a useful procedure when comparing soils, but real pore-size distributions are, of course, quite different from equivalent pore-size distributions in sandy or clayey soil materials. Real distributions can be observed with morphological techniques, as has recently been reviewed elsewhere (e.g., Bouma, 1991).

Bulk density, porosity, and equivalent pore size distributions provide static parameters that can be used to characterize soil structure. Dynamic characteristics can be obtained as well by measuring permeabilities for water and air. Permeabilities are an effective measure for pore continuity patterns, particularly for the larger ones. Methods for measuring hydraulic conductivities are reviewed in Section III.A. For a review of relations between large-pore continuity and permeability, reference has again to be made to Bouma (1991) to avoid duplication.

C. Soil Morphological Techniques

Soil porosity can be studied with morphological techniques at different levels of detail (e.g., Burke et al., 1986). Optical studies by eye or binocular microscopes can use soil lacquer peels and soil monoliths (Van Baren and Bomer, 1979). Soil thin sections or polished, plastic-impregnated soil samples allow more detailed investigations (e.g., Jongerius and Heintzberger, 1975; Miedema et al., 1974; Bullock et al., 1985). Very detailed studies of soil structure, using magnifications of more than 200 x, can be based on application of submicroscopic techniques (e.g., Bisdom, 1983).

The vertical and horizontal continuity of large pores (macropores) is an important soil structure characteristic. It governs infiltration rates, particularly at higher rainfall intensities, and aeration processes in clayey soils that have a soil matrix with fine pores. Macropore-continuity is difficult to study in a two-dimensional thin section, or in hand-specimen of soils in which only relatively small soil volumes are observed. Three approaches can be followed to derive macropore-continuity:

(1) Statistical studies, that translate two-dimensional images into a three-dimensional pattern. Recently, stereology has been applied in this context (e.g., Weibel, 1979; Ringrose-Voase and Bullock, 1984).

(2) Functional characterization of soil macropores obtained by using staining techniques, or other tracers. Thin sections are used in detailed studies. Undisturbed blocks of soil can be percolated with water in which a dye has been dissolved. Also, a gypsum slurry can be used to fill macropores. The necessary equipment for these techniques is minimal.

Several procedures have been followed, depending on the specific objectives of soil structure studies. They include:

(a) Infiltration patterns in dry clay soils with vertical cracks.
Sprinkling irrigation results in the formation of small, 5 mm wide bands of water, along which water moves downwards following the vertical faces of the cracks. These bands are made visible by staining (e.g., Bouma, 1991).

(b) Ponded infiltration in soils with vertical macropores.
A thin gypsum slurry can be used instead of water. After hardening of the gypsum, observations can be made of the patterns of occurrence of the gypsum. Gypsum can only penetrate to a certain level through continuous pores. In turn, presence of gypsum indicates pore continuity (e.g., Bouma et al., 1982; Mackie, 1987; Fitzpatrick et al., 1985).

(c) Formation of horizontal cracks upon drying of clay soils.
Formation of horizontal air-filled cracks impedes the upward, unsaturated flow of water from the water-table to the rootzone. The percentage of horizontal air-filled cracks can be estimated, for a given pressure head, by using a staining test in undisturbed large blocks of soil that are encased in gypsum. The soil block is turned on its side, the upper and lower surfaces are opened, and two sidewalls of the turned cube are closed. Methylene-blue solution is poured into the cube and stains the air-filled cracks. The surface area of these stained cracks is counted after returning the cube to its original position. A separate cube is needed for each (negative) pressure head. The hydraulic conductivity curve for the peds is "reduced" for each pressure head measured in a cube. When, for example, 50% of the horizontal cross-sectional area is stained, K_{unsat} for upward flow is 50% of the K_{unsat} at the same pressure head in the peds (e.g., Bouma, 1984).

(d) Flow patterns in a wet clay soil.
Undisturbed samples of a clay soil that have been close to saturation for a period of several months, are percolated with a 0.1% solution of methylene-blue or another dye that is absorbed by the clay particles. Thin sections are made, in which stained (continuous) and unstained (discontinuous) macropores can be observed and quantified (e.g., Bouma et al., 1979).

Use of the above-mentioned techniques allows the characterization of soil pores in terms of pore type, size, shape, and continuity. Such characterizations are difficult to obtain with soil-physical methods, and they provide an opportunity to more specifically define soil structure.

(3) Use of CT scanning or NMI techniques, allow, in principle, the in situ non-destructive characterization of macropore patterns. These techniques are, however, still experimental (e.g., Warner et al., 1989).

D. Using Soil Structure Data to Predict Soil Hydraulic Properties

In subchapters C and D, and in the associated literature, many methods are described to characterize soil structure. Soil structure in this chapter is considered to have a separate identity and many properties, including hydraulic properties which are being considered here. These can be measured directly in different soil structures, but the question should be raised whether these hydraulic properties can also be predicted by soil structure data. This question is far from academic because direct measurement of hydraulic parameters, such as moisture retention and hydraulic conductivity, is often cumbersome and costly. Many papers have discussed the prediction of hydraulic parameters, using soil texture and bulk densities in regression analysis (e.g., Schuh and Sweeney, 1986; Gupta and Larsen, 1979). Sometimes bulk densities are "significant" variables, sometimes they are not. Regression analysis yields empirical data. Other attempts have been made to use moisture retention data and derived equivalent pore size distributions to predict hydraulic conductivities following a more deterministic approach (e.g., Green and Corey, 1971). These efforts have been marginally effective since quite large "matching factors" were often needed to fit calculated data to measured data. Also, soil structure data derived by morphological methods were used to predict soil hydraulic parameters, but to a much lesser extent. Anderson and Bouma (1973) predicted hydraulic conductivities of a silt loam soil on the basis of staining patterns in thin sections, as did Bouma et al. (1979) for six Dutch clay soils. Earlier, moisture retention curves were also predicted successfully, but only in sands (Bouma and Anderson, 1973). Though successful, these morphological methods are much more laborious than the physical measurement itself. They serve not as an alternative to direct physical measurement but, rather, to increase understanding about the underlying processes. For example, in clay soils very few pores, which occupied a very small volume, determined K_{sat}. Knowing that, it is no surprise that a crude measurement of a bulk density often shows no significant relationship with K.

A more generalized and descriptive approach to predict K_{sat} from morphometric data was followed by Mc Keague et al. (1982). Recently, we have used well-defined soil horizons in well-established soil series as "carriers" of physical information, particularly the unsaturated hydraulic conductivity which is of more interest than K_{sat} (e.g., Wosten et al., 1990). Any soil surveyor can determine the occurrence of a particular soil horizon. Multiple measurements in such horizons provide averages and standard deviations to be used for estimates elsewhere. Sometimes, estimates of hydraulic properties are all that is needed. These properties are often used in simulation models that need many other data regarding weather, crop parameters, water table fluctuations, etc. There is little justification to determine hydraulic parameters in great detail by measurement, while other modeling data are only known in approximate terms. However, when making estimates it is crucial to include a measure for accuracy which should, in turn, be incorporated in the modeling process for all parameters involved.

At this point in time it would seem to be most attractive to make good measurements of hydraulic soil properties. Data obtained should be related to soil parameters such as texture, organic matter content, and bulk density which can be measured directly, but also to soil characteristics such as soil series, horizons, and soil structure to be described in the field. Measured soil hydraulic properties can be expressed in terms of coefficients as defined by, e.g., Van Genuchten (see reference by Wosten et al., 1990). Those coefficients can be related by regression analysis to soil parameters, as discussed, or to more qualitative groupings of soil horizons based on pedological expertise. This, in the opinion of the author of this chapter, is a more promising procedure than attempts to calculate hydraulic parameters directly from physical or morphological soil structure data.

E. Using Soil Morphological Analyses to Improve Physical Measurements

1. Elementary Units of Structure

Many measurement procedures use standard sample sizes, because of fixed dimensions of sampling cylinders or of equipment being used. For example, sampling cylinders with a fixed volume of 100 cm^3 have been used extensively in different laboratories. Equipment, such as the double-ring infiltrometer or the air-permeameter, come in standard sizes. There is good justification to vary sample size as a function of soil structure, as a means to reduce variability among replicate measurements (e.g., Bouma, 1983). Soil structure descriptions as made during soil survey can be used to tentatively define representative elementary volumes of samples (REV's), which are the smallest sample-volumes that can represent a given soil horizon by producing an unbiased population of data. To do so, the elementary units of soil structure (ELUS) have to be distinguished. These are individual sand grains in sandy soils and natural aggregates ("peds") in aggregated soils. Peds can vary in size up to several liters each in very coarse, prismatic subsoil structures. Clods formed by tillage can vary widely in size. Even though emphasis in soil structure descriptions is often placed on the solid phase in terms of soil grains and peds, emphasis should be on the pore space where transport processes take place. By describing grains and peds, information is also provided about the pores between them. In addition, pores that do not result from the packing of grains or peds, should be considered separately. Such pores are, for example, root and worm channels with a cylindrical shape. As a general rule we have proposed that REV's should contain at least 20 ELUS but preferably more, or that any sample taken should have a representative number of channels per unit surface area. Defining REV's as a function of field description of soil structure needs to be further investigated. Data by Anderson & Bouma (1973) clearly illustrate the potential of the procedure

Table 1. Measured hydraulic conductivity of saturated soil (K_{sat}) in soil cores of varying height but with a diameter of 7.5 cm, containing different numbers of elementary units of structure (ELUS)

Sample Volume (cm³)	ELUS (no)	K_{sat} (cm day⁻¹)	S (cm day⁻¹)
230	8	650	350
330	11	329	320
460	15	100	80
780	26	75	30
12000	400	70	20

Measurements were made in a silt loam soil with medium-sized peds with an average volume of approximately 30 cm³. The largest sample was a gypsum-covered column of soil with a diameter of 30 cm.
S = Standard deviation.

(Table 1). Unrealistically high K_{sat} values were measured in soil cores in a silt loam containing fewer than 20 ELUS. Values measured with a gypsum-covered column having a volume of 12 liters, averaged 70 cm day⁻¹. The reason for the high K_{sat} values in the small cores is the unrepresentatively high vertical continuity of cracks between the peds in small samples. Lauren et al. (1988) measured the K_{sat} in a Glossaquic Hapludalf using different horizontal areas of infiltration, varying between 160 x 75 cm and 7 x 6 cm (Table 2). They found that K_{sat} and its variability was a direct function of sample volume which should not be the case, considering general flow theory. Mean values for intermediate

Table 2. K_{sat} values in B_t of a Hapludalf measured in soil columns with five sizes

Column	Dimensions (cm)	Mean	Mode	Median	SD	CV %	No. of sample
A	160x75x20	21.3	10.3	16.6	16.9	79	37
B	120x75x20	13.7	6.4	10.7	11.0	81	36
C	50x50x20	14.4	6.3	10.9	12.5	96	37
D	20(diam)x20	36.6	6.3	20.3	54.9	150	37
E	7(diam)x6	34.5	4.8	16.3	64.0	186	35

(after Lauren et al. 1988)

sizes of the infiltration surface (Size C) were significantly lower than values for smaller cores (D and E) because they contained more peds in cross-section. As a general rule of thumb it was assumed that a representative sample should at least have 20 peds in cross-section.

2. Sample Location in the Soil Profile

Sampling at regular depth intervals is often applied with good results in relatively homogeneous soils with weakly developed soil horizons. When clear soil horizons exist, however, it is preferable to sample by horizon (e.g., Petersen and Calvin, 1965). A sample containing fragments of two adjacent, and as such quite different, soil horizons, will yield data that are hard to interpret. However, it should be realized that pedological horizons as distinguished in soil survey, are not always good "carriers" of data that are relevant for the particular soil structural interpretation being pursued. Some pedological distinctions may be irrelevant in this context, while relevant soil-structural aspects may not be reflected in a particular horizon classification. In general, it is advisable to make a soil structure description before taking soil-structure samples for physical analyses. Such samples should preferably be taken in soil layers with a more or less homogeneous soil structure, as observed in the field. Particular attention should be paid to management-induced boundaries such as the lower boundary of the plow layer, which may occur within a pedological horizon. Another important boundary is the atmosphere-soil interface where crusting and sealing due to rainfall impact may occur in thin layers of only a few centimeters thickness. Selective sampling is desirable here. Selective sampling in soils with clearly expressed morphological features such as macropores or bleached areas around peds can decrease the variability of results obtained. For example, Bouma et al. (1989) showed that hydraulic conductivities inside peds were significantly lower than those in sandy areas between peds, which had formed by soil forming processes. Values were 0.8 m/day and 7.0 m/day respectively. When samples were taken at random, a very high variability of measured K_{sat} resulted because peds and sandy material were present in each sample in a random manner. Selective sampling yielded two distinct populations of data.

III. Effects of Soil Management on Soil Structure and the Associated Hydraulic Properties

A. Methods for Measuring Hydraulic Properties

Numerous methods have been developed to measure soil hydraulic properties (e.g., Klute, 1986). Descriptions in literature are usually confined to a discussion of soil-physical theory, a listing of equipment needs, and operational procedures. Little attention is being paid to the applicability of the various methods in different soils even though this is an important consideration. Bouma (1983), for example, discussed problems encountered when applying the auger-hole method for measuring K_{sat} in clay soils. Puddling the auger-hole walls while boring the hole led to closure of highly conductive planar voids which conducted the water during saturated flow as was demonstrated by using dyes. Unrealistically low K_{sat} values of 0.3 cm/day were therefore obtained with the standard auger-hole method whereas realistic values of 50 cm/day were measured with the column method which used a large, undisturbed volume of soil. In contrast, the auger-hole method yields good, representative results in sandy soils.

Measurements of K_{sat} in a pedal silt loam soil demonstrated that K_{sat} was a function of the volume of measurement (Anderson and Bouma, 1973; Lauren et al., 1988). Relatively small samples contained few peds in cross-section and yielded a significantly higher variability than larger samples. As a general guideline it was suggested to obtain sample volumes containing at least 20 peds in cross-section (Lauren et al., 1987).

Measurements with the double-ring infiltrometer can be made successfully in sandy soils but may yield erroneous results in clayey soils with macropores where the wetting pattern below the infiltration rings is bound to be irregular and quite different from patterns in a sand. Applying a method under unsuited conditions or in the wrong time of the year, such as making measurements of K_{sat} or infiltration rates in dry clay soils, results in erroneous data. Swelling of soil will occur in such soils which results in variable, unstable data. Besides, water will tend to run down air-filled macropores, such as cracks, through an unsaturated soil matrix. This process has been called bypass flow (e.g., Bouma, 1984; White, 1985) and is still not formally being discussed in soil physics textbooks. The process is important in many soils (e.g., Beven and Germann, 1982) and particularly in soils with well developed structures with macropores which are particularly considered in this chapter. Some methods will therefore be reviewed which are suitable for measuring soil hydraulic properties in well structured soils and which are suitable for use at or near the soil surface where soil management affects soil structure most profoundly. Several methods will be briefly reviewed here; reference will be made to publications with a more thorough description. For all other methods reference will be made to Klute (1986).

1. The Suction Crust Infiltrometer

Theoretically conceptualized by Hillel and Gardner (1969), the field crust test was developed in the seventies (Klute, 1986). A recent revision (Booltink et al., 1991) uses one crust and measures a series of fluxes through the crust under different hydraulic heads, as imposed on top of the crust by a Mariotte device in a burette. Each head is associated with a steady flux and subcrust pressure head which, together, represent one point of the K (h) curve. By changing the head on top of the crust, a series of K values can be obtained that may range from saturation (no crust) to pressure heads as low as -20 cm. The particular advantage of the method is the perfect contact of soil and crust, due to the application by hand of the mixed, wet crust material to the soil surface after which it hardens. Hardening occurs rapidly because quick setting cement is used, which is mixed with medium textured sand. A one-dimensional vertical flow system is induced by placing an infiltrometer on top of a column of soil that is carved out in situ. Creation of a one-dimensional system avoids complex calculations of fluxes in a three-dimensional soil volume.

The suction crust infiltrometer is particularly suitable to measure K in soils with macropores, which may have very high K_{sat} values when macropores are continuous. However, upon desaturation, K drops very strongly because macropores fill with air and smaller pores in the soil matrix must conduct the water. Tensiometers may still indicate saturation in terms of zero pressure at unit gradient when water infiltrates into the soil through a crust, while fluxes are much lower than fluxes into soil without a crust at shallow head. The difference is due to flow of water along the walls of macropores, without filling the complete pore.

Infiltration under suction can also be achieved by using a foil, as was proposed in the suction infiltrometer (Perroux and White, 1988). This method works particularly well in sandy soils, but irregular surfaces in clay soils may present contact problems with the foil.

2. The Column Method

The carved-out column with infiltrometer, as described for the crust test above, can be used as well to measure K_{sat} by measuring the steady flux into the column without a crust. Use of a 30 cm diameter infiltrometer and a Mariotte device allows very accurate measurements since low rates of 1 mm/day correspond with 60 cm³/day which can be measured reproducibly within a relatively short period of time. Occurrence of continuous macropores presents a particular problem when measuring K_{sat}. The type of measurement described here characterizes fluxes in situ. In other words, macropores that are vertically discontinuous fill up with water and have to drain through surrounding usually much smaller pores. When, however, a core is sampled and detached from the subsoil measured fluxes are much higher because macropores are now vertically continuous. Lauren et al. (1987) demonstrated these effects for a Typic Hapludalf with worm channels (Table 3).

Presence of continuous macropores in a relatively compact soil matrix creates conditions that deviate strongly from the standard concept of soil which considers soil to be reasonably homogeneous and isotropic. It may, in fact, be advisable to make separate measurements in the matrix and in the macropores and to combine them proportionally (e.g., Bouma et al., 1983, 1989).

Table 3. Differences between K_{sat} values measured in soil columns that are attached and detached

| | K_{sat}(cm/day) | | | | | |
Method	Mean	Median	Mode	SD	CV(%)	No. Samples
Attached	36.6	20.3	6.3	55	150	37
Detached	4296	286.6	1.3	64280	1496	37

Detached columns had higher K_{sat} values due to macropore continuity.

3. Bypass Flow

Flow of free water along macropores through a soil matrix that is unsaturated has been defined as bypass flow (Bouma, 1984). Measurement of bypass flow is relatively simple (Bouma et al., 1981). A large, undisturbed soil core is placed in a cylinder on top of a funnel which allows collection of free water from the core. The cylinder is subjected to rainfall with different intensities and quantities and the amount of outflow is expressed as a percentage of inflow. Illustrations will be provided later in this text when specific case studies are discussed.

B. Examples

A number of case studies will be discussed which cover different soils and climates, as well as different types of soil management.

1. Soil Structure and Hydraulic Properties in a Typic Haplaquent under Grass and after Regular and Deep Tillage

The soils being studied are located in the southwestern part of the Netherlands near Nieuw Vossemeer and are developed in young marine deposits (Kooistra et al., 1985; Van Lanen et al., 1986). They are classified as Typic Haplaquents (Soil Survey Staff, 1975) and as Calcaric Fluvisols (FAO-UNESCO, 1974). These soils have AC profiles and are well drained. They are mostly under cultivation and have a rotation of potatoes, sugar beet, and cereals, mainly winter wheat. Many of these cultivated soils have developed compacted layers (ploughpans) below the tilled topsoil. Some soils occur under permanent grass-

Figure 1. Soil structure under grassland (left picture) and arable land, showing rounded subangular blocky and angular prismatic peds, the latter due to compaction.

land, and these soils have characteristically different structures as demonstrated in Figure 1.

A short description of a representative pedon with a ploughpan is given below and some analytical data are presented in Table 4. The Ap horizon (0-30 cm) is a dark grayish brown sandy loam and has a weak medium subangular blocky structure. The C1 horizon (30-45 cm) is an olive gray sandy loam with a massive ploughpan. The C2 horizon (45-70 cm) is a grayish brown loamy sand with a weak fine subangular blocky structure. This horizon has reddish mottles and gley spots which indicate periods of wetness. Below 70 cm depth the soil consists of laminated sand with reddish mottles. All horizons are calcareous.

Table 4. Analytical data for a Typic Haplaquent with undisturbed ploughpan

Horizon	Depth	Sample Depth (cm)	Particle size distribution (%, w/w)							CaCO₃ %, w/w	pH-Kcl	Organic matter %, w/w
			>150	50-105	50-105	16-50	2-16	<21				
A_p	0-31	0-25	1.6	11.8	41.8	22.1	6.7	16.0	4.7	7.6	1.4
C_1	31-42	38-42	0.6	10.8	50.4	20.4	5.4	12.4	9.2	7.8	0.8
C_2	42-70	58-62	1.8	1.8	47.6	8.9	1.9	6.5	7.6	8.0	0.6

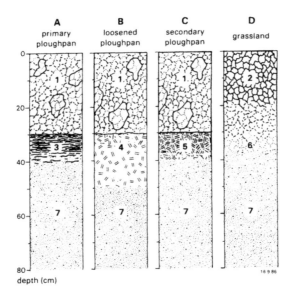

Figure 2. Soil structure in a Typic Haplaquent with a sandy loam texture. The numbers indicate layers with different soil structures and physical properties as specified in Table 5.

Deep loosening by rota-digging had created a mixture of A and C material within a depth of 50 cm. Soil structure in the grassland soil consisted of strongly developed medium, subangular blocky peds. With depth the degree of structure diminished until it was identical to the other soils below a depth of 40 cm. A

Table 5. Physical data for four different types of soil structure in a Typic Haplaquent (see Figure 2)

Layer	Dry (%)	Organic matter (%)	Bulk density (g cm^{-3})
1	14	1.8	1.40
2	12	4.3	1.35
3	10	0.8	1.62
4	9	0.6	1.33
5	9	0.6	1.58
6	11	0.3	1.43
7	5	0.4	1.55

J. Bouma

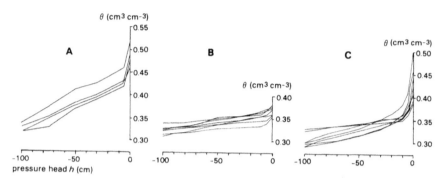

Figure 3. Moisture-retention curves from four samples in pasture land (A), and seven in arable land with a primary ploughpan (B) and a secondary ploughpan formed in three years (C). Depths of samples: 25-45 cm.

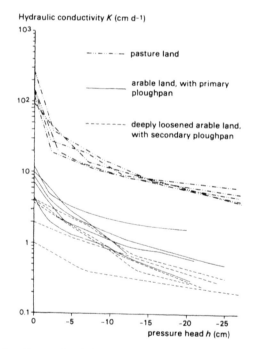

Figure 4. Hydraulic-conductivity curves derived from five samples of each soil: pasture land, arable land with a primary ploughpan, and deeply loosened arable land with a secondary ploughpan formed in three years. Depths of all samples: 25-45 cm.

schematic diagram of structures is presented in Figure 2 and physical data for the different types of land use are summarized in Table 5. All studied fields were tile drained, ground water levels fluctuated between 0.8 m below the surface in winter and 2.0 m below the surface in summer.

Measured hydraulic conductivity and moisture retention curves are shown in Figures 3 and 4. Retention curves are significantly different between saturation and a pressure head of -100 cm, at which pressure moisture contents are identical. Moisture contents at saturation are highest in Type A (Pasture) while the decrease upon decreasing pressure head is most gradual. This indicates the presence of relatively many large- and medium-sized voids which can be seen in thin section (Figure 5). Structure types B and C differ in terms of their moisture contents in the pressure head range from 0 to -10 cm. The deeply tilled soil (Type C) loses much more water than the soil that was regularly plowed (Type B). This loss of water is not associated with emptying of large pores, which are not present (see thin section image in Figure 5), but with water extraction from puddled soil which results in a decrease of the soil volume, as particles are drawn closer together. Samples from the compact ploughpan (Type B) showed no decrease of soil volume upon desaturation because the sand grains were bound together by the fine soil particles providing a relatively high stability. This is illustrated in Figure 6 which shows the microfabric at high magnification. Soil clay and silt particles are present around the sand grains and concentrations occur at the points of contact which provides a stable structure. Extinction patterns, observed with polarizing microscopes, indicate that the clay plates are oriented parallel to the grains. The same soil material looks quite different after puddling, which occurs when energy is applied to soil which has a relatively high moisture content, greater than the lower plastic limit. Then, bonds between grains and clay plates are broken and the clay plates adsorb water which leads to swelling. Grains float in a viscous matrix which fills the spaces between the sand grains. Such a soil material is unstable and does not contain macropores. A picture of puddled soil is shown in Figure 7. The only possibility for restoration of soil stability is by drying. Then, the fine soil particles are drawn towards the grains and concentrate at their points of contact. When rewetted, the bonds remain but only when the soil has been air-dry. If dried to only a moist water content, swelling will take place again and the soil will remain viscous. These processes are schematically represented in Figure 8 from Bouma (1969). Hydraulic conductivity curves (Figure 4) were characteristically different as well. Highest values were measured for the grassland, which is due to the relatively high content of large- and medium-sized pores (Figure 5). Values in the ploughpan were significantly higher than those of the deeply tilled material at the same depth. As stated above, this difference can be attributed to puddling in Type C which results in fewer larger pores.

18   J. Bouma

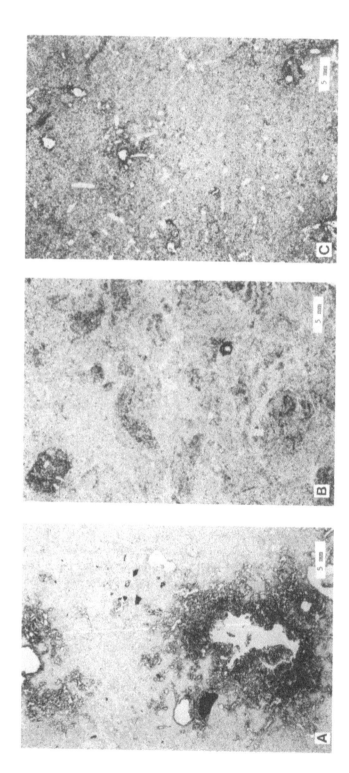

Figure 5. Thin section images of soil structure at 37 cm depth for grassland (A); arable land with primary ploughpan (B) and secondary ploughpan (C). Dark colored pore walls indicate staining, resulting from vertical pore-continuity.

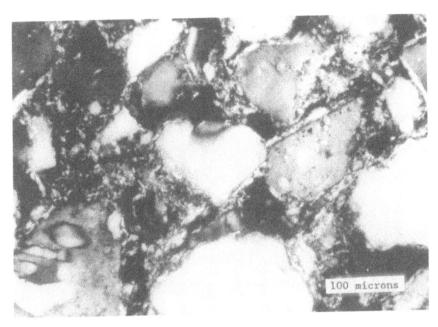

Figure 6. Microfabric of a dry sandy loam soil material which shows clay domains concentrated around sand grains, leaving small packing pores in between. (structure no. III in Figure 8)

2. Soil Structure and Hydraulic Properties in a Typic Haplaquent with Surface and Subsurface Compaction

Effects of different tillage measures on potato growth were studied in an interdisciplinary project in 1976 (Boone et al., 1978; Van Loon and Bouma, 1978). Studies were made in soils that had similar natural properties as soils discussed in the previous subsection. Different treatments (and horizon designations), schematically represented in Figure 9, were studied as follows:

L: Loose, noncompacted soil in which tillage of and traffic on the Ap horizon was made at the appropriate low moisture content to avoid compaction and puddling. Also, soil traffic was minimized by use of special traffic lanes for cultural practices such as herbicide spraying. This treatment was used as a reference or control.
P: Ploughpan. Occurrence of a strongly compacted subsoil below the Ap, achieved by removal of the Ap and by repeatedly driving over the exposed subsoil with a heavy tractor and a loaded wagon, after which the topsoil was redeposited.
C: Compacted surface soil, achieved by driving four times over the soil surface with a heavy tractor.

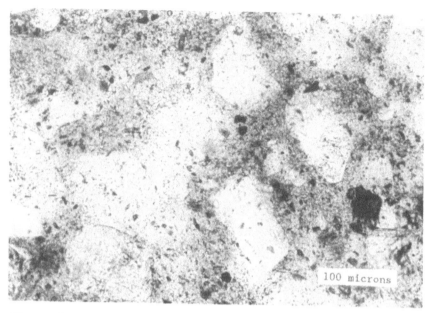

Figure 7. Microfabric of a puddled sandy loam soil material showing dispersed clay and organic matter between sand grains. (structure no. I in Figure 8)

The moisture retention and hydraulic conductivity characteristics, which are of particular interest in the context of this chapter, are presented in Figures 10 and 11. The compacted Ap horizon has a relatively low porosity of 0.38 m³/m³, while the loose Ap had a value of 0.52 m³/m³. However, at a pressure of only 10kPa, moisture retention curves of all horizons show very small differences. Differences among treatments are also strongly reflected in the K- curves. K_{sat} for the Ap-C was only 8 cm/day, while the K_{sat} for Ap-L was 180 cm/day. However, upon desaturation, K drops very strongly in the loose soil because relatively large pores are rapidly filling with air. At a pressure of -10 cm, the two K values are almost identical. At lower pressure heads, K values for the Ap-C are higher than the ones for Ap-L. The K curves for the C22g in which the subsoil ploughpan was formed, showed similar differences even though the range in values was less. The K_{sat} of the C22g-L was 80 cm/day and 20 cm/day for the C22g-P. The difference between surface and subsoil compaction illustrates the importance of the initial condition of the soil material before compaction. Initially disturbed and loose soil, such as Ap material, tends to become more strongly compacted than undisturbed and stabilized soil material in situ, such as is found in the C22g horizon.

The graphs in Figure 11 illustrate important effects of different soil structures, which are a result of differences in soil management, on soil hydraulic properties. Saturated and unsaturated conditions should always be con-

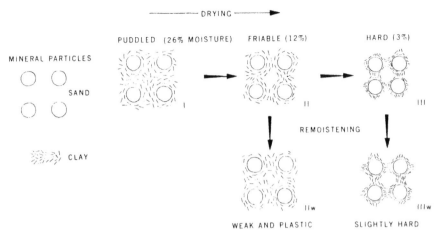

Figure 8. Microstructures in a sandy loam soil material as a function of treatment. When puddled soil (I) is dried, the fine particles concentrate around larger grains (III) leaving fine pores in between. These pores remain after remoistening (IIIw). This material is slightly hard. After slightly drying to a friable consistency (II) remoistening forms a weak and plastic soil (IIw).

Figure 9. Schematic diagram of the soils indicating the positions of the compacted layers.

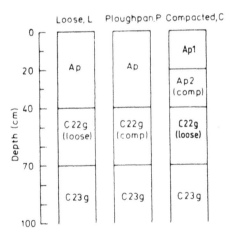

sidered together, because what may be different at saturation may not be different at lower moisture contents. When simulating the soil water regime during the growing season, different moisture contents occur continuously and hydraulic effects of the characteristics as shown in Figures 10 and 11 are quite unpredictable. Simulation results for the growth of potatoes will be presented and discussed in Chapter 4.

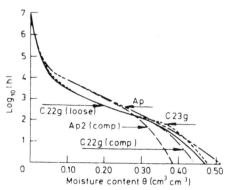

Figure 10. Measured soil-moisture retention curves, h(Θ), of the different layers of the loose, ploughpan, and surface-compacted soils.

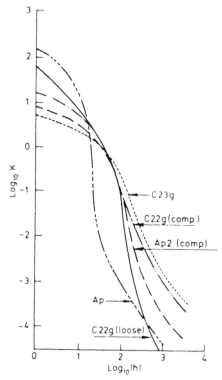

Figure 11. Measured relations between the hydraulic conductivity K(cm d⁻¹) and the pressure head h (cm) of the different layers of the loose, ploughpan, and surface-compacted soils.

3. Soil Structure and Hydraulic Properties of the Surface Horizons of a Typic Pelludert under Range- and Cropland

The study (Smaling and Bouma, 1991) was conducted on a Vertisol in Kenya, which was classified as a Typic Pelludert (Soil Survey Staff, 1975) or Eutric Vertisol (FAO, 1978). The upper 25 cm of the soil had a clay content of 54%, an organic carbon content of 2.4% and a pH-H_2O of 6.2. Between 25 and 50 cm the clay content was 65%. There was no groundwater table and basaltic rotten rock started below 70 cm. Experiments were conducted in rangeland under perennial grasses and in adjacent cropland, manually tilled to a depth of 15 cm. The soils in the rangeland had a strong, angular blocky topsoil with thin, continuous cracks. Rooting was intense, and extended to a depth of 70 cm and more, particularly along ped faces. Below 20 cm, the subsoil had coarse prisms with a width of 8 to 15 cm, separated by cracks with a width of 0.5 to 2 cm. In the cropland topsoils, tillage had destroyed macropores by forming small aggregates. Below the depth of tillage, the soil had coarse prisms with slickensides. Vertical cracks had a width of 0.5 to 1.5 cm and were continuous to a depth of 60 to 80 cm. Rooting patterns of maize were observed after harvest and did not extend significantly beyond a depth of 40 cm.

Particular attention was paid in this study to the process of bypass flow, as defined in Section II.B. Many Vertisols occur in semi-arid climates and occurrence of bypass flow is particularly relevant in these soils with large cracks, because water that flows into the cracks may not be available for the crops if it moves beyond the rootzone. By comparing in this study bypass flow in rangeland with that in cropland, the effect of management on this particular soil hydraulic property can be evaluated.

Only experiments relating to the measurement of bypass flow will be reviewed here. The applied shower had an intensity of 30 mm/hour and lasted for 1 hour on day 1. The same shower was again applied on day 3. Results (Table 6) show bypass percentages for the rangeland topsoil of 47% (first day) and 62% (third day). Corresponding values for the cropland were 19% and 49% respectively. The higher values for the second shower can be explained by wetting of the soil surface layer during the first shower. Infiltration rates in moist soils are lower than those in dry soils. Consequently, higher bypass flow

Table 6. Bypass flow in a Vertisol for two types of land use

Land use	No. of cylinders	Water application (mm)		Bypass flow (%)	
		Day 1	Day 3	Day 1	Day 3
Rangeland	12	30	30	47	62
Cropland	24	30	30	19	49

rates are measured when infiltration rates into the soil matrix are lower. Measured rates of bypass flow are quite substantial, and they are likely to affect the soil water balance significantly because precipitation often occurs in the form of high intensity showers. Water that runs into the cracks will move to their bottoms and will infiltrate from there into the soil matrix. This water may still be available for plant roots if they extend to this depth. For conditions considered here, it is doubtful whether water moving downwards with bypass flow is still available.

The difference in bypass flow observed between rangeland and arable land is due to differences in structure. In cropland, tillage by hand has resulted in the presence of many relatively small aggregates with no major, continuous vertical cracks in between. When a shower is applied, much of the water will infiltrate into peds either at the surface or while moving downwards through the top-soil. However, bypass flow is still significant. In rangeland, on the contrary, large cracks extend to the surface and water that cannot infiltrate into the soil matrix at the soil surface because the infiltration rate is lower than the rate of precipitation, will flow rather easily into the cracks and down into the soil. The observed phenomena are schematically illustrated in Figure 12. As such, the study demonstrates the importance of soil structure in determining rates of bypass flow. So far, only flow of water was considered. Of interest is also movement of dissolved chemicals in the water, such as fertilizers or pesticides. A complete discussion of these aspects is beyond the scope of this text. However, major losses of nutrients can occur when they are applied at a dry soil surface before rainfall. It is preferable to apply fertilizers to a moist surface which allows solution and diffusion into the soil matrix. When the next shower arrives, little fertilizer is left at the surface and there is less leaching as was demonstrated in a Dutch case study for cracking, heavy clay soils (Dekker and Bouma, 1984).

4. Soil Structure and Hydraulic Properties of a Humoxic Tropohumult and an Oxic Humitropept under Forest and Pasture

This study (Spaans et al., 1989) was carried out in the Atlantic zone of Costa Rica. The four selected sites were developed in volcanic mudflows of different age. A relatively old Humoxic Tropohumult and a relatively young Oxic Humitropept (Soil Survey Staff, 1975) were selected under tropical rain forest and under pasture. The pasture on the Humult had been present only for a period of 3 years after clearing of the forest. The other pasture had been present for the last 15 years; before that the land was used for bananas and deforestation occurred 35 years ago. Grazing density was identical at both sites. The yellowish-red Humult had a heavy clay texture and showed a thin A horizon that graded gradually into a deep illuvial B2t horizon. The soil was well drained, quite homogeneous, very friable, and porous to a depth of at least 150 cm, due to high biological activity. Under pasture a clear change occurred in the upper 30 cm from crumb structure to a well developed, medium angular blocky struc-

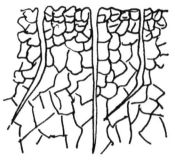 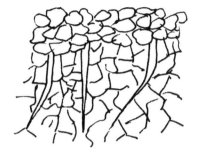

Figure 12. Schematized drawing of soil structures in a Vertisol under rangeland (left) and cropland.

ture with few visible pores inside peds. In the upper 30 cm mottling was observed, which consisted of bleached root channels and ped faces and iron concentrations inside the peds. This type of mottling is indicative of periodic perching of water in surface soil. The dark yellowish brown Oxic Humitropept had a clay loam texture and a thin A horizon on top of a cambia B horizon, containing volcanic tuff and stones. This well-drained, homogeneous soil was porous down to 80 cm depth and had a crumb structure under forest. Just as in the Humult, the Tropept showed mottling patterns under pasture in the upper 30 cm and had strongly developed, angular blocky peds. Results of physical and morphological measurements will be reviewed here only in as much as they relate to the topic of this chapter.

Major differences were observed for measured moisture retention and hydraulic conductivity data when comparing forest and pasture surface soils, but reactions to compaction processes in the two soils were characteristically different (Figures 13 and 14). In the Humult, K_{sat} decreased from 10 m/day to 5 cm/day when comparing forest with pasture. In the Tropept, however, values of 70 cm/day were measured for both types of land use. Porosities in the two Humult soils were virtually identical (Figure 13), but much more water was released in the forest soil when pressure was reduced from saturation to h = -30 cm. This indicates presence of large pores, which is confirmed by the micromorphological analyses which show small aggregates with packing pores in between for the forest soil (Figure 15). Also, the high water filled porosity at saturation for the pasture soil indicates occurrence of puddling as discussed in Section III.B.1, which is also in agreement with the observed, more compact structure. (Figure 15). The two moisture retention curves in the Tropepts are almost identical, which may explain the identical K curves. Use of both physical and micromorphometric techniques allowed a representation of porosities in terms of volumes occupied by three size classes: pores smaller than 30 microns, those between 30 and 5000 microns to be observed and measured in thin sections, and those larger than 5000 microns (Figure 16). Values are presented for a depth range of 5-15 cm which governs the measured conductivity values.

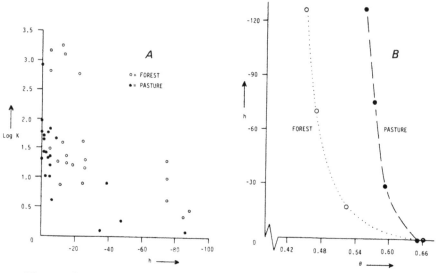

Figure 13. Hydraulic conductivity (A) and moisture retention curves (B) of the Humult topsoils under forest and pasture, showing relations between the hydraulic conductivity K (cm day⁻¹) and the pressure head h (cm) and between h and the volumetric moisture content h (m³ m⁻³), respectively.

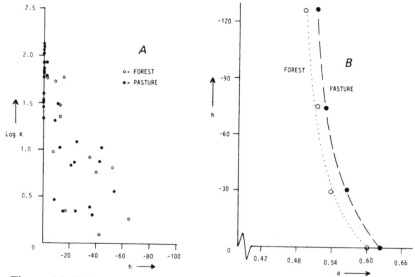

Figure 14. Hydraulic conductivity (A) and moisture retention curves (B) of the Tropept topsoils under forest and pasture.

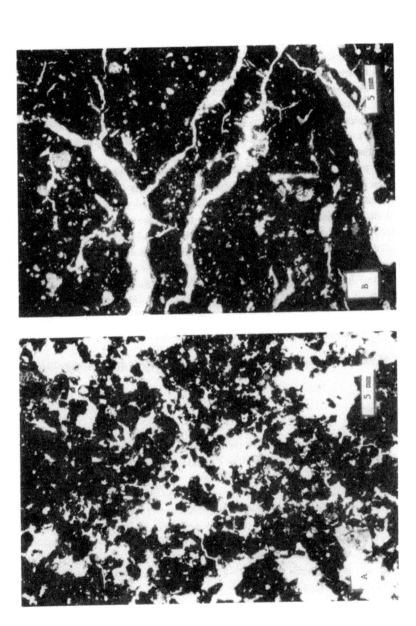

Figure 15. Thin section images of soil structure in a Humoxic Tropohumult under forest (A) and pasture (B).

For the Humult, porosity from 30-5000 microns is reduced from 39 to 6%, while the reduction is less for the Tropept where the change is from 26 to 12%. Measurement of size classes within this range show clear differences between the two soils (Figure 16). No pores between 3000 and 5000 microns occur in the Humult pasture, which shows a clear shift to smaller pore sizes as compared with the forest soil. Changes are much less pronounced for the Tropepts, where the size class between 3000 and 5000 microns is most dominant for both forest and pasture. The measured volumes of pores in the various classes cannot be used directly to calculate K values. However, they allow an explanation of differences obtained among physical measurements which would be difficult to derive without a visualization of pore patterns as can be provided by micromorphological techniques.

This study not only showed differences among hydraulic characteristics due to differences in soil structure following different types of soil management, but it also demonstrated use of combined morphological and physical techniques. In addition, it showed that different soil materials react differently to certain types of soil management. Both soils were subjected to a change from forest to pasture. Even though the old soil had only been in pasture for a period of 3 years, its hydrological properties had deteriorated more than those of the young soil which had been in pasture for 15 years. This difference reflects soil resilience, which appears to be higher in the young soil. Resilience reflects the ability of a soil to bounce back after an event leading to structure deterioration.

IV. Use of Simulation Models to Predict Soil Hydraulic Properties

A. Introduction

Soil hydraulic properties have been interpreted in Section III in terms of moisture retention and hydraulic conductivity data, which are characteristic hydraulic parameters for soils. As such they express static and dynamic properties which are directly related to the pore size distribution and pore continuity. Soil hydraulic properties are generally interpreted in a broader manner, however, as relating to soil water contents in the soil during the seasons. Such data are difficult to obtain because long-term in situ monitoring is necessary, if only because weather conditions play a dominant role in determining water contents in the soil. Fortunately, computer simulation techniques are available now and are being developed to predict soil water contents as a function of weather conditions, crop growth, and water-table fluctuations, if present. To characterize water movement in such models, moisture retention and hydraulic conductivity parameters are needed to predict water contents in space and time as a function of well-defined physical boundary conditions for weather, crop parameters, and water-table dynamics, if relevant. A discussion of simulation models is clearly beyond the scope of this text. Ref-

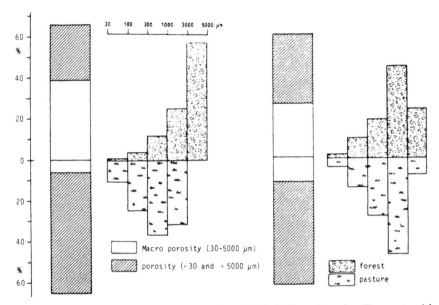

Figure 16. Porosity diagrams for a Humult (left figure) and a Tropept with values obtained by physical (smaller than 30 micron and larger than 5000 micron) and morphometric methods. Numbers above the zero-line characterize forest; below the line numbers represent grassland.

erence is therefore made to Feddes et al., (1988); Bouma (1989); Ritchie and Crum (1989).

Case studies reported in Sections III.B.1 and III.B.2 did include a modeling component, and results will be reviewed in the following subchapter to demonstrate the practical relevance of use of simulation, which is also an effective way to interpret measured physical parameters as discussed in Section III.B.

B. Modeling Soil Water Regimes and Crop Growth in a Typic Haplaquent

Results to be reviewed here were originally reported by Van Lanen et al. (1986) and Feddes et al. (1988).

The model ONZAT was used by Van Lanen et al. (1986) to predict water contents in soil for four soil structure types that were distinguished in Section III.B.1. During the growing seasons of 1982 and 1983 water contents in the soils were periodically measured, and rooting patterns were observed. Simulations for real weather conditions in those years were used to validate the

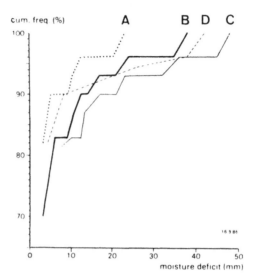

Figure 17. Cumulative frequency distribution of annual moisture deficits for a Typic Haplaquent with four different types of structures, derived from calculations for the period 1955 to 1984. A = grassland; B = primary ploughpan; C = secondary ploughpan; D = loosened ploughpan.

output of the model by comparing measured and calculated water contents. Agreement was considered to be good and after that, the validated model was used to predict water contents and potato growth over a period of 30 years from 1955 to 1984. Simulated data for such a long period allow statistical expressions for important land qualities that are directly related to crop yield. One example is shown in Figure 17 which contains cumulative frequency distributions of annual moisture deficits of potatoes grown on soils with the different types of structure. Use of the grassland structure for simulation involves an exploratory and hypothetical type of procedure, to see what the effect could be of soil structure improvement. The grassland structure is seen in this context as the "best" structure that can be realized in this soil. Soil with the grassland structure shows the lowest deficits, while the secondary ploughpan induces the highest deficits. This is due to better root penetration in the grassland soil than in the soil with the secondary ploughpan which restricts rooting to the upper part of the soil. For example, a deficit of less than 20 mm water has a probability of occurrence of 97% for the grassland structure and 89% for the soil with the secondary ploughpan.

Feddes et al. (1988) made specific simulation runs for the growing season of 1976, which was a dry year. They compared simulated values of soil water contents with measured values and concluded that there was good agreement. The SWACRO model was used to calculate potato yields in loose soil, in soil

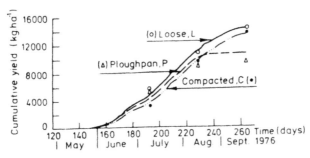

Figure 18. Cumulative dry matter production of tubers for the growing season of 1976 for the three compaction treatments: L, P, and C. Lines represent calculated values by SWACRO; points are measured values.

with a ploughpan, and in soil with a compacted surface. Results (Figure 18) show good agreement between measurements and simulations. Lowest yields were obtained in the ploughpan soil, but initially crop development was good because of relatively high upward unsaturated flow through the ploughpan. The K-curves in Figure 11 indicate that K_{unsat} was indeed higher in the compacted soil than in the loose soil for the relevant pressure head range. However, roots could not penetrate the ploughpan (see Figure 19) and the distance between the bottom of the rootzone and the watertable became too long to allow sufficient upward unsaturated flow. The soil with a compacted surface had slow growth initially due to poor penetration of roots (Figure 18). However, once roots had penetrated the compacted soil, growth increased. Around August 15, the C treatment had a higher growth rate than the P treatment. The C treatment, however, never reached the production level of the L treatment in which roots could well develop from the start of growth. These simulation results demonstrate the influence of both rooting patterns and hydraulic properties of the various soil layers, that were determined by soil management. Differences among treatments are quite significant, as they range from approximately 9000 kg/ha (P) to 13,000 kg/ha (C) and 14,000 kg/ha (L).

This particular case study demonstrates that computer simulation of the soil water regime and associated plant growth can be a useful tool in expressing the effects of soil structure and management on soil hydraulic properties.

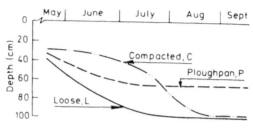

Figure 19. Rooting depths (100% of roots) of the potato crop during the growing season for the three treatments L, P, and C.

V. Conclusions and Recommendations

1. Soil structure can be characterized by physical and morphological measurements. Predictions of basic soil hydraulic properties that are needed for modeling studies, such as moisture retention and hydraulic conductivity, are difficult to derive from such soil-structural measurements. A better procedure would be direct physical measurement of hydraulic conductivity and moisture retention, taking into account soil morphological properties, followed by derivation of relationships between these measured properties and soil characteristics such as texture, bulk density, horizon designations, etc.

2. Combined use of morphological and physical measurement techniques for the study of soil structure is advantageous, because each technique can add unique data to structure characterization. Observed differences in hydraulic conductivity characteristics in studies reviewed here could, for example, be explained by considering pore types, sizes, and continuities as observed with morphological techniques.

3. Differences in soil management have clear effects on soil macrostructure observed in the field within one particular soil series, as was demonstrated for three very different soils in temperate and tropical regions. In turn, management-induced differences in soil structure within a particular soil series resulted in significantly different soil physical properties. When considering soil survey interpretations of certain soil series at detailed farm level, soil structural conditions to be described in the field, should therefore be taken into account to improve interpretations.

4. Many physical methods are available for measuring soil hydraulic properties. However, operational aspects do not always receive sufficient attention. Two methods are reviewed which have been useful in studying hydraulic properties of soil structure in the field. The crust test procedure, as reviewed here, measures unsaturated hydraulic conductivity near saturation in structured soils and is simple to use, and it does not require complicated calculation procedures due to the induced one-dimensional flow system. Bypass flow, which is flow of

free water along macropores through unsaturated soil, is still largely ignored in soil physical theory while it plays a crucial role in well structured soils. A simple field test is described to measure bypass flow.

5. Soil hydraulic properties can be defined in terms of characteristic parameters, such as moisture retention or hydraulic conductivity which are static and dynamic in character. These parameters can be used in simulation models for the soil water regime, when proper physical boundary conditions for the flow system are defined. Different soil structure types in a Haplaquent had characteristically different physical parameters, and simulations showed different physical behavior as well for the soils with the different structures. Thus, the potential of the simulation techniques could be well demonstrated. In a time when extensive field experimentation becomes more difficult because of excessive cost, use of validated simulation models can be an excellent means to generate data on soil water regimes and calculated crop yields for periods of several years. Field experimentation is, however, crucial to allow proper model validation.

6. In the studies that were reviewed we observed and measured different soil structures in different soil series, that were the result of different management. These measurements, in turn, were used to feed simulation models for soil water regimes and the associated crop growth. In the future it would be interesting to "build" different soil structures with the primary soil particles of a given soil material and to explore the effects of the resulting hydraulic parameters on simulated water regimes and crop growth. Thus, soil scientists could define certain "optimal" soil structures by interaction with agronomists or plant physiologists, to be realized in the field by either soil tillage experts or by biological soil management.

7. The different studies reviewed in this chapter illustrate the major importance of climate or weather conditions on structure formation and on soil behavior. Terms, such as soil structure degradation and structure quality, which imply a judgment, should be used with care, because identical structures may behave quite differently in different climates. For example, a compacted layer may be favorable in an arid climate as it increases upward unsaturated flow. In a wet climate, it may be very unfavorable. A grass vegetation and grazing may be favorable for soil structure in a temperate climate where tillage is associated with heavy soil traffic, leading to compaction. However, in a very wet tropical climate grassland and grazing may lead to severe compaction as compared with conditions in the tropical forest. It is therefore advisable to take a process-oriented approach and define soil structure in terms of physical parameters that can be used in simulation modeling. Measurement of these parameters can be improved by morphological characterization of soil structures before measurement: "Look before you measure!".

References

Anderson, J.L. and J. Bouma. 1973. Relationships between hydraulic conductivity and morphometric data of an argillic horizon. *Soil Sci. Soc. Amer. Proc.* 37: 408-413.

Beven, K. and P. Germann. 1982. Macropores and water flow in soils. *Water Resour. Res.* 18: 1311-1325.

Bisdom, E.B.A. 1983. Submicroscopic examination of soils. *Advances in Agronomy.* Vol. 36: 55-96. Academic Press. New York.

Booltink, H.G.W., J. Bouma, and D. Gimenez. 1991. A suction crust infiltrometer for measuring hydraulic conductivity of unsaturated soil near saturation. *Soil Sci. Soc. Amer. J.* 55:566-568.

Boone, F.R., J. Bouma, and L.A.H. de Smet. 1978. A case study on the effect of soil compaction on potato growth in a loamy sand soil. 1. Physical measurements and rooting patterns. *Neth. J. Agric. Sci.* 26: 405-420.

Bouma, J. 1969. Microstructure and stability of two sandy loam soils with different soil management. Agr. Res. Report No. 724, Pudoc, Wageningen. pp. 109.

Bouma, J. 1983. Use of soil survey data to select measurement techniques for hydraulic conductivity. Agr. Water Management 6:177-190.

Bouma, J. 1984. Using soil morphology to develop measurement methods and simulation techniques for water movement in heavy clay soils. In: J. Bouma and P.A.C. Raats (eds.) *Water and Solute Movement in Heavy Clay Soils.* Proceedings of an ISSS Symposium. ILRI, Publ. 37, Wageningen. Neth. 363 pp.

Bouma, J. 1989. Using soil survey data for quantitative land evaluation. In: Stewart, B.A. (Ed.), *Advances in Soil Science,* 9, Springer Verlag, 177-213.

Bouma, J. 1991. Influence of soil macroporosity on environmental quality. *Advances in Agronomy* 46:1-37.

Bouma, J. and J.L. Anderson. 1973. Relationships between soil structure characteristics and hydraulic conductivity. In: *Field Soil Water Regime.* SSSA Special Publication no.5. Soil Science Soc. America, Madison, Wis.

Bouma, J., A. Jongerius, and D. Schoonderbeek. 1979. Calculation of hydraulic conductivity of some saturated clay soils using micromorphometric data. *Soil. Sci. Soc. Am. J.,* 43: 261-265.

Bouma, J., L.W. Dekker, and C.J. Muilwijk. 1981. A field method for measuring short-circuiting in clay soils. *J. Hydrol.* 52 (3/4): 347-354.

Bouma, J., C.F.M. Belmans, and L.W. Dekker. 1982. Water infiltration and redistribution in a silt loam subsoil with vertical worm channels. *Soil Sci. Soc. Am. J.* 46 (5): 917-921.

Brewer, R. 1964. *Fabric and Mineral Analysis of Soils.* John Wiley and Sons. New York.

Brewer, R. 1964. *Fabric and Mineral Analysis of Soils*. John Wiley and Sons. New York.

Bullock, P., N. Fedoroff, A. Jongerius, E. Stoops, T. Tursina, and K. Babel. 1985. *Handbook for Soil Thin Section Description*. Waine Research Publication. Wolverhampton. England.

Burke, W., D. Gabriels, and J. Bouma (Eds.). 1986. *Soil Structure Assessment*. Method Manual Sponsored by the Eur. Comm. Dir. Gen. VI (Agriculture). Balkema. Rotterdam/Boston. 91 p.

Dekker, L.W. and J. Bouma. 1984. Nitrogen leaching during sprinkler irrigation of a Dutch clay soil. *Agr. Water Manag.* 8 (1): 37-47.

Feddes, R.A., M. de Graaf, J. Bouma, and C.D. van Loon. 1988. Simulation of water use and production of potatoes as affected by soil compaction. *Potato Research* 31: 225-239.

Fitzpatrick, E.A., L.A. Mackie, and C.E. Mullins. 1985. The use of plaster of Paris in the study of soil structure. *Soil Use and Manage.* 1 (2): 70-72.

Green, R.E. and J.C. Corey. 1971. Calculation of hydraulic conductivity: A further evaluation of some predictive methods. *Soil Sci.Soc.Amer. Proc.* 35: 3-8.

Gupta, S.C. and W.E. Larson. 1979. Estimating soil water retention characteristics from particle size distribution, organic matter percent and bulk density. *Water Resour.Res.* 15: 1633-1635.

Hillel, D.I. and W.R. Gardner. 1969. Steady infiltration into crust-topped profiles. *Soil Sci.* 107: 137-142.

Jongerius, A. and G. Heintzberger. 1975. Methods in soil micromorphology. A technique for the preparation of large thin sections. Soil Survey Paper no. 10, Netherlands Soil Survey Institute, Wageningen, The Netherlands.

Klute, A. (Ed.) 1986. *Methods of soil analysis*. Part 1: Physical and mineralogical methods, 2nd edn, Agronomy Series, 9, American Society of Agronomy Inc., Madison, Wisconsin.

Kooistra, M.J., J. Bouma, O.H. Boersma, and A. Jager. 1985. Soil structure variation and associated physical properties of some dutch Typic Haplaquents with sandy loam texture. *Geoderma* 36: 215-229.

Lauren, J.G., J.R. Wagenet, J. Bouma, and J.H.M. Wösten. 1987. Variability of saturated hydraulic conductivity in a Glossaquic Hapludalf with macropores. *Soil Science.* 145: 20-28.

Mackie, L.A. 1987. Production of three dimensional representations of soil macropores with a microcomputer. *Geoderma* 40: 275-280.

McKeague, J.A., C. Wang, and G.C. Topp. 1982. Estimating saturated hydraulic conductivity from soil morphology. *Soil Sci. Soc. Amer. J.* 46: 1239-1244.

Miedema, R., Th. Pape, and G.J. van de Waal. 1974. A method to impregnate wet soil samples, producing high quality thin sections, *Neth. J. Agric. Sci.*, 22: 37-39.

Petersen, R.G. and L.D. Calvin. 1965. Sampling. Chapter 5 In: *Methods of Soil Analysis*. Part I. C.A. Black (ed.) Am. Soc. of Agronomy. Agronomy Series no. 9: 54-73.

Ringrose-Voase, A.J. and P. Bullock. 1984. The automatic recognition and measurement of soil pore types by image analysis and computer programs. *J. Soil Sci.* 35: 673-684.

Ritchie, J.T. and J. Crum. 1989. Converting soil survey characterization data into IBSNAT crop model input. In: *Land Qualities in Space and Time*. J. Bouma and A.K. Bregt (Eds.) Pudoc - Wageningen: 155-169.

Schuh, W.M. and M.D.Sweeney. 1986. Particle-size distribution method for estimating unsaturated hydraulic conductivity of sandy soils. *Soil Science* 142: 247-254.

Smaling, E.M.A. and J. Bouma. 1991. Bypass flow and leaching of nitrate in a Kenyan Vertisol at the onset of the major growing season. *Soil Use and Management* 8: 44-48.

Soil Survey Staff. 1951. *Soil Survey Manual*. Agric. Handbook 18. USDA. US Government Printing Office, Washington, D.C., USA.

Soil Survey Staff. 1975. *Soil Taxonomy*. A basic system of soil classification for making and interpreting soil surveys. Agric. Handbook No. 436. USDA. U.S. Governmen Printing Office, Washington, D.C., 754 pp.

Spaans, E.J.A., G.A.M. Baltissen, J. Bouma, R. Miedema, A.L.E. Lansu, D. Schoonderbeek, and W.G. Wielemaker. 1989. Changes in physical properties of young and old volcanic surface soils in Costa Rica after clearing of tropical rain forest. *Hydrological Processes*: 3: 383-392.

Van Baren, J.H.V. and W. Bomer. 1979. Procedures for the collection and preservation of soil profiles. *Technical Paper 1*. ISRIC, Wageningen, The Netherlands. 23 p.

Van Lanen, H.A.J., M.H. Bannink, and J. Bouma. 1986. Use of simulation to assess the effect of different tillage practices on land qualities of a sandy loam soil. *Soil Tillage Research* 10: 347-361.

Van Loon, C.D. and J. Bouma. 1978. A case study on the effect of soil compaction on potato growth in a loamy sand soil. 2. Potato plant responses. *Neth. J. Agric. Sci.* 26: 421-429.

Warner, G.S., J.L. Nieber, I.D. Moore, and R.A. Geise. 1989. Characterizing macropores in soils by computed tomography. *Soil Sci. Soc. Amer. J.* 53: 653-660.

Weibel, E.R. 1979. *Stereological Methods*. Vol. 1. Practical Methods for biological Morphometry. Academic Press, London.

White, R.E. 1985. The influence of macropores on the transport of dissolved and suspended matter through soil. *Advances in Soil Science* 3: 95-121.

Wosten, J.H.M., C.H.J.E. Schuren, J. Bouma, and A. Stein. 1990. Functional sensitivity analysis of four methods to generate soil hydraulic functions. *Soil Sci. Soc. Amer. J.* 54: 827-832.

Physico-Chemical Effects of Salts upon Infiltration and Water Movement in Soils

I. Shainberg and G.J. Levy

I.   Introduction 38
II. Colloidal Properties of Soil Clays: Effect of Soil Solution and
     Exchangeable Cations 38
     A. Effect of Solution Composition on Swelling and
        Dispersion of Clays 38
     B. Swelling and Dispersion in Smectites and Illites 41
     C. Dispersion and Flocculation of Kaolinites 44
     D. Mechanism of Flocculation 45
III. Effect of Soil Exchange Phase Composition and Soil Solution
     Concentration on the Hydraulic Conductivity (HC) 46
     A. Definition and Experimental Background (Saturated Flow) 46
     B. Mechanisms Responsible for Changes in HC 48
     C. Effect of Electrolytes on the HC of Soils 49
     D. Soil Properties Affecting the Response of Sodic Soils to
        Electrolyte Solutions of Medium to High Salinities 54
     E. Response of Sodic Soils to Leaching with Distilled Water
        and Dilute Salt Solution 56
     F. Effect of Exchangeable Magnesium 62
     G. Effect of Exchangeable Potassium 65
     H. Effect of Solution Composition on HC of Kaolinitic Soils 66
     I. Response of Sodic Soils to Readily Dissolving Salts 67
IV. Effect of Soil Exchange Phase Composition and Soil Solution
     Concentration on Infiltration rate (IR) 70
     A. Definition and Experimental Background 70
     B. The Phenomenon of Sealing 71
     C. Effect of Exchangeable Cations on IR of Soils 73
     D. Effect of Electrolyte Concentration on IR of Sodic Soils 77
     E. Effect of Consecutive Rainstorms and Saline Water 79
     F. Effect of Soil Amendments 79
     G. Effect of Electrolyte Concentration on IR of
        Depositional Seals 82
V.   Summary 83
References 84

ISBN 0-87371-889-5

37

I. Introduction

The value and productivity of rainfed or irrigated soils in the arid and semi-arid regions of the world depend on the composition and concentration of the soil solution and the composition of the cations adsorbed on the exchangeable phase of the soil clay. Accumulation of soluble salts in the soil solution imposes a stress on growing crops which may lead to decreased yields and even complete crop failure. Accumulation of dispersive cations such as sodium and, possibly, potassium on the exchange phase may lead to poor soil physicomechanical properties.

The soil fraction responsible for determining the physical behavior of soils is the colloidal clay. This is so because, of the three soil fraction sizes, the clay fraction has the greatest specific surface area and is therefore most active in the physico-chemical processes such as swelling and dispersion. These two processes affect the microstructure of the soil which in turn determines the soils' physical properties, and especially the water transmission properties of the soil which are the concern of this review.

The two common ways to characterize water transmission properties of soil are (i) hydraulic conductivity (HC) and, (ii) infiltration rate (IR). A distinction between these two determinations is essential. HC is usually measured under conditions where the soil surface is not disturbed. Considerable surface disturbance occurs when IR is measured, especially when precipitation or overhead irrigation are involved, leading to seal formation at the soil surface. This, in turn, leads to different water movement in the sealed layer compared with the underlying soil.

The objective of this review is to synthesize information on the physico-chemical effects of salt composition and concentration and exchange phase composition on the water transmission properties of soils.

II. Colloidal Properties of Soil Clays: Effect of Soil Solution and Exchangeable Cations

A. Effect of Solution Composition on Swelling and Dispersion of Clays

The degree of interaction between soil solution and soil particles depends on types and amounts of soil clays. The dominant clay mineral in the semi-arid and arid regions is montmorillonite (a member of the smectite group). Kaolinite is common in the more humid regions while illite is common in both regions. The colloidal properties of clay minerals depend on their specific surface areas (surface area per gram of clay), which depend on the size and shape of the clays. The specific surface areas of montmorillonite, illite, and kaolinite are 750, 120, and 30 $m^2\ g^{-1}$, respectively. This explains why montmorillonite is the most reactive constituent among soil mineral colloids. However, clay reactivity also depends on the cationic composition and on the free electrolyte concentration in the equilibrium solution.

1. Na/Ca Distribution in Smectites

The diffuse double layer at clay surfaces consists of the lattice charge and compensating counterions, which reside in the liquid immediately adjacent to the particles. Counterions are subject to two opposing tendencies: (i) electrostatic attraction to negatively charged clay surfaces and (ii) diffusion from high concentration at the particle surface to low concentration in the bulk solution. The result is an exponentially decreasing counterion concentration from the clay surface to the bulk solution. Divalent ions are attracted to the surface with a force twice as great as that of monovalent ions. Thus, in a divalent ion system, the diffuse double layer is more compressed. With an increase in the electrolyte concentration of the solution, the tendency of counterions to diffuse away from the surface is diminished and the diffuse double layer is further compressed. A complete description of the diffuse double layer theory is presented elsewhere (Bresler et al., 1982; Van Olphen, 1977).

When two clay platelets approach each other, their diffuse counterion atmospheres overlap. Work must be done to overcome the electrical repulsion forces between the two positively charged ionic atmospheres. The electric double layer repulsion force, also called swelling pressure, can be calculated by means of the diffuse double layer theory (Bresler et al., 1982; Van Olphen, 1977). The greater the compression of the ionic atmosphere toward the clay surface, the smaller the overlap of the atmospheres for a given distance between the particles. Consequently, repulsion forces between particles decrease with increasing salt concentration and valence of the adsorbed ions. Because adsorbed Na ions form a more diffuse layer, high swelling pressures develop between Na-montmorillonite platelets, and single platelets tend to persist in dilute solutions (Banin and Lahav, 1968; Shainberg et al., 1971). Conversely, low swelling pressures between Ca-clay platelets and electrical attraction forces between the exchangeable Ca ions and the negative clay surface (forces that are not considered in the diffuse double layer theory) prevent the complete swelling of Ca montmorillonite, even in distilled water, so that Ca platelets aggregate into tactoids or quasi-crystals (Aylmore and Quirk, 1959; Blackmore and Miller, 1961). Each tactoid consists of several (four to nine) clay platelets in parallel array, with an interplatelet distance of 0.9 nm. The exchangeable Ca ions adsorbed on the internal surfaces of the tactoids do not form a diffuse double layer. A diffuse ion layer is present only on the outside of these tactoids (Blackmore and Miller, 1961), and the clay behaves as a system with a much smaller surface area.

In earlier studies of mixtures of monovalent and divalent ions in smectite systems, the two ions were pictured as being mixed at random throughout the exchange complex (Bresler, 1970), which means that for any ratio of Na to Ca

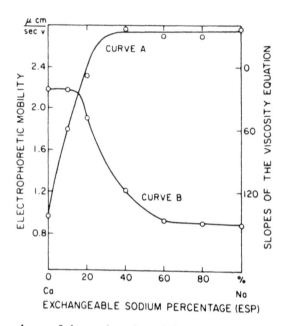

Figure 1. Dependence of electrophoretic mobility (curve A) and the relative size (curve B) of montmorillonite particles on the exchangeable sodium percentage (ESP). (The relative size is expressed in units of the slope in Einstein equation for the viscosity of suspension.) (From Bar-On et al., 1970.)

there are as many Na ions inside the tactoids as on their external surfaces. Mering and Glaeser (1954) hypothesized that "demixing" of the cations occurs so that a few interlayer spaces contain mainly Na ions and others mainly Ca ions. McAtee (1961) also inferred, from X-ray diffraction patterns of clay saturated with mixtures of mono- and divalent ions, that "demixing" had occurred.

Using viscosity and light transmission measurements, Shainberg and Otoh (1968) studied the size and shape of montmorillonite particles saturated with a mixture of Na and Ca ions in the adsorbed phase. They found (Figure 1, curve B) that the introduction of a small percentage of Na onto the exchange complex of Ca tactoids was not sufficient to break the tactoid apart, but more Na (>20%) resulted in tactoid breakdown. The platelets were completely separated when 50-60% of the adsorbed Ca was replaced by Na. The location of the adsorbed ions in montmorillonite saturated with a mixture of mono- and divalent cations was estimated by Bar-On et al. (1970) through the measurement of the electrophoretic mobility of clay particles in suspension (Figure 1, curve A). The addition of a small amount of exchangeable Na to Ca saturated clay had a considerable effect on the electrophoretic mobility. The fact that the size of the Ca tactoids was not affected by a low exchangeable sodium percentage (ESP)

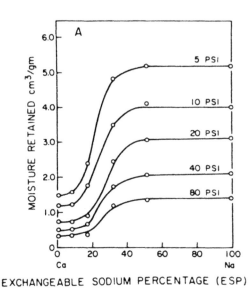

Figure 2. Water retention as a function of ESP and pressure applied on montmorillonite. (From Shainberg et al., 1971.)

(Figure 1, curve B), whereas the electrophoretic mobility increased rapidly, supports the "demixing" model. At low ESP values, the adsorbed Na concentrates on the external surfaces of the tactoids and their electrophoretic mobility increases rapidly. As the ESP of the system increases, Na penetrates into the tactoids and causes their disintegration.

B. Swelling and Dispersion in Smectites and Illites

1. Na-Ca Systems

The volume of water retained by montmorillonite as a function of ESP for various consolidation pressures is presented in Figure 2. Clay swelling is affected only slightly by an increase in ESP at values less than 15. These results compare favorably with those of McNeal and Coleman (1966) on soil clays, in which replacing about 15% of adsorbed Ca with Na had a little effect on clay swelling. With a further increase in ESP, a very sharp increase in macroscopic swelling was found for two montmorillonitic soils.

In a stable clay suspension, dispersed particles frequently collide because of Brownian motion, but separate again because of diffuse double layer forces. When salt is added to the clay suspension, particles adhere upon collision, forming flocs that settle. The suspension is then separated into bottom sediment and particle-free supernatant liquid. The minimum electrolyte concentration that causes flocculation is referred to as the flocculation value, which depends on counterion valency. Flocculation values for Na and Ca montmorillonite are 7-20 $mmol_c$ l^{-1} NaCl and 0.25-1.09 $mmol_c$ l^{-1} $CaCl_2$, respectively (Goldberg and Glaubig, 1987 and references cited therein), and for Na and Ca illites they are 40-50 $mmol_c$ l^{-1} NaCl and 0.25 $mmol_c$ l^{-1} $CaCl_2$, respectively (Arora and Coleman, 1979).

Oster et al., (1980) found that a small increase in ESP of a Na/Ca montmorillonite had a considerable effect on the flocculation value. Similar effects of exchangeable Na on the electrophoretic mobility of Ca montmorillonite were reported (Figure 1, curve A). According to Verwey and Overbeek's stability theory (Van Olphen, 1977), these two properties are interrelated and both are explained by the "demixing" model, which postulates that Na is the predominant cation on the external surfaces.

The effect of exchangeable Na on the flocculation value of Ca-illite suspensions follows more closely a simple linear relationship (Oster et al., 1980), which suggests that "demixing" in Na/Ca illite is not as pronounced as in Na/Ca montmorillonite. However, illite at low ESP was more dispersed than the corresponding montmorillonite (Oster et al., 1980). Illite particles consist of a number of platelets stacked to a thickness of 10 nm with a specific surface area of 120 m^2 g^{-1} (for Fithian illite) as compared with a specific surface area of 750 m^2 g^{-1} for montmorillonite. Electron micrographs reveal that illite particles have irregular surfaces and that the planar surfaces are terraced. Upon close approach, the unavoidable mismatch of the terraces leads to poor contact between the edges and the surfaces, leading to smaller edge-to-face attraction forces and, consequently, a higher flocculation value for Na-illite compared with Na-montmorillonite (Oster et al., 1980).

2. Effect of Magnesium

There are conflicting opinions regarding the importance of exchangeable Mg on swelling and dispersion of soil clays. Magnesium is a more hydrated ion than calcium (Norrish, 1954) and its presence causes a thicker diffuse layer (Shainberg and Kemper, 1967). Thus, it would be expected that a Na-Mg soil would deteriorate more than a Na-Ca clay with the same ESP, when subjected to similar solution concentrations. Rahman and Rowell (1979) found for montmorillonite that swelling depended solely on the ESP and electrolyte concentration. Conversely, illite, and kaolinite showed a specific effect of Mg (Bakker and Emerson, 1973; Emerson and Smith, 1970). Mg-soil (kaolinite) dispersed in water when remolded at a water content of 15 % by weight, whereas the Ca-soil started to disperse at 20 % (Figure 3). For Mg-illites a lower ESP was required to initiate dispersion of the dry clay when immersed in water than for Ca-illites. By comparing the Ca-Mg and Na-Ca forms it was deduced that the percentage of exchangeable Mg that caused the same degree of dispersion in water was ten times that of exchangeable Na (Figure 4).

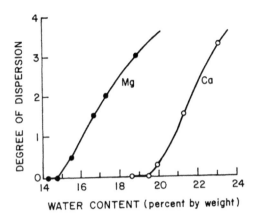

Figure 3. The visual dispersion in water of surface aggregates of a red soil remolded at various water contents. (From Emerson and Smith, 1970.)

3. Effect of Potassium

There is some evidence that K-saturated montmorillonite does not behave as predicted by the double-layer theory. According to Rowell (1963) Na-montmorillonite is fully dispersed and upon swelling forms a homogeneous gel, while K-montmorillonite forms a non-homogeneous gel with many regions where the particles do not swell but behave in a manner similar to illitic clays.

In a mixture of cations, and particularly where the exchangeable potassium percentage (EPP) is low, potassium ions are preferentially adsorbed at regions with high charge density. The fixation effect is apparent and potassium does not behave as a monovalent dispersive cation (Shainberg et al., 1987a).

4. Effect of Hydroxy-Al and -Fe

Aluminum and Fe are common in soils and their hydroxides and oxides occur as coatings on clay minerals (Rich, 1968). The interactions between the Al and Fe species and the clay are complex because of the tendency of these species to hydrolyze, polymerize, and eventually precipitate in the presence of clays. It is generally accepted that the various forms of Al and Fe present in the soil promote clay flocculation and reduce clay swelling and dispersion under sodic conditions (Deshpande et al., 1968; El Rayah and Rowell, 1973; Frenkel and Shainberg, 1980, Goldberg and Glaubig, 1987, etc.). Oades (1984) studied the interactions between Al and Fe species and clays, and the effects thereof on clay flocculation and water uptake. Both types of polycations showed similar efficacy in flocculating various clays and soil clay. However, Al polycations were significantly more efficient than Fe polycations in decreasing water uptake and swelling in montmorillonite. Oades (1984) suggested that this difference was a result of polymer morphology. Al polycations have a planar shape (~ 5 Å thick)

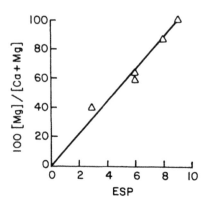

Figure 4. A comparison of the percentage of exchangeable magnesium and sodium causing the same degree of dispersion in water of dry calcium-magnesium and calcium-sodium Muloorima clay aggregates, respectively. (From Emerson and Chi, 1977.)

and can therefore neutralize clay layer charge more completely than can the Fe polycations, being spheres 30-100 Å in diameter. An additional mechanism by which Hydroxy-Al and -Fe increase clay stability was proposed by Keren (1980) and Shainberg et al. (1987c). They observed that in the presence of hydroxy-Al and -Fe the affinity of the clay for Ca increased, thereby reducing the amount of adsorbed Na and its deleterious effect on clay swelling and dispersion.

C. Dispersion and Flocculation of Kaolinites

Schofield and Samson (1954) studied the behavior of kaolinite in suspension and found that a pure Na-kaolinite flocculated at pH < 7 under conditions that dispersed illite and montmorillonite. Acidified kaolinite retained exchangeable Na, which showed that the crystals were negatively charged. When shaken in dilute NaCl solutions, Cl⁻ was positively adsorbed, which proved that the edges were positively charged. The attraction of the positive charges on the edges for the negative charges on the planar surfaces was regarded as the cause for flocculation, which occurred in the absence of salt. Deflocculation of the salt-free suspension occurred on the addition of NaOH. At pH > 8, the crystals negatively adsorbed Cl, which showed that the edges were no longer positively charged. Thus, eliminating the positive charge at the edges dispersed the clay.

Deflocculation of kaolinite suspension could also be brought about by adding small amounts of sodium oxalate, sodium pyrophosphate, sodium polymetaphosphate, sodium alginate, and sodium bentonite (Schofield and Samson, 1954;

Durgin and Chaney, 1984). In a recent study, Levy et al. (1991) found that by adding small amounts of Na-hexametaphosphate (10 mg l⁻¹) to kaolinite the response of the clay to sodic conditions was similar to that of smectite. The negatively charged polyanions are adsorbed on the positively charged edges, and thus prevent edge-to-face flocculation. A high increase in the flocculation value of Na-kaolinite was observed with increasing amounts (up to 30%) of Na-smectite (Arora and Coleman, 1979; Schofield and Samson, 1954). They attributed this strong effect to the disposition of smectite platelets on positively charged edges of kaolinite. Similarly, Frenkel et al. (1978) found that the HC of a kaolinitic soil from North Carolina (with 10.6% clay) was not affected by 20% Na on the exchange complex. However, when the soil was mixed with 2% montmorillonite (about 20% of the clay content), the mixture was very susceptible to dispersion. Chiang et al. (1987) reached a similar conclusion.

The effect of natural organic molecules on kaolinite dispersion was studied by Durgin and Chaney (1984) and Shanmuganathan and Oades (1983a). Durgin and Chaney (1984) studied the organic constituents of water extracts from Douglas fir roots that cause kaolinite dispersion. They found that aliphatic and aromatic carboxylic acids and organics with molecular sizes greater than 10^4 nominal molecular weight were most effective in causing dispersion. Shanmuga-nathan and Oades (1983a,b) studied the physical properties of kaolinitic soils with a net charge of zero and soils with a net positive charge (manipulated by poly Fe(III)-OH cations). They distinguished three groups of anions in order of effectiveness with respect to dispersion of the soil: phosphate = fluvate > citrate = oxalate = tartrate > salicylate = lactate. This order was the same as that for the amount of anion adsorption by the soil, which is controlled by the energy of binding of the anions to the Fe(III) associated with the clay surfaces (Shanmuganathan and Oades, 1983a, and references cited therein). The specific adsorption of the anions on the clay edges increases the surface charge density. It can also be regarded as decreasing the zero point of charge (ZPC) of the clay, i.e., make the clay more negative and thereby promote dispersion (Shanmuga-nathan and Oades, 1983a).

D. Mechanism of Flocculation

Association between the positive edge surface of one particle and the flat negative oxygen surface of another was the mechanism proposed by Schofield and Samson (1954) to explain and describe the flocculation of kaolinite. Van Olphen (1977) postulated similar positively charged sites on montmorillonite edges and edge-to-surface attraction to explain montmorillonite flocculation. The edge-to-face attraction hypothesis is supported by the following observations. First, the open structure of flocculated Na-montmorillonite gel contains about 250 g of water per gram of clay, which corresponds to a film thickness of 330 nm. This thickness exceeds the range of forces of the diffuse double layer and/or van der Waals attraction forces. Thus, only edge-to-face attraction forces are possible. Second, addition of small amounts of Na polymetaphosphate to Na-montmorillonite suspensions increased very sharply the flocculation value of the suspension. Chemisorption of the polyanion at the edges, and reversal of the positive edges charge to negative, may explain the effect of the dispersant on flocculation. Third, the flocculation value of Na-montmorillonite increases with increasing pH (Goldberg and Glaubig, 1987). With increase in pH, the charge on the edges becomes negative, and the suspension becomes more stable.

Several experimental observations oppose and contradict the edge-to-face mechanism in smectites. If edge-to-face association is prevented, the open structure of the gel should collapse. Thus, the gel volume of flocculated Na-montmorillonite containing polymetaphosphate or at high pH should be small. Experimentally, the open structure of the clay gel was maintained (Keren et al., 1988; Oster et al., 1980). Hence, another association mode was suggested by Keren et al. (1988), based on the fact that the measured charge density at a smectite surface is only an average value. The actual charge density is quite heterogeneous (Stul and Mortier, 1974) with regions on the clay where the charge density is twice the average value and is similar to the charge density on vermiculite. On surfaces with high charge density (vermiculite), there is a net electrostatic attraction force between the platelets (Norrish, 1954). Because the montmorillonite platelets are long, flexible, and have the capability of bending (Shomer and Mingelgrin, 1978), one platelet may simultaneously form a cohesive junction point with other platelets at several locations of high charge densities to form an open network with high gel volume. Thus, at neutral and low pH, edge-to-face association is the dominant mode of association in Na-montmorillonite gel. At high pH or in the presence of polyanions that neutralize the charge on the edge, face-to-face association in regions with high charge densities is taking place (Keren et al., 1988). To enable the electrostatic attraction between regions of high charge densities, the diffuse double layer must be further compressed, and higher electrolyte concentration is needed to coagulate the clay.

III. The Effect of Soil Exchange Phase Composition and Soil Solution Concentration on the Hydraulic Conductivity (HC)

A. Definition and Experimental Background (Saturated Flow)

One of the most common parameters used to quantify changes in the physical properties of the soil and especially water flow through soil is the hydraulic conductivity. Water flow in soil takes place in accordance with Darcy's equation (Hillel, 1980a):

Physico-Chemical Effects of Salts

$$q = \frac{V}{At} = K \frac{\Delta H}{L} \tag{1}$$

where q = flux (mm h^{-1}), V = volume of water passing through a soil column (mm^3), a = cross sectional area of the soil column (mm^2), t = time elapsedΔH = hydraulic head difference (mm), L = length of soil column (mm). Thus ΔH/L is the hydraulic head gradient and K = hydraulic conductivity (mm h^{-1}). (h),HC of the soil depends on the properties of the soil as well as the properties of the percolating fluid. With the objective of eliminating the fluid properties from the conductivity parameter, the intrinsic soil permeability, k, is defined (Hillel, 1980a):

$$k = \frac{K\eta}{\rho g} \tag{2}$$

where k = intrinsic permeability (m^2), K = hydraulic conductivity (m s^{-1}), η = fluid viscosity (N s m^{-2}), ρ = fluid density (kg m^{-3}), and g = gravitational acceleration (m s^{-2}).

A number of models have been proposed for describing the relations between the structure of a porous material and its permeability. In 1927, Kozeny (according to Hillel, 1980a) used the mean hydraulic radius (cross-section divided by wetted perimeter) as given by e / S where e is the porosity and S is the surface area of the particles per unit volume of porous material. His equation is:

$$k = a^{-1} e^3 s^{-2} \tag{3}$$

The empirical factor a was considered to be a constant that combined shape and tortuosity factors and was stated to be approximately equal to 5. With the development of methods for measuring the size distribution of pores, equations describing the effects of size of pores on permeability were introduced (Childs and Collis George, 1950; Marshall, 1958).

These types of models, however, do not take into account the effect of fluid properties on the soil matrix geometry. Swelling and dispersion of clay in response to electrolyte composition and concentration in the percolating solution change the size of the conducting pores and hence affect HC. A semi-empirical approach along these lines was proposed by McNeal (1968) who described saturated HC with the equation:

$$1-r = fx^n / (1+fx^n) \qquad (4)$$

where r is the ratio of HC when leaching the soil with a given solution to HC of divalent cation-saturated soil leached with a high salt concentration solution, f is an empirical constant characteristic of the soil, x is a colloid swelling factor (based upon a demixed ion swelling model), and n is a factor which varies with the estimated soil ESP.

The most common measure of the effect of exchangeable cations and salinity on soil permeability is the determination of relative HC values in the laboratory. High salt concentration solutions are used to measure the inherent permeability of the soil matrix, and to assure exchange equilibrium between the solution and the adsorbed phase. Thereafter, the soil column is leached with a series of percolating solutions of decreasing concentrations, but of appropriate composition to maintain a given exchangeable cation composition (e.g., a given sodium adsorption ratio, SAR). The measurements are carried out on fragmented samples which have been dried, ground, passed through a standard sieve, and packed in a uniform manner. Samples are also commonly treated with CO_2 or wetted under suction and treated with a biological inhibitor to remove most of the influence of entrapped air and microbial activity on time-dependent soil HC changes. A filter paper, or a thin layer of washed sand, is placed on the soil surface to avoid possible disturbance when solutions are added to or taken from the soil column.

B. Mechanisms Responsible for the Changes in HC

Two main mechanisms have been proposed to explain the decrease in soil permeability. Quirk and Schofield (1955) suggested that the swelling of clay particles, which increases with an increase in clay sodicity, could result in blocking or partial blocking of the conducting pores. Rowell et al. (1969) showed that the initial reductions in HC could be attributed to swelling. McNeal et al. (1966) found a linear relationship between HC reduction and macroscopic swelling of extracted soil clays with correlation coefficients varying from 0.94 to 0.97. They were able to relate clay swelling measurement to theoretically calculated interlayer swelling values.

Deflocculation and dispersion were proposed as the second main mechanism (Quirk and Schofield, 1955). This mechanism operates when the charged plates have separated enough so that the attraction forces are no longer strong enough to oppose repulsion forces and hence the clay platelets detach by an external force. Opinions differ, however, concerning how the dispersed clay platelets affect the HC. Studying micrographs obtained from a scanning electron microscope, Chen and Banin (1975) concluded that the dispersed clay particles were locally rearranged to create a continuous network in the pores, thus decreasing the HC. In other studies, macroscopic movement of the dispersed clay platelets and their plugging the soil pores was considered as the mechanism controlling the reduction in HC (Felhendler et al., 1974; Pupisky and Shainberg, 1979). The importance of dispersion in affecting soil permeability was also recognized in other studies (Frenkel et al., 1978; Rhodes and Ingvalson, 1969; Shainberg et al., 1981a,b).

The differences between swelling and dispersion processes are important. Swelling of reference clays (Shainberg et al., 1971) and clay in soil (Emerson and Bakker, 1973) is not greatly affected by low ESP values (below 10 to 15) but increases markedly as the ESP increases above 15 (see previous section). Conversely, dispersion of clay is very sensitive to low levels of sodicity, and increases markedly at the low ESP range (Figure 1). For soil clay, dispersion occurred at low ESP, in the range of 3-5 (Emerson and Bakker, 1973). Deflocculation and dispersion of clay are possible only at solution concentrations below the flocculation value, but swelling is a continuous process and decreases gradually with increases in solution concentration. Since the flocculation values of montmorillonite clay with ESP 10 is 4 mmol$_c$ l^{-1}, it may be concluded that the dispersion mechanism can operate only at very dilute salt solutions, and under these conditions it is operative even at low ESP values. Swelling is essentially a reversible process and adding electrolytes or divalent cations should recover the permeability of a soil. Dispersion and particle movement are essentially irreversible and may lead to the formation of an impermeable clay layer in the soil profile.

Figure 5 shows the importance of particle dispersion and movement as compared with swelling. In this experiment, a loamy soil of 20% montmorillonitic clay and a clay soil of 50% montmorillonitic clay were equilibrated by leaching with 0.02 mmol$_c$ l^{-1} solution of SAR 10. Following equilibration, the soils were leached with distilled water (DW). The loamy soil was completely sealed upon leaching with DW, but the clay soil maintained a relative HC of about 20 percent. In the calcareous clay soil (Grumusol) leached with DW salt concentration in the soil solution was maintained at 4 mmol$_c$ l^{-1} by dissolution of carbonates. At ESP 10, this concentration prevented clay dispersion, and the HC decrease was due only to clay swelling. Conversely, in the noncalcareous loamy soil (Netanya), soil solution concentration was lowered below 1 mmol$_c$ l^{-1}; thus clay dispersion occurred and the soil was totally sealed.

C. Effect of Electrolytes on the HC of Soils

Permeability of soil to water depends on the ESP of the soil and on the salt concentration (Quirk and Schofield, 1955; McNeal et al., 1966, 1968; Yaron and Thomas, 1968; Frenkel et al., 1978). The higher the proportion of exchangeable Na and the lower the electrolyte concentration of the percolate, the larger the reduction in HC. To describe the relationship between HC and solution composition, Quirk and Schofield (1955) developed the concept of

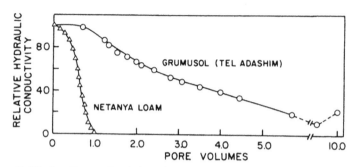

Figure 5. Hydraulic conductivity of two montmorillonitic soils having an ESP of 10 and leached with distilled water. (From Shainberg and Oster, 1978.)

"threshold concentration". This is the concentration of salts in the percolating solutions which causes a 10 to 15% decrease in the soil permeability at a given ESP value. Soil permeability can be maintained, even at high ESP values, provided that the salt concentration of the water is above a critical (threshold) level. For their particular soil, a noncalcareous silty loam, the threshold concentration values were approximately 0.6 mmol$_c$ l^{-1} for the Ca-soil, 2.3 and 9.5 mmol$_c$ l^{-1} for soils with ESP of 5.8 and 21, respectively, and 250 mmol$_c$ l^{-1} for Na-soil. According to Quirk and Schofield (1955), even Ca-soil may show a reduction in HC, provided that the salt concentration is below 0.6 mmol$_c$ l^{-1}. This has been verified by Emerson and Chi (1977), who observed dispersion of soil saturated with divalent cations at salt concentrations below 0.5 to 2 mmol$_c$ l^{-1}. When water of salinity below 1 mmol$_c$ l^{-1} (rain or snow water) is applied to a soil, even Ca-soil and soils of low ESP may disperse and lose some permeability.

The basic approach of Quirk and Schofield (1955) has been extended to a large number of additional soils by workers such as McNeal and Coleman (1966) and Cass and Sumner (1982a,b). In most of these studies, the electrolyte concentrations of the percolating solutions were maintained above 3 mmol$_c$ l^{-1} (e.g., McNeal and Coleman, 1966; McNeal et al., 1966, 1968; Yaron and Thomas, 1968; Rhoades and Ingvalson, 1969) and the swelling mechanism controlled the HC of the soils. Typical effects of salt concentration and ESP on soil HC are demonstrated in Figure 6. In their studies, McNeal and Coleman (1966), showed that properties of the solution (i.e., salt concentration and SAR) alone were not sufficient to characterize the HC of different soils upon leaching with a given solution. McNeal and Coleman (1966) further found that agreement among soils with similar clay mineralogy was improved when similar changes in the HC of soils were characterized by different combinations of salt concentration and ESP, rather than SAR (Figure 7). Equivalent salt solution series, defined as solutions with combinations of SAR and salt concentration

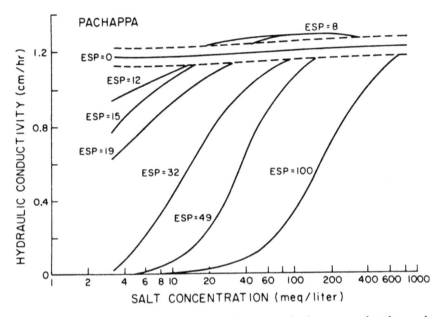

Figure 6. Hydraulic conductivity of Pachappa sandy loam as related to salt concentration and ESP. (From McNeal and Coleman, 1966.)

producing the same extent of clay swelling in a given soil, was used for predicting changes in the soil's HC (Jayawardane, 1977, 1979). In a recent study, Jayawardane and Blackwell (1991) showed a high positive correlation between the log transformation of salt concentration and SAR of equivalent salt solutions from the following equation:

$$\log E_c = \log a_1 + b_1 \log SAR \tag{5}$$

where E_c is salt concentration (mmol$_c$ l^{-1}), and a_1 and b_1 are constants for each equivalent salt solution series for a given soil. Jayawardane and Blackwell (1991) also found that the same set of equivalent salt series (i.e., the same constants a_1 and b_1 from the series) was satisfactorily applied to two soils from Tasmania and three soils from the USA, which varied in their clay content and mineralogy. Thus, these investigators suggested that a single set of equivalent salt solutions values could be applied to a large group of soils and enable predicting changes in their HC. Since this concept considers clay swelling the cause for changes in HC, its possible validity is limited to solutions where E_c is >3 mmol$_c$ l^{-1}.

Russo and Bresler (1977a, 1980) tested the effects of mixed Na-Ca salt solutions on soil water diffusivity and unsaturated HC for a loamy soil from Gilat, Israel. When Ca-saturated, the hydraulic properties of this soil were

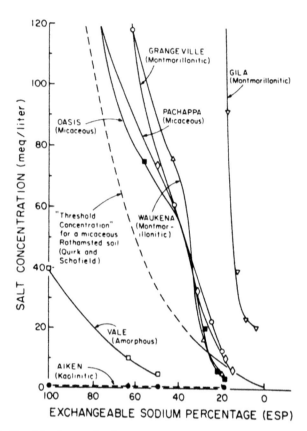

Figure 7. Combinations of salt concentration and ESP required to produce a
25 % reduction in hydraulic conductivity for selected soils. (From McNeal and
Coleman, 1966.)

independent of solution concentration. However, for Na-Ca systems, the
diffusivity and HC decreased either as the solution concentration decreased or
as the SAR increased. The negative effect of high SAR and low solution
concentration decreased with decreasing soil water content (Russo and Bresler,
1977a, 1980). A similar study was recently carried out by Lima et al. (1990) on
a Yolo loam, a moderately fine-textured soil from California. They too,
observed that as the SAR increased and the solution concentration decreased, the
HC, diffusivity, and sorptivity decreased, whereas water retention increased.
Lima et al. (1990) showed the increase in nonlinearity of the log HC function
with the increase in salinity, especially for larger water content values (Figure
8). They ascribed this phenomena to the predominance of salt's effect on size
of larger pores. Extrapolation of the log HC curves in Figure 8 to values of HC
at saturation, yielded values greater than those obtained in independent measurements. The smaller the SAR and the more concentrated the leaching solution,

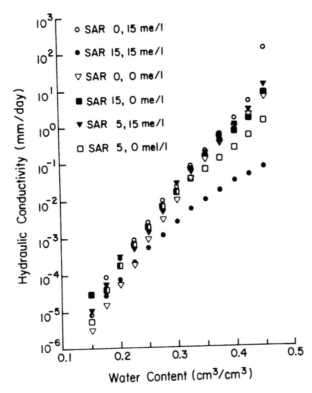

Figure 8. Salinity effects on hydraulic conductivity. (From Lima et al., 1990.)

the larger the discrepancy between the extrapolated and measured HC values. Lima et al. (1990) suggested that this difference was due to a lack of chemical equilibrium, between the soil and the leaching solution in the diffusivity studies, which the calculated unsaturated HC curves were based on. Conversely, a chemical equilibrium was attained in the saturated HC measurements.

In the aforementioned studies, soil properties significantly affecting the response of soil permeability to sodicity and electrolyte concentration seemed to be clay content, clay mineralogy, iron or aluminum oxide content, organic matter content, and bulk density. However, other soil properties are very important in determining the response of sodic soils when the soils are leached with DW (simulating rain and snow water). In the following discussion a separation is therefore made between percolating solutions of low concentration (below 3 mmol$_c$ l^{-1}) and those having concentrations above 3 mmol$_c$ l^{-1}. This separation is justified in terms of the mechanisms responsible for the HC decreases as discussed in the previous chapter.

D. Soil Properties Affecting the Response of Sodic Soils to Electrolyte Solutions of Medium to High Salinities

1. Effect of Soil Texture

McNeal et al. (1968) showed that for soils having variable clay content but nearly uniform clay mineralogy fraction (a clay fraction consisting of 42% montmorillonite, 29% mica, 16% quartz plus feldspars, and 13% of other species including chlorite, vermiculite, and amorphous minerals), HC reductions caused by increases in exchangeable sodium or decreases in total electrolyte concentration were greater for soils having higher clay content. This effect was in addition to the decreases in absolute HC associated with increases in clay content. Frenkel et al. (1978) also reported that the susceptibility of soils to sodic conditions increased with an increase in clay content.

2. Effect of Clay Mineralogy

McNeal and Coleman (1966) and Yaron and Thomas (1968) concluded that the most labile soils were those high in 2:1 layer silicates, especially montmorillonite, and the least labile were those high in kaolinite and sesquioxides. McNeal et al. (1966) found a good correlation between relative soil permeability and the swelling of extracted soil clays, which in turn was affected by soil solution concentration and composition. Several workers have used montmorillonite swelling behavior to predict HC response to solutions of various salt concentrations and ionic compositions. Lagerwerff et al. (1969) coupled predictions of clay swelling with the Kozeny equation (Equation 3) to estimate HC decreases during expansion of the soil matrix. However, being unable to calculate *a priori* the electrical potential at the midplane between swelling soil particles in mixed-cation systems, they used the midplane electrical potential as an empirical index which was allowed to vary until a reasonable fit was provided between theoretical and experimental values. A semi-empirical approach was taken by McNeal et al. (1968), who described saturated permeability as a function of salt composition and concentration, and of colloid swelling (Equation 4).

Russo and Bresler (1977b) developed a model for montmorillonitic soils to predict the dependence of HC on moisture content, SAR, and solution concentration. The model was based on the diffuse double layer theory for swelling and on the Marshall equation (Marshall, 1958) for estimating HC as a function of moisture content. The model was sensitive to the relationship between number of platelets per clay particle (N) and the ESP of the soil. Russo and Bresler (1977b) used N-ESP relationships of Shainberg and Otoh (1968), Shainberg and Kaiserman (1969) and Dufey et al. (1976) and obtained relatively good agreement between model calculations and experimental data.

Frenkel et al. (1978), studying the response of kaolinitic soils to sodic conditions, found that the HC values of these soils were less affected by the ESP and the electrolyte concentration than were those of montmorillonitic soils.

Similarly, Rhoades and Ingvalson (1969) concluded that a much higher ESP was needed to appreciably reduce HC for vermiculitic as compared to montmorillonitic soils.

It is evident that swelling was considered the main mechanism for HC decreases as the ESP of the soil increased and the salt concentration decreased. The models and experimental results showed that the effect was normally greatest for soils having high contents of swelling minerals (such as smectites) and for soils having high clay contents.

3. Effect of Aluminum and Iron Oxides

Studying the effect of sodicity and electrolyte concentration on a group of soils from the Hawaiian Islands, McNeal et al. (1968) found that despite the high textural clay contents in these highly weathered and hydrous oxide-stabilized soils, the effect of ESP and electrolyte concentration on the HC was essentially negligible until iron-and aluminum-containing hydrous oxides were chemically removed. El Swaify (1973), studying tropical soils, arrived at similar conclusions. McNeal's (1968) conclusion was that the cementing action of iron oxides prevented swelling and dispersion. Deshpande et al. (1968) concluded that aluminum oxides, rather than iron oxides, were more effective in preventing soil swelling and dispersion during leaching with sodic water. Many studies with montmorillonite clay supported this observation (Alperovitch et al., 1985; El-Rayah and Rowell, 1973; Oades, 1984; Keren and Singer, 1989). Alperovitch et al. (1985) proposed that the high efficiency of hydroxy-Al in maintaining high HC (i.e., preventing clay swelling and dispersion) was due to its high charge density and planar structure. The spherical shape and low charge of hydroxy-Fe reduced its ability to prevent swelling and dispersion (Oades, 1984).

Changes in the pH were found to have similar effects on the efficiency of hydroxy-Al and -Fe polymers in maintaining high HC in sand-smectite mixtures under sodic conditions (Keren and Singer, 1990). For both systems higher HC was observed at pH 5.5 than at pH 7.5. Keren and Singer (1990) ascribed the latter results to the fact that the net electrostatic charge for hydroxy polymers increases with a decrease in pH at the range lower than their point of zero charge, which is 9.4 and 8.1 for hydroxy-Al and -Fe, respectively (Kinniburgh et al., 1975). Shainberg et al. (1987c) compared the effect of $FeCl_3$ to that of $AlCl_3$ on the HC of a sandy loam soil under sodic conditions and observed, opposite to the aforementioned studies that Fe was better than Al in maintaining high HC. They hypothesized that both Al and Fe may act in coordination with organic matter to stabilize clays as well as directly on the clay. Iron may be more effective than aluminum in the former case (Shainberg et al., 1987c). The increase in the HC, relative to the control, of a smectitic clayey soil treated with different species of Al was in the following decreasing order: $Al^{3+} > Al(OH)^{2+} > Al(OH)^+_2 > Al(OH)_3$ (Wada and Beppu, 1989). However, Wada and Beppu (1989) observed that treatment with Al^{3+} resulted in a higher acidity and a larger decrease in exchangeable bases compared with the polymeric hydroxy-Al

I. Shainberg and G.J. Levy

cations. From their results, Wada and Beppu (1989) concluded that the number of electric charges on the Al atom is more important than the size of the Al and hydroxy-Al cations on the bonding between clay particles.

4. Effect of Soil Density

Frenkel et al. (1978) studied the effect of soil bulk density on the HC decreases resulting from sodic and electrolyte concentration, and concluded that soil sensitivity to excessive ESP and low electrolyte concentration increased with increase in bulk density.

E. Response of Sodic Soils to Leaching with Distilled Water and Dilute Salt Solution

1. Effect of Leaching with Distilled Water

Despite the obvious effects of electrolytes on soil permeability, the main emphasis has been given to the exchangeable sodium percentage and electrical conductivity of the irrigation water as an index to the degree to which sodium will occupy exchange positions on the clay surface. Soils in arid and semiarid regions often have high ESP before being irrigated and for this reason the quantity of electrolyte in the irrigation water should be given greater consideration.

Although the above paragraph was written in 1955 (Quirk and Schofield, 1955), it is still pertinent today. Even in arid regions where irrigation is essential for maintaining agricultural production, intermittent rain may lower the electrolyte concentration below the flocculation value determined by the soil properties. In Israel, where water having a SAR value of 26 and an electrical conductivity (EC) of 4.6 dS m^{-1} is used for commercial irrigation, no permeability problems during the irrigation season (summer) were observed (Frenkel and Shainberg, 1975). The electrolyte concentration in the irrigation water was enough to prevent the dispersive effect of the sodium. However, upon applying distilled water (to simulate rain water in winter), the soil dispersed and its permeability decreased to 26% of the initial 9-12 mm h^{-1} value (Frenkel and Shainberg, 1975). Similarly, in some areas in the Central Valley of California waters of very low salinity (0.05 to 0.2 dS m^{-1}) are used for irrigation. The electrolyte concentration of these waters is below the threshold concentration typical to the soils and permeability problems often arise (Mohammed et al., 1979).

Felhendler et al. (1974) measured the HC of two montmorillonite soils (a sandy loam and a silt loam) as function of the SAR and salt concentration of the percolating solution, and found that both soils were only slightly affected by the SAR of the percolating solution up to SAR 20, as long as the concentration of the percolating solution exceeded 10 mmol$_c$ l^{-1} (Figure 9). However, when the percolating salt solution was displaced by distilled water simulating rainfall, the

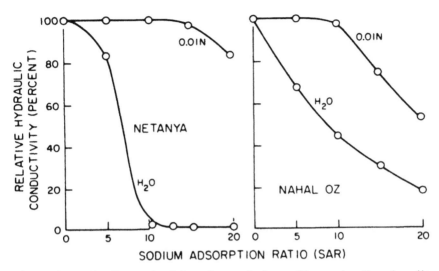

Figure 9. Hydraulic conductivity of a sandy loam (Netanya) soil and a silty loam (Nahal-Oz) soil as a function of the SAR and the concentration of the leaching solutions.(From Felhender et al., 1974.)

response of the two soils differed drastically. The HC of the silty soil dropped to 42 and 18% of the initial value for soils having ESP values of 10 and 20, respectively. The HC of the sandy loam soil dropped to 5 and 0% of the initial value for the same conditions, respectively. It was also noted that the clays in the sandy loam soil were mobile and appeared in the leachate, whereas no clay dispersion took place in the silty loam soil. Felhendler et al. (1974) postulated that the HC response was associated with the potential of clay to disperse when the soil was leached with distilled water. Clay in the sandy loam soil dispersed strongly when the soil was leached with distilled water, but clay in the silty loam soil did not. However, they presented no hypothesis to explain why clay particles of similar mineralogy will disperse in one soil and not in the other.

2. Effect of Low Electrolyte Concentration on the HC of Sodic Soils

The effect of displacing 0.01 N solutions of SAR 10, 15, 20, and 30 with distilled water or salt solutions of 1, 2, or 3 $mmol_c\ l^{-1}$ on the relative HC of, and clay dispersion from Fallbrook soil-sand mixture was studied by Shainberg et al. (1981a,b). From the results presented in Figure 10 it is evident that when leached with distilled water, even a low ESP was enough to appreciably reduce the HC of the Fallbrook soil. Electrolyte concentration of 2 $mmol_c\ l^{-1}$ in the percolating solution prevented the adverse effect of ESP 10 on the HC of this soil. The adverse effect of 15% Na in the exchange complex was prevented by a solution of 3 $mmol_c\ l^{-1}$.

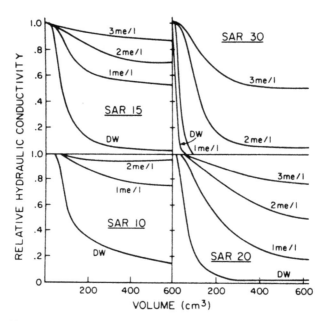

Figure 10. Relative HC of Fallbrook soil-sand mixture, equilibrated with 0.01 N solutions of SAR 10, 15, 20, or 30, and leached with distilled water (DW) or salt solutions of 1, 2, or 3 mmol$_c$ l^{-1}. (From Shainberg et al., 1981a.)

Figure 11 shows the effect of the solution concentration on the concentration of clay in the Fallbrook soil effluent. It is clear that clay dispersiveness and movement are very sensitive to the electrolyte concentration in the percolating solution. For example, in soil equilibrated with SAR 15 solution, leaching with distilled water or 1 or 2 mmol$_c$ l^{-1} solutions resulted in peak clay concentrations of 1.0, 0.1, and 0.02% clay, respectively, in the effluent.

The response of the Fallbrook soil to a combination of sodicity and electrolyte concentration is summarized in Figure 12. When the percolating solution concentration was maintained at >3 mmol$_c$ l^{-1}, no reduction in HC took place until the SAR exceeded 12. For solution concentrations of 2, 1, and 0.5 mmol$_c$ l^{-1}, threshold SAR values were 9, 6, and 4, respectively. When Fallbrook soil was leached with distilled water, the detrimental effect of exchangeable sodium started at an ESP value of about 1%. Once the threshold ESP was exceeded, the HC was always very sensitive to further increases in exchangeable sodium. The more dilute the leaching solution, the more sensitive the soil to an incremental increase in the ESP beyond the threshold value.

The combined effect of soil swelling and dispersion on the HC of montmorillonite-sand mixture when leached with dilute solutions was studied by Keren and Singer (1988). They concluded that the degree of clay swelling before replacing the solution with electrolyte concentration below the flocculation value (FV) determines whether clay will leave the system and hence the changes in the

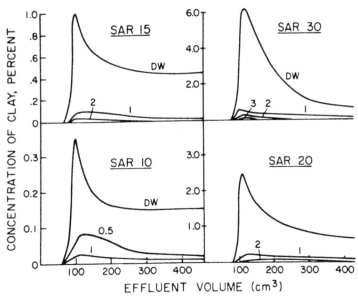

Figure 11. Concentration of clay in effluent of the low salt concentration and distilled water (DW) leachates from the Fallbrook soil mixture equilibrated with 0.01 N solutions of SAT 10, 15, 20, and 30. (From Shainberg et al., 1981a,b.)

HC. They found that for solutions of SAR 10 and 20 subsequent introduction of a solution with electrolyte concentration below the FV causes a further decrease in pore radii and part of the dispersed clay particles are trapped in the narrow pores. Thus, the HC further decreased but no clay appeared in the leachate. For the same conditions but with solutions of SAR 5, an increase in HC was observed following the washing out of clay. Keren and Singer (1988) ascribed these results to the fact that clay swelling is limited at SAR 5.

The effect of ESP on the HC of soils leached periodically with rainwater should be reemphasized; even a low ESP may enhance loss of HC of the soil. Five percent sodium on the exchange complex of a soil like Fallbrook reduced the HC of the soil to 20% of its initial value. The high sensitivity of soils to low levels of exchangeable sodium when leached with "high quality water" (0.7 $mmol_c$ l^{-1}) can explain the results of McIntyre (1979), who concluded, following the measurement of the HC of 71 Australian soils, that an ESP value of 5 should be accepted for Australian soils as separating sodic from normal soils. The U.S. Salinity Laboratory Staff (1954) suggested an ESP value of 15 to designate sodic soils because they used electrolyte concentration >3 $mmol_c$ l^{-1} in determining soils HC.

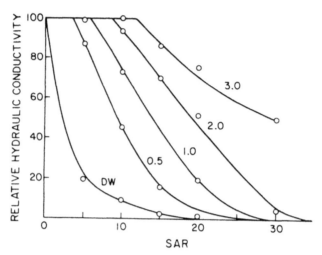

Figure 12. Effects on Fallbrook soil-sand mixture's HC of the ESP of distilled water (DW) and various solution salt concentrations. (From Shainberg et al., 1981a,b.)

3. Effects of Weathering of Soil Minerals on the Soils' Response to Leaching with Distilled Water

Arid land soils release 3 to 5 $mmol_c$ l^{-1} of Ca and Mg to solution as a result of the dissolution of plagioclase, feldspars, hornblende, and other minerals (Rhoades et al., 1968). The solution composition of a calcareous soil at a given ESP when brought in contact with distilled water can be calculated (Shainberg et al., 1981b). The $CaCO_3$ dissolves and replaces sodium from exchange sites until the solution is in simultaneous equilibrium with exchange sites and with the $CaCO_3$ solid phase. Assuming that the apparent $CaCO_3$ solubility product is 10^{-8} and in equilibrium with a CO_2 pressure of 10-3.5 atm, the ECs of the solutions in equilibrium with soils having ESP 5, 10, and 20 are 0.4, 0.6, and 1.2 dS m^{-1}, respectively (Shainberg et al., 1981b). These concentrations are enough to prevent the deleterious effect of exchangeable Na even when the soil is leached with rain water.

The sensitivity of sodic soils to solutions of low electrolyte concentrations led Shainberg et al. (1981a,b) to hypothesize that a major factor causing differences among various sodic soils in their susceptibility to HC decreases when leached with low-electrolyte water was their rate of salt release by mineral dissolution. Mineral dissolution determines the electrolyte concentrations of percolating solutions. They postulated that sodic soils containing minerals ($CaCO_3$ and a few primary minerals) that readily release soluble electrolytes will not readily disperse when leached with distilled water (simulating rain water) at moderate

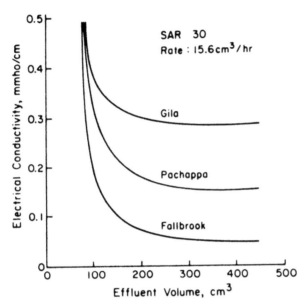

Figure 13. Effect of effluent volume on the EC of effluents from columns of Gila, Pachappa, and Fallbrook soils during leaching with distilled water. (From Shainberg et al., 1981b.)

ESPs, because they will maintain high enough salt concentrations in the soil solution (~ 3 mmol$_c$ l^{-1}) to prevent clay dispersion. Additionally, the ESPs in these soils will be reduced, because most of the released salts are Ca and Mg. Conversely, salt concentration in the soil solution of soils lacking readily weatherable minerals will be below a threshold concentration (the flocculation value), and the soils will be more susceptible to clay dispersion. Consequently, lodgement of dispersed clay particles in the conducting pores may markedly decrease the HC of such soils.

The data presented in Figures 5 and 6 for the two Israeli soils supported the hypothesis of Shainberg et al (1981a,b). Soil that was more stable chemically (Netanya sandy loam) was also more susceptible to sodic conditions when leached with distilled water. This hypothesis is further supported by measurements of clay dispersivity, HC, and the mineral weathering properties of three California soils (Figures 13, 14, and 15). The chemically stable Fallbrook soil was the more sensitive soil to exchangeable Na effects on clay dispersion and loss in HC in spite of the presence of sesquioxides and kaolinite in its clay fraction. Gila and Pachappa soils, which have higher mineral dissolution rates, were less affected by exchangeable Na.

Clay will be in the leachate only in sandy soils (Figure 4). When the soil's clay content is high, the conducting pore size is usually small (resulting in low HC), and the dispersed clay moves only short distances before it clogs the

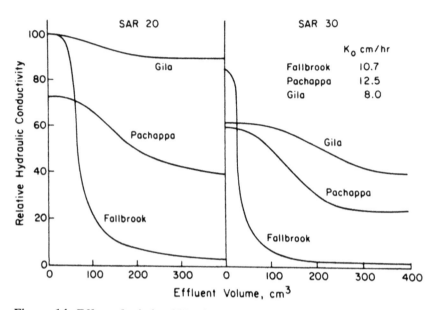

Figure 14. Effect of relative HCs during leaching of ESP 20 and ESP 30 soil mixtures with distilled water. (From Shainberg et al., 1981b.)

pores. Thus, in loamy and clay soils the dispersion mechanism still operates, but no macroscopic movement of the clay particles is observed. When the clay soil is mixed with sand, the dispersion mechanism is evident.

F. Effect of Exchangeable Magnesium

Some of the irrigation waters in numerous regions in the world contain a high concentration of Mg. The use of such waters for irrigation, coupled with an increase in irrigation efficiency, will result in large increases in exchangeable Mg in the soil. In some soils the exchangeable Ca/Mg molar ratio is already less than one.

At present, the issue concerning the effect of adsorbed Mg on the hydraulic properties of soil is controversial. The U.S. Salinity Laboratory Staff (1954) grouped Ca and Mg together as similar ions, beneficial in developing and maintaining soil structure. On the other hand, Van der Merwe and Burger (1969) found that Na-Mg saturated soil was structurally less stable than Na-Ca soil. From these results they concluded that exchangeable Mg can have a deleterious effect on the structure and permeability of the soil. McNeal et al. (1968) also showed that soils high in exchangeable Mg had a lower HC relative to soils high in exchangeable Ca. In other studies a distinction was made between the direct effect of exchangeable Mg in causing decreases in HC, which

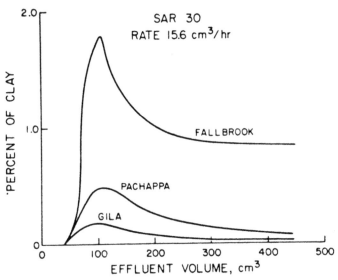

Figure 15. Effluent clay concentration with passage of distilled water at the rate of 15.6 cm^2 hour^{-1} for the three soil mixtures at ESP=30. (From Shainberg et al., 1981b.)

has been termed a "specific effect", and the inability of Mg in irrigation waters to counter the accumulation of exchangeable Na in the soil (Chi et al., 1977; Emerson and Chi, 1977; Rahman and Rowell, 1979). In the above-mentioned studies Mg was shown to have a specific effect in soils dominated by illite, but not in montmorillonitic soils. It was also found that exchangeable Mg appeared to have an adverse effect on the HC of a sodic sandy loam soil dominated by smectite and kaolinite only when leached with distilled water (Rowell and Shainberg, 1979). Alperovitch et al. (1981) found that in a calcareous soil exchangeable Mg had no specific adverse effect on the HC whereas in noncalcareous soils Mg caused a decrease in the HC of the soil (Figure 16). They suggested that the presence of Mg enhances the dissolution of CaCO$_3$ in the calcareous soil, and the electrolytes prevent clay dispersion and HC losses in the Na/Mg calcareous soil. The observed favorable effect of exchangeable Ca, compared to Mg, on the HC of the noncalcareous soils could result from the fact that exchangeable Ca maintains higher salt concentration in the soil solution compared to exchangeable Mg in chemically stable soils (Alperovitch et al. 1981). Furthermore, Shainberg et al. (1988) concluded that the lower HC values of the Na-Mg smectite system compared to the Na-Ca system are related to the effect of Mg on the hydrolysis of these clays. These conclusions were based on the findings of Kreit et al. (1982) who showed that the presence of a high concentration of Mg at the clay surface slowed down the release of octahedral

64

I. Shainberg and G.J. Levy

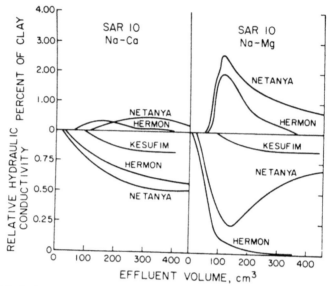

Figure 16. Relative HC and concentration of clay in the effluent of six Israel soils equilibrated with 0.01 solutions of SAR 10 and leached with distilled water. Where no data for clay concentration are given, no clay appeared in the effluent. (Alperovich et al., 1981.)

Mg from the lattice (Kreit et al., 1982), and lowered the EC of the Na-Mg clay solution (Shainberg et al., 1988).

In a study conducted with 3 soils from South Africa, the HC measurements showed that exchangeable Mg resulted in lower HC values compared to exchangeable Ca (Levy et al., 1988b). This specific effect of Mg was observed in the three soils and at all ESP levels (Figure 17), despite the fact that the electrolyte concentrations of the soil solution following leaching with distilled water were similar to, or higher in the Na-Mg systems than in the Na-Ca systems.

A further study on the hydrolysis of kaolinitic soils (Levy et al., 1989) supported the latter findings. The rate of hydrolysis of kaolinitic soils was not affected by the cationic composition of the exchangeable phase (i.e., Ca or Mg). Levy et al. (1989) postulated that exchangeable Mg did not inhibit hydrolysis in the soils studied because of the limited isomorphic substitutions in kaolinite which lead to small amounts of octahedral Mg in the crystal.

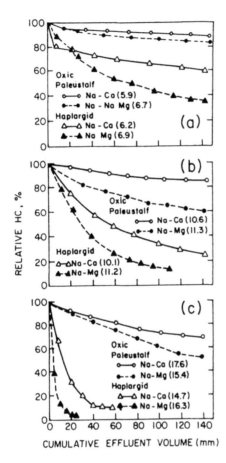

Figure 17. The relative HC of the Haplargid and Oxic paleustalf, initially equilibrated with 0.01 N solutions of (a) SAR 5, (b) SAR 10, and (c) SAR 20, while leached with distilled water. Numbers in parentheses indicate ESP. (From Levy et al., 1988.)

G. Effect of Exchangeable Potassium

Of the four major cations found in the soil (Ca, Mg, Na, and K), the effect of exchangeable potassium on soil permeability is probably the least clear since results vary or conflict, possibly due to differences in clay mineralogy and sample preparation procedures. In some studies it has been noted that exchangeable K and exchangeable Na had the same deleterious effect on the HC of the soil (Quirk and Schofield, 1955). Other researchers have reported that the effect of exchangeable K on the permeability of the soil was not as favorable as that of the divalent cations (Ca and Mg) but not as negative as that of Na (Reeve et al., 1954; Gardner et al., 1959; Ahmed et al., 1969). For instance in K-Ca kaolinite, Elgabaly and Elghamry (1970) found that with increasing K as complementary ion to Ca on the clay, water permeability was lowered but always remained higher than that of Na-Ca clay systems. At an EPP of 15, water permeability was reduced to about only one-fourth of that in Ca saturated clay, while it was 21 times lower when K was replaced by Na at the same Ca level. The decrease in permeability was even more pronounced when homoionic clays were tested. The permeability of the K-saturated kaolinite system was 14 times lower than that of the Ca-saturated system, whereas the Na-saturated clay was 950 times lower.

Levy and van der Watt (1990) also observed that increasing the amount of K in the exchangeable phase resulted in a decrease in HC. They observed that the magnitude of this phenomena depended on the clay minerals present in the soil. The smallest effects were found in kaolinitic soils and the biggest in an illitic soil.

On the other hand, K-saturated soils were found to have larger aggregates and greater stability with regard to percolating solutions, suggesting that adsorbed K has a favorable effect on soil permeability (Cecconi, et al., 1963; Ravina, 1973). In addition, Chen et al. (1983), reported that an EPP in the range of 10-20 improved the HC of some Israeli soils. However, at high EPP values (58-76) HC of the soils decreased to ~20% of that of EPP=0. Ravina and Markus (1975) also found that the permeability of a smectitic grumusol increased when the EPP was increased from 4 to 26. In a recent study on the effect of exchangeable K on the HC of smectite-sand mixtures, Shainberg et al. (1987a,b) concluded that the effect of K depended on the charge density of the smectitic clay. They found that the higher the charge density of the clay the more favorable the effect of exchangeable K on the HC.

The above results indicate that in some soils K takes an intermediate position (between Na and Ca), whereas in others K even improves permeability (in comparison with Ca-soil). It seems that K fixation is a possible mechanism that affects permeability. This hypothesis is supported by the fact that the cation exchange capacity of the soils decreased upon saturation with K (Steemkamp, 1965), the c-spacing of K-saturated smectite decreased (in comparison with Na-clay) and part of the smectite was transformed to an illite like clay. This transformation may significantly affect the physical properties of the soils.

H. Effect of Solution Composition on the HC of Kaolinitic Soils

Differences of opinion can be found in the literature on the effect of kaolinite clay on the HC of soils leached with sodic water. McNeal and Coleman (1966) and Yaron and Thomas (1969) concluded that the most and least labile soils were those high in 2:1 layer silicates (especially montmorillonite) and those high in kaolinite and sesquioxides, respectively. El-Swaify and Swindale (1969) found a negligible effect of exchangeable Na, even in the absence of salinity, on the HC of tropical soils whose clay fractions were dominated by kaolins, iron oxides, amorphous silicates, and gibbsite. McNeal et al. (1968) found that the "stability" of the HC of Hawaiian soils under high-Na and low salt conditions was greatly reduced by partial removal of free iron oxides, pointing to the cementing action of iron oxides in preventing dispersion. Velasco-Molina et al. (1971), concluded that, in the virtual absence of electrolytes, the order of soil dispersion with respect to the dominant clay mineral at a given ESP was montmorillonite > Halloysite > mica. Elgabaly and Elghamry (1970) found that the HC of ground and sieved kaolinitic systems with ESP > 10 decreased rapidly when leached with distilled water. Frenkel et al. (1978) concluded that although the HC of a kaolinitic soil was less affected than that of a montmorillonitic soil when leached with a solution of low salt concentration, the HC of the former soil was reduced markedly, even at an ESP of 10, when leached with distilled water. Frenkel et al. (1978) and Chiang et al. (1987) concluded that soil behavior depended on the type of mineral that is present in combination with kaolinite. Soil kaolinites containing small amounts of montmorillonite and mica were dispersive (Arora and Coleman, 1979; Frenkel et al., 1978; Schofield and Samson, 1954). This phenomenon has been ascribed to the breakup of the edge-to-face particle association of kaolin structure by the adsorption of negatively charged montmorillonite faces on the positively charged kaolin edges.

In kaolinitic soils, pH also affects dispersion because of the variable nature of the positive charge. Peele (1936) showed that permeability and aggregate size of a Cecil series soil decreased as lime was added. Arora and Coleman (1979) found that raising the pH of a Georgia kaolinite from 7 to 8.3 created more dispersion than in any of the other samples including smectites, illites, and vermiculites. Studying the effect of EC, ESP, and soil pH on the HC of three kaolinitic soils from the southeastern United States, Chiang et al. (1987) showed that the HC of the Cecil soil was very sensitive to changes in EC, Na content, and pH, while the Davidson and Iredell soils were well flocculated and unaffected by changes in EC at ESP <3. These researchers suggested, that the observed differences in structural stability may be related to parent material in that sandy soils derived from granite rock (Cecil) are more prone to dispersion--related problems than more clayey, oxidic soils derived from mafic material (Davidson and Iredell). Thus, kaolinitic soils, being dispersive, will also respond to increased EC in the soil solution.

I. Response of Sodic Soils to Readily Dissolving Salts

1. Gypsum Application

Gypsum is the most commonly used amendment for sodic soil reclamation, primarily because of its low cost, availability, and handling (Oster, 1982). Its value has been known for a long time (Hilgard, 1906; Kelley and Arany, 1928). The amount of gypsum required to reclaim a sodic soil depends on the amount of exchangeable Na in the soil profile (U.S. Salinity Laboratory Staff, 1954).

Gypsum added to a sodic soil can initiate permeability increases due to both electrolyte concentration and cation exchange effects (Loveday, 1976). The relative significance of the two effects is of interest for several reasons. If the electrolyte effect is sufficiently great to prevent dispersion and swelling of soil clays, surface application of gypsum may be worthwhile. In this case the amount required depends on the amount of high-quality water applied and the rate of gypsum dissolution (Oster, 1982), and it is somewhat independent of the amount of exchangeable Na in the soil profile. Conversely, in soils where the electrolyte concentration effect is insignificant and the main effect results from cation exchange, the amount of gypsum required depends on the quantity of exchangeable Na in any given depth of soil. The cation exchange process has formed the basis of several "gypsum requirement" tests (U.S. Salinity Laboratory Staff, 1954; Oster and Frenkel, 1980; Keren and O'Connor, 1982; Oster, 1982). The relative significance of the electrolyte effect was estimated recently (Shainberg et al., 1982) by comparing the effects of equivalent amounts of gypsum and $CaCl_2$ applied at the soil surface on the hydraulic conductivity of three Israeli soils leached with distilled water. The exchange reclamation was similar with both amendments, but there was a difference in the electrolyte effect. The slow dissolution and the relatively long-term electrolyte effect of gypsum were very important in maintaining high hydraulic conductivity in a chemically stable soil which did not release salt into the soil solution. For a corresponding $CaCl_2$ treatment, complete sealing of the soil took place. Conversely, the efficiency of the two amendments was similar for a calcareous soil. Even without gypsum, this soil released enough electrolytes into the soil solution to prevent clay dispersion and hydraulic conductivity losses. The ability of gypsum to maintain a moderate concentration of electrolyte in the percolating water is very important in preventing seal formation at the soil surface under rainfall conditions.

2. Effect of Lime on the Response of Soils to Sodic Conditions

The presence of fine $CaCO_3$ particles in soils is known to improve the physical conditions of sodic soils (U.S. Salinity Laboratory Staff, 1954). Because of low $CaCO_3$ solubility in soils having pH greater than 7.5, lime is not effective in the exchange mechanism by which exchangeable Na is replaced by Ca, and thus sodic soils containing $CaCO_3$ are common in the semiarid and arid regions of the world. To explain the effect of $CaCO_3$, the U.S. Salinity Laboratory Staff (1954) and Rimmer and Greenland (1976) suggested that lime in soils acts as a cementing agent which stabilizes soil aggregates and prevents clay dispersion. Another mechanism which explains the beneficial effect of $CaCO_3$ is its potential for dissolving and maintaining the concentration of soil solution at concentration levels above the flocculation value of the soil clays, thus preventing their dispersion. According to the dissolution mechanism, exchange reclamation is still negligible because of the low concentration of Ca ions in the soil solution, and clay dispersion is prevented mainly because of the electrolyte concentration effect (Shainberg et al., 1981a). The rate of $CaCO_3$ dissolution in controlling

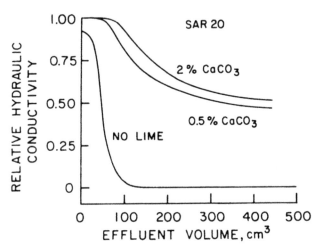

Figure 18. Effect of lime on the HC of noncalcareous soil. (From Shainberg and Gal, 1982.)

HC of sodic soils was determined by Shainberg and Gal (1982) by mixing lime-free soils with low percentages of powdered lime (0.5 and 2.0 percent). Although the HC of the lime-free soils dropped sharply when 0.01 N solutions of SAR 20 were displaced with distilled water, mixing the soils with powdered lime prevented both HC losses and clay dispersion (Figure 18). The increase in electrolyte concentration in the soil solution due to $CaCO_3$ dissolution was suggested by Shainberg and Gal (1982) as the mechanism responsible for the beneficial effect of lime.

The relative weight of each of the two aforementioned mechanisms by which $CaCO_3$ stabilizes soil structure, can be estimated from studies on its effect on the infiltration rate and crust formation under rainfall conditions (e.g. Agassi et al., 1981; Ben-Hur et al., 1985). If $CaCO_3$ acts as a cementing material, calcareous soils should not be as sensitive to crust formation as will noncalcareous soils be. However, if the dissolution mechanism is dominant, the concentration of the electrolytes at the soil's surface exposed to rain will be insufficient, and both types of soils will be sensitive to crusting (see further discussion in Section IV.C).

IV. Effect of Soil Exchange Phase Composition and Soil Solution Concentration on Infiltration Rate (IR)

A. Definition and Experimental Background

When water is supplied to the soil surface, whether by precipitation or irrigation, some of the water penetrates the surface and flows into the soil, while some may fail to penetrate and instead accumulate at the surface or flow over it. Infiltration is the term applied to the process of water entry into the soil, usually by downward flow through the soil surface. Hence, "infiltration rate" (IR) is defined as the volume flux of water flowing into the profile per unit of soil surface area under any set of circumstances. IR and the variation thereof with time are known to depend upon initial water content and suction, texture, structure, and uniformity of the soil profile (Hillel, 1980b). Generally, IR is high during the early stages of infiltration, particularly when the soil is initially quite dry, but decreases monotonically to approach a constant rate asymptotically, due to a decrease in the matric suction gradient which occurs as infiltration proceeds.

The decrease in IR from an initially high rate can also result from a gradual deterioration of soil structure and consequent partial sealing of the profile by the formation of a dense surface seal. It is well known that seal formation at the soil surface predominates in the decrease of infiltration during rain (Chen et al., 1980; Duley, 1939; Epstein and Grant, 1973; Morin and Benyamini, 1977). The IR values of sealed soils depend on the HC of the seal formed at the soil surface, hence some researchers tried to measure the HC of the seal. McIntyre (1958) found that the HC of the upper and lower layers of the seal of a fine sandy loam soil were 2000 and 200 times lower than the HC values recorded for the undisturbed soils. Bresler and Kemper (1970) measured a HC value of 1.5 X 10^{-4} mm s^{-1} in the upper 2-3 mm of a sealed clay loam soil. However, since the crust is a very thin layer and it is very difficult to separate it from the rest of the soil profile, it is extremely difficult to determine its HC. Consequently, IR is widely used to characterize water penetration into the soil, particularly when seal formation is involved.

Numerous formulas, some entirely empirical and others theoretically based, have been proposed in attempts to express IR as a function of time or total volume of water infiltrated into the soil (i.e., cumulative infiltration). The theoretical equations (Green and Ampt, 1911; Philip, 1957) arise out of mathematical solutions to physically based theories of infiltration. The empirical expressions (Horton, 1940; Holtan 1961) are not so restrictive as to the mode of water application, since they do not imply surface ponding from time zero on, as do the Green-Ampt and Philip equations (Hillel, 1980b). An equation commonly used is a Horton-type equation developed by Morin and Benyamini (1977) which describes the infiltration process as a function of cumulative rain rather than cumulative time:

$$I_t = I_c + (I_i - I_c) \exp^{(-ypt)} \qquad \qquad (6)$$

where I_t = instantaneous infiltration rate (mm h^{-1}); I_c = asymptotical final infiltration rate (mm h^{-1}); I_i = initial infiltration rate (mm h^{-1}); y = a constant related to the stability of the soil surface aggregates (mm^{-1}); p = rain intensity (mm h^{-1}); and t = time elapsed from the beginning of the storm (h).

Calculated IR values derived from this equation were in good agreement with values measured with rainfall simulators such as those developed by Morin et al. (1967) and Miller (1987b). Soil samples are packed in trays, which are then placed at a predetermined slope and subjected to simulated rain. Rain intensity, drop impact energy, and "rain" chemistry can be controlled according to the experimental design. When rain water is to be simulated, water of electrical conductivity of <0.01 dS m^{-1} is used. Volume of the percolating water and that of runoff are measured and the IR is calculated.

B. The Phenomenon of Sealing

1. Seal Formation

When soil is exposed to a rainstorm, the raindrop impact causes severe changes in the structure of the soil surface and a reduction in soil permeability is observed (Duley, 1939). It has been suggested (Epstein and Grant, 1973) that seal formation due to the beating action of raindrops results from the mechanical breakdown of aggregates, at the soil surface, resulting in the formation of primary particles and microaggregates, which in turn, reduce the porosity of the soil surface. Furthermore, raindrop impact causes direct compaction of the soil surface (Epstein and Grant, 1973).

Morin et al. (1981) attributed seal formation mainly to raindrop impact which destroyed the aggregates at the soil surface thus forming a continuous thin layer with very low permeability. These workers postulated that the sealing efficiency of the seal results from suction forces at the seal-soil interface which hold the dispersed soil particles together in a continuous dense layer. The suction forces are a result of the large differences in hydraulic conductivity between the seal and the underlying soil.

Agassi et al. (1981) suggested that seal formation is due to two mechanisms: (i) physical disintegration of soil aggregates and their compaction caused by the impact of the raindrops and (ii) chemical dispersion and movement of clay particles into a region at 0.1-0.5 mm depth, where they lodge and clog the conducting pores. Both of these mechanisms act simultaneously as the first enhances the latter.

2. Structure of Seals

In order to explain the genesis and properties of seals, several studies have been conducted on the structure of soil seals resulting from rainfall. Duley (1939) studied micrographs of seals, obtained with an optic microscope with a magnification of X15, and found that the seal was a very thin layer, closely packed and with a higher density than the profile underneath. McIntyre (1958) found that seals consist of two distinct parts: an upper skin seal, 0.1 mm thick, attributed to compaction by raindrop impact, and a "washed in" region, 2 mm thick and of decreased porosity, attributed to the accumulation of fine particles. The "washed in" zone was formed only in easily dispersed soils (McIntyre, 1958). Onofiok and Singer (1984) studied the morphology of three soils using a scanning electron microscope (SEM), and reported of a decreased apparent porosity at the 0.3-0.8 mm depth measured directly on the micrographs, which they ascribed to the presence of the "washed in" zone. Studying micrographs of seals obtained from a petrographic microscope with magnification of X100, Evans and Buol (1968) observed well-oriented particles immediately under the seal lying parallel to the soil surface. Chen et al. (1980) and Tarchitzky et al. (1984) examined, using SEM micrographs, seals of Israeli soils, and also observed a thin skin seal about 0.1 mm in thickness. They did not, however, find an accumulation of fine particles in the 0.1-2.8 mm region as was observed by McIntyre (1958) and Onofiok and Singer (1984). Similar calculated clay percentages in the 0.1-2 mm zone and the > 2 mm depth were reported in support of the SEM observations (Tarchitzky et al., 1984). These researchers did, however, observe an increase of the bulk density of the upper 4 mm from 1.35-1.48 Mg m^{-3} in undisturbed soil samples to 1.74-1.88 Mg m^{-3} in the crusted samples. The increase in bulk density was ascribed to the breakdown of aggregates and the subsequent consolidation of primary particles (Tarchitzky et al., 1984). In a further study, Gal et al. (1984) showed that for distilled water (simulation of rain water), the presence of the "washed in" zone depended on the ESP of the soil and hence on the susceptibility of the soil to dispersion. In soils with moderate ESP levels the "washed in" zone was clearly noticeable, whereas when the ESP was < 1 there was no sign of the "washed in" zone. In the studies of Chen et al. (1980) and Tarchitzky et al. (1984) tap water (EC ~0.6 dS m^{-1}) and soil samples with low ESP were used. These conditions are not favorable for clay dispersion and hence can explain why the "washed in" zone was not observed in these two studies.

Levy et al. (1988a) argued that the aerial structure of the seal cannot be assumed homogeneous as there are large differences in the scale of measurements when sealed areas were studied by either infiltration or SEM micrographs. In their study, Levy et al. (1988a) observed that seals formed in soils consisted of two distinct microtopographical features. The first were mounds which exhibited relatively high permeability and structure bearing some resemblance to that of unsealed soil. The second were plains with low permeability and a seal that was clearly visible in the SEM micrographs. The permeability of the mounds and plains as well as the area occupied by the mounds decreased with an increase in the ESP of the soil.

It is evident that seal formation in soils exposed to rain is associated with clay dispersion and movement in the soil. Soil surfaces are especially susceptible to the chemistry (electrolyte concentration and cationic composition) of the applied water because of the mechanical action of the falling drops and the relative freedom of particle movement at the soil surface (Oster and Schroer, 1979).

C. Effect of Exchangeable Cations on the IR of Soils

1. Na-Ca Systems

The effect of soil sodicity (i.e., ESP) on the IR and seal formation of four soils varying in texture, clay mineralogy, and $CaCO_3$ content was studied by Kazman et al. (1983) using distilled water in a rain simulator. The IR of the four soils was very sensitive to low levels of ESP. The IR of two soils, the same soils on which the HC studies were made (Figure 9), are presented in Figures 19 and 20. The final IR of the Netanya sandy loam of ESP 1.0 was 7.5 mm h^{-1}, whereas the final IR of the same soil but having ESP values of 2.2 and 11.6 dropped to 2.3 and 0.6 mm h^{-1}, respectively. An ESP value of 2.2 was enough to cause a drop of 70% in the final IR. Similar effects of ESP were observed in the Nahal-Oz soil. The depth of rain required to approach the final IR was also ESP dependent (Figures 19 and 20).

Upon comparing the IR data (Figures 19 and 20) with the HC data (Figure 9), it is evident that the IR is more sensitive to low ESP than is the HC. Furthermore, the IRs of the calcareous and noncalcareous soils are equally sensitive to low levels of ESP (Ben-Hur et al., 1985; Kazman et al., 1983). The HC of calcareous soils, on the other hand, is not sensitive to this condition. These two phenomena are explained by the same mechanism. The soil solution electrolyte concentration determines the response of soils to low ESP values. When a sodic soil is leached with distilled water, $CaCO_3$ dissolution maintains electrolyte concentration in the soil solution at a value close to 3 mmol$_c$ l^{-1} which is sufficient to prevent clay dispersion with subsequent decreased HC. However, the soil solution concentration at the soil surface is determined solely by the concentration of the applied water (i.e., nearly distilled water for rain water), thus aggregate dispersion takes place at the soil surface at low ESP levels even in calcareous soils. The higher susceptibility of the surface of soils to low ESP levels is supplemented by two other factors: (i) mechanical impact of the raindrops, which enhances dispersion; and (ii) absence of soil matrix (sand particles), which slows clay dispersion and clay movement (Oster and Schroer, 1979). In the studies of Kazman et al. (1983), rain energy was the same for all the experiments, hence differences in the IR curves for the various soil samples were due to the potential of the soil to disperse as determined by soil sodicity.

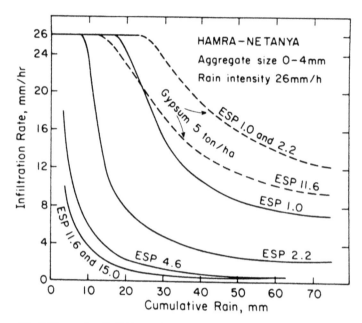

Figure 19. Effect of soil ESP (and phosphogypsum) on the infiltration rate of Hamra (Netanya) soil. (From Kazman et al., 1983.)

The drop in the IR of soils with ESP < 1 (Figure 19) was then mainly due to the mechanical action of the raindrops.

It is, however, very difficult to separate the mechanical and chemical mechanisms as they are complementary. The kinetics of the chemical clay dispersion depends markedly on the intensity of the mechanical mixing of the raindrops. Agassi et al. (1985) studied the interaction between the physical effect of the raindrops and the chemical effects of the composition and concentration of the applied water. They found that in a situation where both mechanisms (i.e., rain with energy and distilled water causing dispersion) were in operation, seals with low permeabilities were formed (3 mm h[-1]), even in a soil with low ESP. When the chemical effect was diminished by using saline water, they obtained seals with a relatively high permeability (~8 mm h[-1]). On the other hand, when rain with very low energy (fog-type rain) was used together with distilled water, a limited reduction in the permeability of the soil was observed. This reduction was related to changes in the HC of the soil profile in accordance with the ESP of the soil since no seal was evident at the soil surface. On the basis of these results, Agassi et al. (1985) concluded that in the absence of the physical mechanism the chemical one does not come into effect at low ESP levels. Evidently, the chemical mechanism needs some activation energy for it to start operating at the soil surface, which in this case is provided by the impact energy of the raindrops.

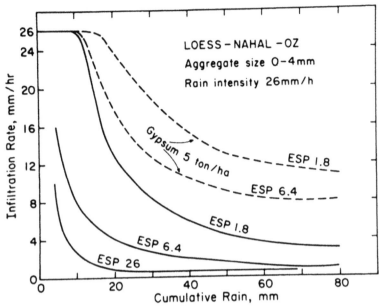

Figure 20. Effect of soil ESP (and phosphogypsum) on the infiltration rate of loess soil. (From Kazman et al., 1983.)

Highly weathered kaolinitic soils in the southeastern United States also suffer from dispersion-related degradation of physical properties following the introduction of Na to the exchangeable complex. The effect of surface application of $NaNO_3$ at fertilizer N rates on infiltration, runoff, and soil loss was studied by Miller and Scifres (1988). Runoff from untreated soil commenced only after 20 mm of rain and the final IR was 10 mm h[-1]. The $NaNO_3$ treatment resulted in an immediate surface sealing and a final IR of 2.5 mm h[-1]. Although the effect of exchangeable Na on the HC of kaolinitic soils is limited (Chiang et al., 1987; Frenkel et al., 1978), its effect on seal formation and clay dispersion under conditions of rain is significant. Emerson (1967) postulated that significant changes in the HC will only follow spontaneous dispersion of aggregates. Kaolinitic aggregates do not spontaneously disperse and thus will not affect the HC. However, aggregates from kaolinitic soils can be mechanically dispersed, by the beating action of raindrops, to form a seal with low permeability. Kaolinitic soils from the south of the United States are comparable to montmorillonitic soils (Kazman et al., 1983) with respect to dispersivity and seal formation when exposed to rain. This resemblance in the response to raindrop impact suggests that the phenomena of soil sealing and the correlation between sealing and soil dispersivity (Miller and Baharuddin, 1986) are of a general nature.

2. Effect of Exchangeable Magnesium

Exchangeable Mg can cause a deterioration in soil structure and develop a "magnesium solonetz" (Ellis and Caldwell, 1935). In addition, Mg enhances dispersion in montmorillonite and illite compared to Ca (Bakker and Emerson, 1973; Shainberg et al., 1988). Levy et al. (1988b) compared the effect of Mg to that of Ca, as complementary cation to Na, on the HC and IR of three soils from South Africa. The results showed that exchangeable Mg had a similar effect to that of exchangeable Ca on the IR. By contrast, Mg was not as efficient as Ca in maintaining high HC under sodic conditions. These researchers concluded that the adverse effect of Mg on clay dispersion was pronounced in the HC determinations and not in the IR studies because in the latter chemical clay dispersion is not the dominant mechanism controlling soil permeability.

Keren (1989) compared the effect of exchangeable Mg to that of Ca (with no Na in the exchangeable complex) on the IR of two montmorillonitic soils from Israel. The IR of the Mg soils were always lower than those for the Ca soil whether the soil was calcareous or not. Keren (1989) attributed this difference to the larger width of hydration shell of the adsorbed Mg compared to that of the Ca ion. The width of hydration shell determines the strength of the link between the clay tactoids (the wider the shell the weaker the links), and hence affects aggregate stability. In a further study, using the same two soils, Keren (1990) compared the effect of Mg to that of Ca as the complementary cation to Na under rain kinetic energy range of 3.2-22.9 kJ m^{-3}. At the highest kinetic energy (similar to that used by Levy et al., 1988b) the effect of Mg on the IR was comparable to that of Ca. However, in the middle range of rain kinetic energy studied (8.0 and 12.5 kJ m^{-3}), the IR was lower in the Na-Mg treated soils than in the Na-Ca soils. Keren (1990) suggested that the soil aggregates in the Na-Ca system are more stable than those in the Na-Mg system because of the larger width of the hydration shell of the adsorbed Mg. However, this difference can be noticed, as previously proposed by Levy et al. (1988b), only under conditions where chemical dispersion is dominant in controlling the IR (i.e., raindrops with low to medium kinetic energy).

3. Effect of Exchangeable Potassium

Although the effect of exchangeable K on the water transmission properties of the soil is controversial (see discussion in the section on HC), its effect on the IR and seal formation has not received much attention. Levy and van der Watt (1990) compared the effect of K on both the HC and IR of three South African soils to that of Ca and Na. Their results showed that when K was the complementary cation to Ca, increasing the EPP resulted in the decrease in the IR. However, when compared to Na, exchangeable K had a more favorable effect on the IR than Na at all the ESP/EPP levels tested. Thus, Levy and van der Watt (1990) concluded that like in the HC measurements, K has an intermediate effect on the IR soils exposed to rain, between those of Ca and Na.

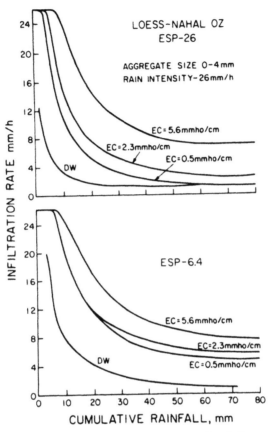

Figure 21. Effect of electrolyte concentration in rain-simulation experiments on the infiltration rate of Hamra soil. (From Agassi et al., 1981.)

D. Effect of Electrolyte Concentration on the IR of Sodic Soils

Agassi et al. (1981) studied the effect of electrolyte concentration and soil sodicity on the IR of two loamy soils in Israel with the aid of a laboratory rain simulator. Comparing their results (Figures 21 and 22) to the results of Felhendler et al. (1974) who studied the HC of the same soils (Figure 9) revealed that the IR was by far more sensitive than the HC to the electrolyte concentration of the applied water. For example, the HC of the Natanya soil equilibrated with SAR 10 solution was not affected by the electrolyte concentration in the percolating solution provided it was above 5 mmol$_c$ l^{-1} (Figure 9). Thus, from the results obtained with these Israeli soils and some soils from California (Shainberg et al., 1981a,b) it is evident that concentration >5 mmol$_c$

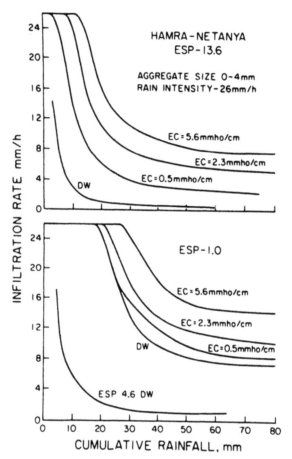

Figure 22. Effect of electrolyte concentration in rain-simulation experiments on the infiltration rate of loess soil. (From Agassi et al., 1981.)

l[-1] is enough to prevent the adverse effect of sodium in the exchange complex on the HC of soils with ESP levels in the range of 0-15.

By contrast, the IRs of the Israeli soils in the above ESP range were affected by electrolyte concentration up to 50 mmol$_c$ l[-1] (Figures 21 and 22). Similarly, Oster and Schroer (1979), who studied a Heimdal soil from North Dakota, found that cation concentration greatly affected IR even at low SAR levels. They observed an increase in final IR from 2 to 28 mm h[-1] as the cation concentration in the applied solutions having SAR levels between 2 and 4.6 increased from 5 to 28 mmol$_c$ l[-1].

The results presented in Figures 21 and 22 show that salt concentration had an effect not only on the final IR, but also on the rate at which the IR dropped from the initial to its final value. The lower the concentration of the electrolyte, the faster the rate at which the IR decreased. Similarly, for the same electrolyte concentration, increasing soil ESP results in a sharper decrease in the IR and a lower final IR.

E. Effect of Consecutive Rainstorms and Saline Water

Alternate application of rain and saline water predominate in many semiarid regions of the world. The stability and reversibility of the crusts formed was found to depend on the period of drying and water quality (Hardy et al., 1983; Levy et al., 1986). The term crust is used to distinguish a dried seal from a wet seal. In their experiments, Hardy et al. (1983) studied the effect of subsequent saline water (SW) sprinkling on the stability of crusts formed under distilled water (DW) rain. They found that application of SW in the second storm caused an increase in the final IR of the Netanya soil to 5.6 mm h^{-1} compared to 2 mm h^{-1} in the first storm with DW (Figure 23). They observed that in this soil the raindrop impact caused a breakdown of the old crust and a formation of a new, more permeable one due to the use of SW. In the loess soil from Nahal-Oz only a slight increase in the final IR was observed in the consecutive SW storm (Figure 23). Hardy et al. (1983) concluded that in the loess soil the structure of the crust formed in the first storm was stable and did not break during the second storm. In a further study on crust stability, Levy et al. (1986) observed for the loess soil from Nahal-Oz that if the crust was sufficiently dry, applying a second storm with SW increased the final IR to 6.4 mm h^{-1} compared to 2.5 mm h^{-1} at the end of the former DW storm. In the case where only partial drying occurred, changing from distilled water in the first storm to saline water in the consecutive storm did not increase the final IR (Levy et al., 1986). It was then concluded that sufficient drying of the crust weakens its structure and causes its breakdown by the beating action of the raindrops in the subsequent storm. The quality of the water in that storm will determine the characteristics of the newly formed crust.

F. Effect of Soil Amendments

1. Application of Phosphogypsum

Chemical dispersion during the process of seal formation can be prevented by spreading phosphogypsum (PG) at the soil surface (Agassi et al., 1982; Kazman et al., 1983; Keren and Shainberg, 1981). PG is a byproduct of the phosphate fertilizer industry and is available in large quantities. Application of PG at 5 Mg ha^{-1} prior to the rain on two Israeli soils (Kazman et al., 1983) prevented the rate at which the IR dropped as well as the final IR (Figures 19 and 20). For instance, the IR of the sandy loam from Netanya with ESP 1.0 increased from 7.5 mm h^{-1} in the control to 12 mm h^{-1} after PG application. In the case where the ESP was 11.6, the IR increased from 0.6 to 10 mm h^{-1} (Figure 19). PG

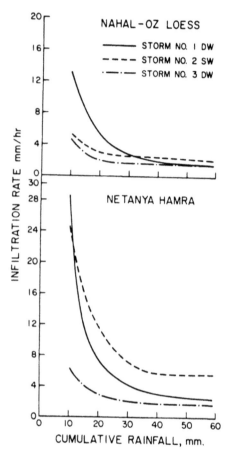

Figure 23. Infiltration rates of Nahal-Oz and Netanya soils subjected to three consecutive storms of sequence DW-SW-DW. (From Hardy et al., 1983.)

application showed similar efficiency in maintaining high IR values in Ultisols from Georgia (Miller, 1987a).

Keren and Shainberg (1981) found that PG was more effective in maintaining IR than mined gypsum because its rate of dissolution was much higher. The dissolution rate is very important in the case of IR and seal formation because of the short contact time between rainwater and gypsum at the soil surface. The EC of the soil solution from soil surface treated with mined gypsum was apparently not sufficient to prevent clay dispersion and seal formation due to slow dissolution of this type of gypsum.

The main mechanism by which PG affects the IR of soils exposed to low electrolyte source of water is by dissolution and release of electrolytes into the soil solution. Agassi et al. (1986) compared the IR of a loess soil treated with 5 Mg ha^{-1} powdered PG and exposed to DW rain to that of the same soil

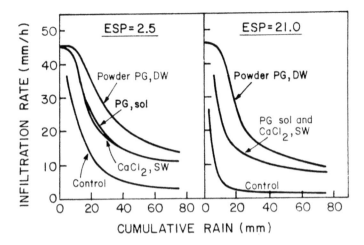

Figure 24. The effect of phosphogypsum (PG) powder and distilled water (DW) rain, compared with CaCl$_2$ solution (SW) and PG solution, on the IR of loess soil with high and low ESP. (From Agassi et al., 1986.)

exposed to saline "rainwater" of 0.01 M CaCl$_2$ or saturated PG solution. Although the salinity of the percolating water was similar (= 2 dS m^{-1}) in all three treatments, the IR curves obtained with the saline solutions were lower than those obtained with the PG treatment (Figure 24). It was evident that PG affects the IR of the soils not only through its effect on the EC of the percolating water (Agassi et al., 1986). These researchers postulated that powdered PG at the soil surface may affect seal formation by two additional physical mechanisms: (i) by interfering mechanically with the structure of the seal and thus disturbing the formation of a continuous seal; and (ii) by serving as a mulch and partially protecting the soil surface from the beating action of the raindrops.

2. Application of Polymers

Preventing seal formation and thereby maintaining high IR values could be achieved by improving soil structure and aggregate stability at the soil surface. Interest in organic polymers as soil conditioners and stabilizers was enhanced in the early 1950s (Chepil, 1954). However, most of the early studies used high rates of polymers which proved economically infeasible for agriculture.

There has recently been a renewed interest in organic polymers, mainly polyacrylamides (PAM) and polysaccharides in general (*Soil Sci.*, Special Issue, Vol. 141, No. 5, 1985) and particularly in improving soil structure, aggregate stability, infiltration rate, and soil erosion (Ben-Hur et al., 1989; Helalia and Letey, 1988; Shainberg et al., 1990; Shaviv et al., 1986; Smith et al., 1990). Under optimal conditions, treatment with PAM increased the final IR from 2.0 to 23.5 mm h^{-1} and rain intake from a storm of 80 mm from 12.3 to 64.2 mm

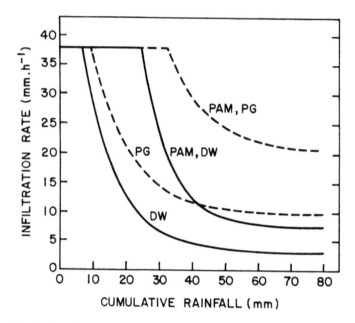

Figure 25. Infiltration rate of grumusol as a function of cumulative rainfall, phosphogypsum treatment (PG, 5 ton-ha[-1]), and PAM application (20 kg-ha[-1]). (From Shainberg et al., 1990.)

(Shainberg et al., 1990). However, optimal conditions included addition of electrolytes to the soil by either spreading PG or using electrolyte solution. When the PAM was not supplemented with electrolytes the final IR was only 6.5 mm h[-1] (Figure 25). Shainberg et al. (1990) concluded that in order for the PAM to be effective in improving soil IR, it is essential that the soil colloids be in a flocculated state. This can be obtained by supplying a source of electrolytes to the soil surface. Similar results but with low energy rain to simulate irrigation conditions were obtained by Smith et al. (1990).

G. Effect of Electrolyte Concentration on the IR of Depositional Seals

The previous sections dealt with the IR of seals formed by the impact energy of waterdrops from rain or overhead irrigation. Such types of seals are also known as structural seals. Another type of seal is that formed by translocation of fine soil particles and their deposition at a certain distance from their original location (Arshad and Mermut, 1988; Chen et al., 1980; Eisenhauer, 1984). When the velocity of running water exceeds the shear strength of the soil surface, the water erodes the soil and entrains sediments. As water velocity decreases the sediments in the water deposit and a depositional seal is formed. Formation of a depositional seal is important in surface (furrow and basin) irrigation where there is no water drop impact (Kemper et al., 1985) and may lead to intake rate as low as 1 mm h^{-1} (Oster and Singer, 1985).

Shainberg and Singer (1985) studied in the laboratory the effect of electrolyte concentration on the HC of depositional seals. They found that the penetration rate of soil suspensions of approximate concentration of 1% was drastically affected by low levels of electrolytes in the solution. The HC of the depositional seal was two to three orders of magnitude lower than that of the bulk soil for solution with EC <0.3 dS m^{-1}. Increasing the EC of the suspension either by directly adding electrolytes to it, or by spreading PG on the soil surface resulted in the formation of a depositional seal with HC of only one order of magnitude lower than that of the bulk soil. Shainberg and Singer (1985) suggested that the HC of the depositional seal depended on whether or not the clay in the suspended solution was in a flocculated state, which in turn depended on the EC of the solution. They further hypothesized that when the EC exceeded the flocculation value of the suspended clay, the seal formed consisted of flocculated particles deposited randomly in an open structure resulting in high permeability.

The latter hypothesis was supported by results obtained from a direct study of the micromorphology of depositional seals (Southard et al., 1988). These researchers observed that the density and micromorphology of these seals depended strongly on the EC of the soil solution. Seals formed from soil material suspended in DW had a high bulk density and had oriented birefringent layers of clay. When the soil material was suspended in solutions containing some electrolytes, less dense and more porous seals with no birefringent oriented clay layers were formed (Southard et al., 1988).

V. Summary

The swelling and dispersion of clays are affected by the composition of the adsorbed cations and the electrolyte concentration in the soil solution. Adsorbed Na ions form a thick diffuse double layer, high swelling and dispersed clay particles which tend to persist in dilute solution. The low swelling pressure between Ca-clay platelets prevents their dispersion, and the Ca-saturated platelets stack into tactoids or quasi crystals. In a mixed Na/Ca system, introducing a small amount of Na (ESP < 15) to a largely Ca saturated clay has little effect on the swelling. A larger amount of Na (ESP > 15) brings about a breakdown of the tactoid and intensive swelling. Conversely, clay dispersion takes place at low ESP values, but only when the electrolyte concentration is below the flocculation value. When Na is introduced into Ca-montmorillonite, the Na ions will concentrate on the external surfaces of the tactoids (demixing) leading to a higher ESP at the external surfaces compared with the ESP at the internal surfaces. This phenomenon explains the pronounced effect of Na even in clays and soils of low sodicity. Swelling and/or dispersion of soil colloids alters the geometry of soil pores and thus greatly affects the rate of water flow in soils.

The HC of a soil depends on both the exchangeable cation composition and the salt concentration and composition of the percolating solution. High HC can be maintained, even at high ESP levels, provided the solution concentration is above a critical (threshold) level. Leaching with salt solutions at concentrations exceeding the flocculation value prevents clay dispersion and can cause a reduction in the HC by clay swelling. At this concentration range, susceptibility of soils to sodic conditions is enhanced by increasing clay content and the presence of expandable clay minerals (smectites). At low ESP levels little soil swelling is expected and marked reduction in HC may occur by clay dispersion in dilute salt solutions. Soil susceptibility to sodic conditions when leached with very dilute solutions, depends on its potential to release electrolytes through dissolution of primary minerals and/or $CaCO_3$. Soils that release enough electrolytes to prevent clay dispersion may maintain relatively high HC.

Water infiltration into soils exposed to rain or overhead irrigation is determined by the permeability of the seal formed at the soil surface. Seal formation results from two complementary processes: breakdown and compaction of aggregates at soil surface by impact of water drops and chemical dispersion of clay particles. Drop impact increases the susceptibility of soil surface to quality water and exchangeable cation composition. Increasing soil ESP from 1 to 2.5 is enough to cause a drop of 70% in the final IR. PG application controls seal formation and maintains high IR values because it dissolves and prevents clay dispersion at the soil surface. Surface application of polymers and PG stabilizes the structure of the soil surface and prevents seal formation. Soil management that increases rain infiltration contributes to water and soil conservation.

References

Agassi, M., J. Morin, and I. Shainberg. 1982. Infiltration and runoff control in the semi-arid region of Israel. *Geoderma* 28:345-355.

Agassi, M., J. Morin, and I. Shainberg. 1985. Effect of raindrop impact energy and water salinity on infiltration rates of sodic soils. *Soil Sci. Soc. Am. J.* 49:186-190.

Agassi, M., I. Shainberg, and J. Morin. 1981. Effect of electrolyte concentration and soil sodicity on infiltration rate and crust formation. *Soil Sci. Soc. Am. J.* 45:848-851.

Agassi, M., I. Shainberg, and J. Morin. 1986. Effect of powdered phosphogypsum on the infiltration rate of sodic soils. *Irrig. Sci.* 7:53-61.

Ahmed, S., L.D. Swindale, and S.A. El-Swaify. 1969. Effects of adsorbed cations on physical properties of tropical black earths. 1. Plastic limit, percentage stable aggregates and hydraulic conductivity. *J. Soil Sci.* 20:255-268.

Alperovitch, N., I. Shainberg, and R. Keren. 1981. Specific effect of magnesium on the hydraulic conductivity of sodic soils. *J. Soil Sci.* 32:543-554.

Alperovitch, N., I. Shainberg, R. Keren, and M.J. Singer. 1985. Effect of clay mineralogy and aluminum and iron oxides on the hydraulic conductivity of clay-sand mixtures. *Clays Clay Miner.* 33:443-450.

Arora, H.S. and N.T. Coleman. 1979. The influence of electrolyte concentration on flocculation of clay suspensions. *Soil Sci.* 32:543-554.

Arshad, M.A. and A.R. Mermut. 1988. Micromorphological and physicochemical characteristics of soil crust types in Northwestern Alberta, Canada. *Soil Sci. Soc. Am. J.* 52:724-729.

Aylmore, L.A.G. and J.P. Quirk. 1959. Swelling of clay-water systems. *Nature* 183:1752-1753.

Bakker, A.C. and W.W. Emerson. 1973. The comparative effect of exchangeable calcium, magnesium and sodium on some physical properties of red-brown earth subsoils. III. The permeability of Shepperton soil and comparison methods. *Aust. J. Soil Res.* 11:159-165.

Banin, A. and N. Lahav. 1968. Particle size and optical properties of montmorillonite in suspension. *Isr. J. Chem.* 6:235-250.

Bar-On, P., I. Shainberg, and I. Mochaeli. 1970. The electrophoretic mobility of Na/Ca montmorillonite particles. *J. Colloid Interface Sci.* 33:471-472.

Ben-Hur, M., J. Faris, M. Malik, and J, Letey. 1989. Polymer as soil conditioner under consecutive irrigation and rainfall. *Soil Sci. Soc. Am. J.* 53:1173-1178.

Ben-Hur, M., I. Shainberg, D. Bakker, and R. Keren. 1985. Effect of soil texture and $CaCO_3$ content on water infiltration in crusted soil as related to water salinity. *Irrig. Sci.* 6:281-294.

Blackmore, A.V. and R.D. Miller. 1961. Tactoid size and osmotic swelling in Ca montmorillonite. *Soil Sci. Soc. Am. Proc.* 25:169-173.

Bresler, E. 1970. Numerical solution of the equation for interacting diffuse layer in mixed ionic system with non-symmetrical electrolytes. *J. Colloid Interface Sci.* 3:278-2283.

Bresler, E., and W.D. Kemper. 1970. Soil water evaporation as affected by wetting methods and crust formation. *Soil Sci. Soc. Am. Proc.* 34:3-8.

Bresler, E., B.L. McNeal, and D.L. Carter. 1982. *Saline and sodic soils.* Springer-Verlag, New York.

Cass, A. and M.E. Sumner. 1982a. Soil pore structural stability and irrigation water quality. I. Empirical sodium stability model. *Soil Sci. Soc. Am. J.* 46:503-506.

Cass, A. and M.E. Sumner. 1982b. Soil pore structural stability and irrigation water quality. II. Sodium stability data. *Soil Sci. Soc. Am. J.* 46:507-512.

Cecconi, S., A. Salazar, and M. Martelli. 1963. The effect of different cations on the structural stability of some soils. *Agrochimica* 7:185-204.

Chen, Y. and A. Banin. 1975. Scanning electron microscope (SEM) observations of soil structure changes induced by Sodium-Calcium exchange in relation to hydraulic conductivity. *Soil Sci.* 120:428-436.

Chen, Y., A. Banin, and A. Borochovitch. 1983. Effect of potassium on soil structure in relation to hydraulic conductivity. *Geoderma* 30:135-147.

Chen, Y., J.T. Tarchitzky, J. Brouwer, J. Morin, and A. Banin. 1980. Scanning electron microscope observation of soil crusts and their formation. *Soil Sci.* 130:49-55.

Chepil, W.S. 1954. The effect of synthetic conditioners on some phases of soil structure and erodibility by wind. *Soil Sci. Soc. Am. Proc.* 18:386-390.

Chi, C.L., W.W. Emerson, and D.G. Lewis. 1977. Exchangeable calcium, magnesium and sodium and the dispersion of illites in water. I. Characterization of illites and exchange reactions. *Aus. J. of Soil Res.* 15:243-253.

Chiang, S.C., D.E. Radcliffe, W.P. Miller, and K.D. Newman. 1987. Hydraulic conductivity of three southeastern soils as affected by sodium, electrolyte concentration and pH. *Soil Sci. Soc. Am. J.* 51:1293-1299.

Childs, E.C. and N. Collis-George. 1950. The permeability of porous materials. Proc. R. Soc. Ser. A. 201:392-405.

Deshpande, T.L., D.J. Greenland, and J.P. Quirk. 1968. Changes in soil properties associated with the removal of iron and aluminum oxides. *J. Soil Sci.* 19:108-122.

Dufey, J.E., A. Banin, H. Laudelout, and Y. Chen. 1976. Particle shape and sodium self-diffusion coefficient in mixed Na-Ca montmorillonite. *Soil Sci. Soc. Am. J.* 40:310-314.

Duley, F.L. 1939. Surface factors affecting the rate of intake of water by soils. *Soil Sci. Soc. Am. Proc.* 4:0-64.

Durgin, P.B. and J.G. Chaney. 1984. Dispersion of kaolinite by dissolved organic matter from Douglas-Fir roots. *Can. J. Soil Sci.* 64:445-455.

Eisenhauer, D.E. 1984. Surface sealing and infiltration with surface irrigation. Unpublished Ph.D. Dissertation, Colorado State Univ. Fort Collins, Colorado.

Elgabaly, M.M. and W.A. Elghamry. 1970. Water permeability and stability of kaolinite systems as influenced by adsorbed cation ratio. *Soil Sci.* 110:107-110.

Ellis, J.H. and O.G. Caldwell. 1935. Magnesium clay solonetz. Trans. 3rd Congr. *Soil Sci.* Vol. I, p. 348-350.

El-Rayah, H.M.E. and D.L. Rowell. 1973. The influence of Fe and AL hydroxides on the swelling of montmorillonite and the permeability of a Na-soil. *J. Soil Sci.* 24:137-144.

El-Swaify, S.A. 1973. Structural changes in tropical soils due to anions in irrigation water. *Soil Sci.* 115:64-72.

El-Swaify, S.A. and L.D. Swindale. 1969. Hydraulic conductivity of some tropical soils as a guide to irrigation water quality. *Ninth Int. Congr. Soil Sci. Trans.* 1:381-389.

El-Swaify, S.A., S. Ahmed, and L.D. Swindale. 1970. Effects of adsorbed cations on physical properties of tropical red and tropical black earths. II. Liquid limit, degree of dispersion and moisture retention. *J. Soil Sci.* 21:188-198.

Emerson, W.W. 1967. A classification of soil aggregates based on their coherence in water. *Aust. J. Soil Res.* 5:47-57.

Emerson, W.W. and A.C. Bakker. 1973. The comparative effects of exchangeable calcium, magnesium, and sodium on some physical properties of red-brown earthsubsoils. *Aust. J. Soil Res.* 11:151-157.

Emerson, W.W. and C.L. Chi. 1977. Exchangeable Ca, Mg and Na and the dispersion of illites. II. Dispersion of illites in water. *Aust. J. Soil Res.* 15:255-263.

Emerson, W.W. and B.H. Smith. 1970. Magnesium, organic matter and soil structure. *Nature* (London) 228:453-454.

Epstein, E. and W.J. Grant. 1973. Soil crust formation as affected by raindrop impact. In A. Hadas et al. (ed.) *Ecological studies, 4*, Physical aspects of soil water and salts in ecosystems, p. 195-201. Springer Verlag, Berlin-Heidelberg-New York.

Evans, D.D. and S.W. Buol. 1968. Micromorphological study of soil crusts. *Soil Sci. Soc. Am. Proc.* 32:19-22.

Felhendler, R., I. Shainberg, and H. Frenkel. 1974. Dispersion and hydraulic conductivity of soils in mixed solution. *Trans. of the 10th Intern. Cong. of Soil Sci.* (Moscow) I:103-112. Moscow:Nauka Pub. House.

Frenkel, H. and I. Shainberg. 1975. Chemical and hydraulic changes in soils irrigated with brackish water. In: *Irrigation with Brackish Water*, Intern. Symp. Beersheva, Israel.

Frenkel, H. and I. Shainberg. 1980. The effect of hydroxy Al and Fe polymers on montmorillonite particle size. *Soil Sci. Soc. Am. J.* 44:626-629.

Frenkel, H., J.O. Goertzen, and J.D. Rhoades. 1978. Effects of clay type and content, exchangeable sodium percentage, and electrolyte concentration on clay dispersion and soil hydraulic conductivity. *Soil Sci. Soc. Am. J.* 48:32-39.

Gal, M., L. Arcan, I. Shainberg, and R. Keren. 1984. The effect of exchangeable sodium and phosphogypsum on the structure of soil crusts. *Soil Sci. Soc. Am. J.* 48:872-878.

Gardner, W.R., M.S. Mayhugh, J.O. Goertzen, and C.A. Bower. 1959. Effect of electrolyte concentration and ESP on diffusivity of water in soil. *Soil Sci.* 88:270-274.

Goldberg, S. and R.A. Glaubig. 1987. Effect of saturating cation, pH and aluminum and iron oxides on the flocculation of kaolinite and montmorillonite. *Clays Clay Miner.* 35:220-227.

Green, W.H. and G.A. Ampt. 1911. Studies on soil physics: I. Flow of air and water through soils. *J. Agr. Sci.* 4:1-24.

Grim, R.E. 1968. *Clay Mineralogy.* 2nd ed. New York: McGraw-Hill.

Hardy, N., I. Shainberg, M. Gal, and R. Keren. 1983. The effect of water quality and storm sequence upon infiltration rate and crust formation. *J. Soil Sci.* 34:665-676.

Helalia, A.M. and J. Letey. 1988. Cationic polymer effects on infiltration rates with a rainfall simulator. *Soil Sci. Soc. Am. J.* 52:247-250.

Hilgard, E.W. 1906. *Soils their formation, properties composition and relation to climate and plant growth in the humid and arid regions.* Macmillan, London.

Hillel, D. 1980a. *Fundamentals of soil physics.* Academic Press, New York.

Hillel, D. 1980b. *Application of soil physics.* Academic Press, New York.

Holtan, H.N. 1961. A concept for infiltration estimates in watershed engineering. *U.S. Dept. Agr., Agr. Res. Service Publ.* p. 41-51.

Horton, R.E. 1940. An approach toward a physical interpretation of infiltration-capacity. *Soil Sci. Soc. Am. Proc.* 5:399-417.

Jayawardane, N.S. 1977. The effect of salt composition of groundwater on the rate of salinization of soils from a water table. Unpublished Ph.D. dissertation, Univ. of Tasmania.

Jayawardane, N.S. 1979. An equivalent salt solutions method for predicting hydraulic conductivity of soils for different salt solutions. *Aust. J. Soil Res.* 17:423-428.

Jayawardane, N.S. and P.S. Blackwell. 1991. Relationship between equivalent salt solution sereis of different soils. *J. Soil Sci.* 42:95102.

Kazman, Z., I. Shainberg, and M. Gal. 1983. Effect of low levels of exhangeable Na (and phosphogypsum) on the infiltration rate of various soils. *Soil Sci.* 135:184-192.

Kelley, W.P. and A. Arany. 1928. The chemical effect of gypsum, sulphur, iron sulphateand alum on alkali soil. *Hilgardia* 3:393-420.

Kemper, W.D., T.J. Trout, M.J. Brown, and R.C. Rosenau. 1985. Furrow erosion and water and soil management. *Trans. ASAE.* 28:1564-1572.

Keren, R. 1980. Effect of titration rate, pH, and drying process on cation exchange capacity reduction and aggregate size distribution of montmorillonite hydroxy-Al complexes. *Soil Sci. Soc. Am. J.* 44:1209-1212.

Keren, R. 1989. Water-drop kinetic energy effect on water infiltration in calcium and magnesium soils. *Soil Sci. Soc. Am. J.* 53:1624-1628.

Keren, R. 1990. Water-drop kinetic energy effect on infiltration in sodium-calcium-magnesium soils. *Soil Sci. Soc. Am. J.* 54:983-987.

Keren, R. and G.A. O'Connor. 1982. Gypsum dissolution and sodic soil reclamation as affected by water flow velocity. *Soil Sci. Soc. Am. J.* 46:726-732.

Keren R. and I. Shainberg. 1981. Effect of dissolution rate on the efficiency of industrial and mined gypsum in improving infiltration in a sodic soil. *Soil Sci. Soc. Am. J.* 47:1001-1004.

Keren, R. and M.J. Singer. 1988. Effect of low electrolyte concentration on hydraulic conductivity of sodium/calcium-montmorillonitesand system. *Soil Sci. Soc. Am. J.* 52:368-373.

Keren, R. and M.J. Singer. 1989. Effect of low electrolyte concentration on hydraulic conductivity of clay-sand-hydroxy polymers systems. *Soil Sci. Soc. Am. J.* 53:349-355.

Keren, R. and M.J. Singer. 1990. Effect of pH on permeability of clay-sand mixture containing hydroxy polymers. *Soil Sci. Soc. Am. J.* 54:1310-1315.

Keren, R., I. Shainberg, and E. Klein. 1988. Settling and flocculation value of sodium-montmorillonite particles in aqueous media. *Soil Sci. Soc. Am. J.* 52:76-80.

Kinniburgh, D.G., J.K. Syers, and M.L. Jackson. 1975. Specific adsorption of trace amounts of calcium and strontium by hydrousoxides of iron and aluminum. *Soil Sci. Soc. Am. J.* 39:464-470.

Kreit, J.E., I. Shainberg, and A.J. Herbillon. 1982. Hydrolysis and decomposition of hectorites in dilute solutions. *Clays Clay Miner.* 30:223-231.

Lagerwerff, J.V., F.S. Nakayama, and M.H. Frere. 1969. Hydraulic conductivity related to porosity and swelling of soil. *Soil Sci. Soc. Am. Proc.* 33:3-11.

Levy, G.J. and H.v.H. van der Watt. 1990. Effect of exchangeable potassium on the hydraulic conductivity and infiltration rate of some South African soils. *Soil Sci.* 149:69-77.

Levy, G.J., N. Alperovitch, A.J. van der Merwe, and I. Shainberg. 1989. The hydrolysis of kaolinitic soils as affected by the type of the exchangeable cation. *J. Soil Sci.* 40:613-620.

Levy, G.J., P.R. Berliner, H.M. du Plessis, and H.v.H. van der Watt. 1988a. Microtopographical characteristics of artificially fromed crusts. *Soil Sci. Soc. Am. J.* 52:784-791.

Levy, G.J., H.v.H. van der Watt, and H.M. du Plessis. 1988b. Effect of sodiummagnesium and sodium-calcium systems on soil hydraulic conductivity and infiltration. *Soil Sci.* 146:303-310.

Levy, G.J., I. Shainberg, and J. Morin. 1986. Factors affecting the stability of soil crusts in subsequent storms. *Soil Sci. Soc. Am. J.* 50:196-201.

Levy, G.J., I. Shainberg, N. Alperovitch, and A.J. van der Merwe. 1991. Effect of Na-hexametaphosphate on the hydraulic conductivity of kaolinite-sand mixture. *Clays Clay Miner.* 39:131-136.

Lima, L.A., M.E. Grismer, and D.R. Nielsen. 1990. Salinity effects on yolo loam hydraulic properties. *Soil Sci.* 150:451-458.

Loveday, J. 1976. Relative significance of electrolyte and cation exchange effects when gypsum is applied to a sodic clay soil. *Aust. J. Soil Res.* 14:361-371.

Marshall, T.J. 1958. A relation between permeability and size distribution of pores. *J. Soil Sci.* 9:1-8.

McAtee, J.L. 1961. Heterogeneity in montmorillonites. *Clays Clay Miner.* 5:279-288.

McIntyre, D.S. 1958. Permeability measurements of soil crusts formed by raindrop impact. *Soil Sci.* 85:185-189.

McIntyre, D.S. 1979. Exchangeable sodium, subplasticity and hydraulic conductivity of some Australian soils. *Aust. J. Soil Res.* 17:115-120.

McNeal, B.L. 1968. Prediction of the effect of mixed-salt solutions on soil hydraulic conductivity. *Soil Sci. Soc. Am. Proc.* 32:190-193.

McNeal, B.L. and N.T. Coleman. 1966. Effect of solution composition on soil hydraulic conductivity. *Soil Sci. Soc. Am. Proc.* 20:308-312.

McNeal, B.L., D.A. Layfield, W.A. Norvell, and J.D. Rhoades. 1968. Factors influencing hydraulic conductivity of soils in the presence of mixed-salt solutions. *Soil. Sci. Soc. Am. Proc.* 32:187-190.

McNeal, B.L., W.A. Norvell, and N.T. Coleman. 1966. Effect of solution composition on soil hydraulic conductivity and on the swelling of extracted soil clays. *Soil Sci. Soc. Am. Proc.* 30:308-315.

Mering, J. and R. Glaeser. 1954. On the role of the valency of exchangeable cations in montmorillonite. *Bull. Soc. Fr. Miner. Cristallogr.* 77:519-530.

Miller, W.P. 1987a. Infiltration and soil loss of three gypsumamended ultisols under simulated rainfall. *Soil Sci. Soc. Am. J.* 51:1314-1320.

Miller, W.P. 1987b. A solenoid-operated variable intensity rainfall simulator. *Soil Sci. Soc. Am. J.* 51:832-834.

Miller, W.P. and M.K. Baharuddin. 1986. Relationship of soil dispesivity to infiltration and erosion of southeastern soils. *Soil Sci.* 142:235-240.

Miller, W.P. and J. Scifres. 1988. Effect of sodium nitrate and gypsum on infiltration and erosion of a highly weathered soil. *Soil Sci.* 148:304-309.

Mohammed, E.T.Y., J. Letey, and R. Branson. 1979. Sulphur compounds in water treatment. *Sulphur in Agriculture* 3:7-11.

Morin, J., S. Goldberg, and I. Seginer. 1967. A rainfall simulator with a rotating disk. *Trans. Am. Soc. Agric. Engrs.* 10:74-79.

Morin, J. and Y. Benyamini. 1977. Rainfall infiltration into bare soils. *Water Resour. Res.* 13:813-817.

Morin, J., Y. Benyamini, and A. Michaeli. 1981. The dynamics of soil crusting by rainfall impact and the water movement in the soil profile. *J. Hydrol.* 52:321-335.

Norrish, K. 1954. The swelling of montmorillonite. *Disc. Faraday Soc.* 18:120-134.

Oades, J.M. 1984. Interactions of polycations of Al and Fe with clays. *Clays Clay Miner.* 32:49-57.

Onofiok, O. and M.J. Singer. 1984. Scanning electron microscope studies of surface crusts formed by simulated rainfall. *Soil Sci. Soc. Am. J.* 48:1137-1143.

Oster, J.D. 1982. Gypsum usage in irrigated agriculture: A review. *Fertilizer Res.* 3:73-89.

Oster, J.D. and H. Frenkel. 1980. The chemistry of the reclamation of sodic soils with gypsum and lime. *Soil Sci. Soc. Am. J.* 44:41-45.

Oster, J.D. and F.W. Schroer. 1979. Infiltration as influenced by irrigation water quality. *Soil Sci. Soc. Am. J.* 43:444-447.

Oster, J.D. and I. Shainberg. 1979. Exchangeable cation hydrolysis and soil weathering as affected by exchangeable sodium. *Soil Sci. Soc. Am. J.* 43:70-75.

Oster, J.D. and M.J. Singer. 1984. *Water penetration problems in California soils.* Univ. of California, Davis California.

Oster, J.D., I. Shainberg, and J.D. Wood. 1980. Flocculation value and gel structure of Na/Ca montmorilonite and illite suspensions. *Soil Sci. Soc. Am. J.* 44:955-959.

Peele, T.C. 1936. The effect of calcium on the erodibility of soils. *Soil Sci. Soc. Am. Proc.* 1:47-58.

Philip, J.R. 1957. The theory of infiltration: 4. Sorptivity and algebraic infiltration equations. Soil Sci. 84:257-264.

Pupisky, H. and I. Shainberg. 1979. Salt effects on the hydraulic conductivity of a sandy soil. *Soil Sci. Soc. Am. J.* 43:429-433.

Quirk, J.P. and R.K. Schofield. 1955. The effect of electrolyte concentration on soil permeability. *J. Soil Sci.* 6:163-178.

Rahman, A.W. and D.L. Rowell. 1979. The influence of Mg in saline and sodic soils: a specific effect or a problem of cation exchange. *J. Soil Sci.* 30:535-546.

Ravina, I. 1973. The mechanical and physical behaviour of Ca-Clay soil and K-Clay soil. *Ecol. Stud.* 4:131-140.

Reeve, R.C., C.A. Bower, R.H. Brooks, and F.B. Gschwend. 1954. A comparison of the effects of exchangeable Na and K upon the physical conditions of soils. *Soil Sci. Soc. Am. Proc.* 18:130-132.

Remley, P.A. and J.M. Bradford. 1989. Relationship of soil crust morphology to interrill erosion parameters. *Soil Sci. Soc. Am. J.* 53:1215-1219.

Rhoades, J.D. and R.D. Ingvalson. 1969. Macroscopic swelling and hydraulic conductivity properties of four vermiculite soils. *Soil Sci. Soc. Am. Proc.* 33:364-369.

Rhoades, J.D., D.B. Kruger, and M.J. Reed. 1968. The effect of soil mineral weathering on the sodium hazard of irrigation waters. *Soil Sci. Soc. Am. Proc.* 32:643-647.

Rich, C.I. 1968. Hydroxy interlayers in expansible layer silicates. *Clays Clay Miner.* 16:15-30.

Rimmer, D.L. and D.J. Greenland. 1976. Effect of $CaCO_3$ on the swelling of a soil clay. *J. Soil Sci.* 27:129-139.

Rowell, D.L. 1963. Hydraulic conductivity of clays during shrinkage. *Soil Sci. Soc. Am. Proc.* 30:289-292.

Rowell, D.L. and I. Shainberg. 1979. The influence of magnesium and easily weathered minerals on hydraulic conductivity changes in a sodic soil. *J. Soil Sci.* 30:719-726.

Rowell, D.L., D. Payne, and N. Ahmad. 1969. The effect of the concentration and movement of solutions on the swelling, dispersion and movement of clay in saline and alkali soils. *J. Soil Sci.* 20:176-188.

Russo, D. and E. Bresler. 1977a. Effect of mixed Na/Ca solutions on the hydraulic properties of unsaturated soils. *Soil Sci. Soc. Am. J.* 41:713-717.

Russo, D. and E. Bresler. 1977b. Analysis of the saturatedunsaturated hydraulic conductivity in a mixedNa/Ca soil system. *Soil Sci. Soc. Am. J.* 41:706-710.

Russo, D. and E. Bresler. 1980. Soil-warer-suction relationship as affected by soil solution compositionand concentration. In: *Agrochemicals in soils.* A.Banin and U. Kafkafi (eds.). Pergamon Press, England, pp. 287-296.

Schofield, R.K. and H.R. Samson. 1954. Flocculation of kaolinite due to the attraction of oppositely charged crystal faces. *Disc. Faraday Soc.* 18:138145.

Shainberg, I. and M. Gal. 1982. The effect of lime on the response of soils to sodic conditions. *J. Soil Sci.* 33:489-498.

Shainberg, I. and A. Kaiserman. 1969. Kinetics of the formation and breakdown of Ca-montmorillonite tactoids. *Soil Sci. Soc. Am. Proc.* 33:547-551.

Shainberg I. and W.D. Kemper. 1967. Electrostatic forces between clay and cations as calculated and inferred from electrical conductivity. *Proc. 14th Nat. Conf. Clays Clay Miner.* p. 117-132.

Shainberg, I. and H. Otoh. 1968. Size and shape of montmorillonite particles saturated with Na/Ca ions. *Isr. J. Chem.* 6:251-259.

Shainberg, I. and J.M. Singer. 1985. Effect of electrolyte concentration on the hydraulic properties of depositional crust. *Soil Sci. Soc. Am J.* 49:1260-1263.

Shainberg, I., E. Bresler, and Y. Klausner. 1971. Studies on Na/Ca montmorill-onite systems. I. The swelling pressure. *Soil Sci.* 111:214-219.

Shainberg, I., J.D. Rhoades, and R.J. Prather. 1981a. Effect of low electrolyte concentration on clay dispersion and hydraulic conductivity of a sodic soil. *Soil Sci. Soc. Am. J.* 45:273-277.

Shainberg, I., J.D. Rhoades, D.L. Suarez, and R.J. Prather. 1981b. Effect of mineral weathering on clay dispersion and hydraulic conductivity of sodic soils. *Soil Sci. Soc. Am. J.* 45:287-291.

Shainberg, I., R. Keren, and H. Frenkel. 1982. Response of sodic soils to gypsum and $CaCl_2$ application. *Soil Sci. Soc. Am. J.* 46:113-117.

Shainberg, I., N. Alperovitch, and R. Keren. 1987a. Charge density and Na/K/Ca exchange on smectites. *Clays Clay Miner.* 35:68-73.

Shainberg, I., R. Keren, N. Alperovitch, and D. Goldstein. 1987b. Effect of exchangeable potassium on the hydraulic conductivity of smectite-sand mixtures. *Clays Clay Miner.* 35:305-310.

Shainberg, I., M.J. Singer, and P. Janitzky. 1987c. Effect of aluminum and iron oxides on hydraulic conductivity of sandy loam soil. *Soil Sci. Soc. Am. J.* 51:1283-1287.

Shainberg, I., N. Alperovitch, and R. Keren. 1988. Effect of magnesium on the hydraulic conductivity of sodic smectite-sand mixtures. *Clays Clay Miner.* 36:432-438.

Shainberg, I., D.N. Warrington, and P. Rengasamy. 1990. Water quality and PAM interactions in reducing surface sealing. *Soil Sci.* 149:301-307.

Shanmuganathan, R.T. and J.M. Oades. 1983a. Influence of anions on dispersion and physical properties of the A horizon of a red brown earth. *Geoderma* 29:257-277.

Shanmuganathan, R.T. and J.M. Oades. 1983b. Modifications of soil physical properties byaddition of calcium compounds. *Aust. J. Soil Res.* 21:285-300.

Shaviv, A., I. Ravina, and D. Zaslavsky. 1986. Surface application of anionic surface conditioners to reduce crust formation. p.286-293. In F. Callebaut et al. (ed.) *Int. Worksh. Assessment of soil surface sealing and crusting.* *Proc.* Ghent, Belgium.

Shomer, I. and U. Mingelgrin. 1978. A direct procedure for determining number of platesin tactoids of smectites: The Na/Ca montmorillonite case. *Clays Clay Miner.* 26:135-138.

Smith, H.J.C, G.J. Levy, and I. Shainberg. 1990. Waterdroplet energy and soil amendments: Effect on infiltration and erosion. *Soil Sci. Soc. Am. J.* 54:1084-1087.

Steemkamp, C.J. 1965. Potassium fixation by black sub-tropical clay and changes in mineral lattice of clay on fixation. *S. Afr. Agric. Sci.* 8:535-542.

Stul, M.S. and W.J. Mortier. 1974. The heterogeneity of the charge density in montmorillonites. *Clays Clay Miner.* 22:391-396.

Southard, R.J., I. Shainberg, and M.J. Singer. 1988. Influence of electrolyte concentration on the micromorphology of artificial depositional crust. *Soil Sci.* 145:278-287.

Tarchitzky, J., A. Banin, J. Morin, and Y. Chen. 1984. Nature, formation and effects of soil crusts formed by water drop impact. *Geoderma* 33:135-155.

U.S. Salinity Laboratory Staff. 1954. Diagnosis and improvement of saline and alkali soils. *U.S. Dep. of Agric. Handbook No. 60.* Washington, D.C.

Van der Merwe, A.J. and R. Burger. 1969. The influence of exchangeable cationson certain physical properties of a saline-alkali soil. *Agrochemophysica* 1:63-66.

Van Olphen. 1977. *An introduction to clay colloid chemistry.* 2nd. ed. Wiley, New York.

Velasco-Molina, H.A., A.R. Swoboda, and C.I. Godfrey. 1971. Dispersion of soils of different mineralogy in relation to sodium adsorption ratio and electrolyte concentration. *Soil Sci.* 111:282-287.

Wada, K. and Y. Beppu. 1989. Effect of aluminium treatments on permeability and cation status of a smectite clay. *Soil Sci. Soc. Am. J.* 53:402-406.

Yaron, B. and G.W. Thomas. 1969. Soil hydraulic conductivity as affected by sodic water. *Water Resour. Res.* 4:545-552.

Transport of Inorganic Solutes in Soil

S.E.A.T.M. van der Zee and G. Destouni

I.   Introduction 95
II. Chemical Interaction in Relation with Transport 96
    A. Chemical Equilibria 97
    B. Kinetics 105
    C. Effect of Chemical Behavior on Transport 107
III. Soil Heterogeneity 118
    A. Non-Reactive Solute 119
    B. Sorptive Solute 124
    C. Nitrate Case Study 126
IV. Conclusions 132
References 133

I. Introduction

The water unsaturated (or vadose) zone is a main resource for agricultural production. It is important to maintain a good soil quality, because otherwise the yield or its quality may be adversely affected. Moreover, it is necessary to protect the biosphere in a wider sense against undesirable changes. Evolving nutrient deficiency or contamination may cause shifts in the type or activity of soil biota, and the dynamics of essential elements and pollutants at a local scale (leading to changes in vegetation or potential use of soil), as well as at a global scale.

For a proper management of soils we need to understand the processes that concern the behavior of nutrients and contaminants in soil. This understanding is also needed to manage our ground water resources. After all, in temperate regions most ground water recharge has moved through the vadose zone. The aspects of solute behavior in soil that we are often interested in is their bio-availability and fluxes to other compartments (ground water, atmosphere, surface water). In both aspects transport plays a dominant role. Therefore, many con-

ISBN 0-87371-889-5
© 1992 by Lewis Publishers

ceptual, physico-chemical as well as mathematical transport models have been developed.

Considering inorganic nutrients and contaminants we can identify two major problems that complicate transport modeling. The first problem is that most inorganic solutes must be considered to be reactive in the unsaturated soil/plant system, due to interactions with the solid phase, with soil biota, and the vegetation root system. In fact, the capacity of the soil matrix to chemically interact with solutes is of major importance for the biosphere. The soil matrix, by retaining and releasing solutes, is able to attenuate "short term" fluctuations in the solute concentrations at the soil boundaries. Such an attenuation (or buffering) is crucial because it protects the (local) soil biosphere against adverse effects of time-dependent changes. Examples of time-dependent changes that involve different time scales are inundation of river flood plains by polluted water (years), or acid deposition (decades). The buffering mechanisms are based on (bio)chemical reactions between the matrix and the solution. The diversity of reactions that can lead to either retention or release of solutes may profoundly affect the transport behavior. A second complication with regard to transport is the intrinsic heterogeneity of soil. Soil is heterogeneous with regard to physical and chemical properties as well as biological activity. Consequently, solute transport behavior varies from place to place. However, in most cases the properties of the soil system from place to place are inadequately known, and must be considered to be random.

The effects of (bio)chemical interactions and soil heterogeneity on solute transport in soil may be modeled (conceptually as well as mathematically) with different approaches. In this chapter we discuss the various approaches that have been used for the description of (bio)chemical interactions in and their combination with transport, and for the description of transport in heterogeneous media.

II. Chemical Interaction in Relation with Transport

Different charged solid phase constituents are responsible for the chemical reactivity of soil. Examples are primary minerals, clay colloids, oxides, and organic colloids. With each of these comprehensive terms in fact constituents are grouped that may reveal different behavior. Well-known clay mineral subgroups are, for instance, the kaolinites common to strongly weathered soils, mica minerals (e.g., illites) which are more often found in desert soils or in temperate regions, and smectites (e.g., montmorillonite). The kaolinites (1:1 layer minerals) show no swelling and do not fix K like the illites (1:2 layer minerals). The montmorillonites (1:2 layer minerals) do not show K-fixation either, but may swell significantly in water.

The oxides in contact with the aqueous phase are usually hydroxylated. Besides true hydroxides (e.g., gibbsite) hydrous oxides or oxyhydroxides characterized with a hydroxylated surface also occur. Because most interactions occur at the mineral surface of oxides, which has much in common with hydroxides, no distinction is made in this text. Important hydroxides are those of Al, Fe, and Mn.

Perhaps most difficulty is encountered in describing soil organic colloids, due to the large diversity. A rough division may distinguish recognizable compounds (e.g., polysaccharides) and amorphous compounds such as humic substances.

At the interface of the mentioned solid phase constituents and the aqueous soil solution interaction via sorption processes occurs. By sorption processes we denote the range of low energetic "physical" adsorption to relatively high energetic chemical precipitation and surface precipitation. To be able to interpret and predict solute behavior under varying conditions a detailed understanding and knowledge of sorption processes is needed. The available models for sorption summarize the current understanding.

A. Chemical Equilibria

In this part we will discuss the various approaches that have been used to model sorption processes for ionic inorganic compounds by soil and soil constituents. The electrical charge of the solid/solution interface due to dissolved ions and the charged solid surface implies that electrostatic effects are always relevant. However, not all models account for these effects explicitly. An example of such a group of models that account for electrostatic effects implicitly is the (cat)ion-exchange approach.

1. Cation Exchange Models

The cation exchange models were among the earliest adsorption models used in soil chemistry. Usually they are based on a constant exchange capacity or constant electric charge of the solid phase. This approximation is reasonable for many clay colloids, with the exception of, e.g., kaolinites (which do not have a charge caused by isomorphic substitution). Assumptions other than a fixed number of exchange sites are, for instance, that only exchange of ions of the same charge (positive or negative) is treated and that the activity of adsorbed ions can be approximated by their mole fraction (Vanselow, 1932). For ion exchange on a constant charge surface, a rigorous thermodynamic treatment was given by Gaines and Thomas (1953), who made no assumptions except that the adsorbed phase components have to be specified. The thermodynamic mass action model by Gaines and Thomas (1953) is closely related with the exchange models as developed by Vanselow (1932). Neither of these models deals explicitly with the electric potential, thereby assuming that all counterions are located in the same plane. The electric charge of the solid phase is completely neutralized in this plane.

The results obtained by Gaines and Thomas (1953) by an analysis of the reaction

$$z_A^{-1} \, A_s + z_B^{-1} \, B \rightleftharpoons z_B^{-1} \, B_s + z_A^{-1} A \tag{1}$$

yielded an exchange constant given by (K_N)

$$K_N \equiv \frac{N_B^{(1/z_B)}}{N_A^{(1/z_A)}} \, \frac{a_A^{(1/z_A)}}{a_A^{1/z_B}} \tag{2}$$

where in (1) and (2), z_i is the valency of compound i ($i = A, B$), the subscript s defines the adsorbed state, N_i and a_i are the equivalent mole fraction and solution activity of i, respectively. The exchange constant (K_N) is an empirically measurable selectivity coefficient based on equivalent fractions (N). The selectivity coefficient may be related with the thermodynamic equilibrium constant of the exchange reaction.

In Table 1 various other exchange models that can be related to the Gaines-Thomas model are presented. A special case is the well-known Gapon model for mono-divalent exchange (Gapon, 1933). The Gapon equation is a predecessor of the statistical exchange models of, e.g., Davis and Rible (1950) and Krishnamoorthy and Overstreet (1950). The Gapon equation has been often used in saline soil problems (Bresler et al., 1982) and theoretical soil chemistry studies (Bruggenwert and Kamphorst, 1982). This model infers a one-to-one competition for the sites available to the adsorbed ions of different valency (Bolt and van Riemsdijk, 1991) and should be considered as convenient rather than theoretically well founded. Observe that the Gapon constant (K_G) as usually defined is inversely related to (K_N) and "K(Gapon)" of Table 1 (Bruggenwert and Kamphorst, 1982). The relationships between the exchange approach and electrochemical (Gouy) models was discussed by Bolt and van Riemsdijk (1991).

2. Electrochemical Models

The presence of ions in a porous medium of which the solid phase is also charged means that electrostatic forces play an important role. The electrochemical models take these forces explicitly into consideration. Their differences are mainly in the description of the specific interactions. Several examples will be summarized.

The simplest electrochemical model is usually referred to as the Gouy-model, which considers only electrostatic (and no specific) effects. This leads to a charged surface in contact with a solution layer that balances the solid phase

Table 1. Cation exchange coefficients

Coefficient	Expression
K(Kerr)	$\equiv N_B a_A/(N_A a_B)=K_N^z \ for \ z_A=z_B$
K(Vanselow)	$\equiv 4K_N^{-2}/(2-N_B)$
K(Eriksson)	$\equiv N_A^2 a_B/(N_B a_A^2)=K_N^{-2}$
K(DKO)	$\equiv 4K_N^{-2}/(2-N_B/2)$
K(Gapon)	$\equiv N_A \ a_B^{0.5}/(N_B a_A)=(K_N\sqrt{N_B})^{-1}=K_G^{-1}$

A and B are cations, N and a are equivalent mole fractions and solution activity, respectively, and z is the valency. K_N is defined by Equation (2).

charge (i.e., the electrical double-layer system). In the solution layer, counter ions with the same charge as the solid phase and co-ions with an opposite chargeare present in different quantities. In case of freely diffusing ions in the electrical field we speak of a diffuse double layer (DDL). Because the electrochemical potential of all ions is equal throughout the DDL, the concentration depends on the distance from the surface charge according to the Boltzmann equation

$$c_i = c_{0,i} \ \exp(-z_i \ F\Psi/RT) \tag{3}$$

Here, i denotes the ion, c_0 is the concentration at infinite distance, z is the valency of the ion, F is Faraday's constant, Ψ is the electric potential, R is the gas constant, and T is absolute temperature. Due to the presence of charged ions, the electric potential changes as a function of position perpendicular to the solid surface. The Poisson equation describes how Ψ changes with position.

Combination of the Poisson equation with (3) yields the Poisson-Boltzmann equation for the DDL, which is a second-order differential equation in Ψ. For various cases that include truncated and extended DDL with symmetric or asymmetric electrolytes, details were provided by Bolt (1982). A discussion for a porous solid was given by Lyklema (1968), and a further examination of ion exclusion was performed by Edwards and Quirk (1962), Edwards et al. (1965), and De Haan (1964). Because in a purely electrostatic description no specific interactions are taken into account, the Gouy-model has been used mainly for clay-systems, where solid charges were due to isomorphic substitution. In such situations, where we deal with a constant charge of the solid and usually with a chemically unreactive solid surface, the electrostatic effects may be dominant.

More complicated electrochemical models take specific interactions into consideration. Whereas clay mineral planes may be rather unreactive in a chemical sense, this is not the case of (hydr)oxides, organic matter, and, e.g., clay edges with (hydr)oxide type of behavior. These latter surfaces may react with solutes which may lead to both ion specific effects and a variable surface charge. A variable surface charge arises due to the reaction of potential determining ions (PDI) with solid surface sites. An example is surface hydrolysis of (hydr)oxides by the protonation reaction

$$S-OH + H^+ \rightarrow S-OH_2^+ \tag{4a}$$

This reaction implies that the surface charge depends on the solution-pH. Besides surface hydrolysis two other types of specific interaction may be worthwhile to mention. These are surface complexation, e.g., of a metal ion (M^{2+})

$$S-OH + M^{2+} \rightarrow S-OM^+ + H^+ \tag{4b}$$

and surface ligand exchange

$$S-O-H + (M-OH)^+ \rightarrow (S-O-M)^+ + H_2O \tag{4c}$$

where S refers to the surface. In principle it is possible that reactions occur with more than one hydroxyl group as well as with surface groups that are not singly coordinated.

The reactions (4) are ion as well as surface specific and can be described as a chemical equilibrium that depends on the reactant activities. This leads to a chemical equilibrium coefficient K. It is also worthwhile to mention that reactions such as (4) may be rewritten in an alternative fashion. By smearing out the charge over the different surface groups, Van Riemsdijk et al. (1987) were able to consider only one reaction constant instead of two which are usually needed to describe both the release and the adsorption of a proton by the S-OH group. Therefore, they called this approach the one pK-model. The versatility of this one pK-model was illustrated by Hiemstra et al. (1987, 1989). With respect to the change of potential as a function of distance to the surface, it is essential how close compounds approach the surface as this determines specificity and, e.g., the importance of the DDL for adsorption. Both inner and outer sphere complexes (the first being closest to the surface) may be distinguished in the Inner Helmholtz Zone (IHZ). This zone is bounded by the Outer Helmholtz Plane (OHP) where the DDL begins (Bolt and van Riemsdijk, 1991).

Reviews of electrochemical models based on the mentioned processes were given by Westall and Hohl (1980) and Bolt et al. (1991). One of the well known models is the Constant Capacitance model (Stumm et al., 1976, Schindler and Gamsjäger, 1972). In this model all surface charge is countered by adsorbed ions positioned at a particular distance from the surface. Therefore, the potential decreases almost linearly to zero from the surface to the plane where the adsorbed ions reside. Resembling the Constant Capacitance model was the model by Stumm et al. (1970) and Huang and Stumm (1973). In the latter cases the constant capacitance was not an adjustable parameter, but follows from Gouy-Chapman theory that was extended to have a fixed number of surface sites. Such a fixed number of sites is implicit for charge development due to surface complexation. The mentioned constant capacitance models describe a condensed Stern layer, which is exactly the opposite from a freely extended Gouy layer or DDL. In addition to these two extremes, several intermediate models have been developed. One of the corrections to the Gouy-model that was made by Stern (1924) was that ions in the DDL have a distance of closest approach to the surface (equal to, e.g., hydrated ionic radius). The resulting empty Stern layer (or IHZ) may be "filled" when dehydrated counter ions penetrate into the IHZ. As long as we deal only with electrostatic effects, distinguishing a Stern layer (or IHZ) has limited effects. However, large double layer capacitances may result that suggest that the counter charge approaches very close to the surface. One hypothesis to deal with this result is the assumption of a porous solid (Lyklema, 1968).

The other approach to deal with apparently large capacitances is to allow pair formation between counter ions and surface sites, e.g., according to (4b) or (4c). In the Basic Stern model, specifically adsorbing ions reside in the IHZ close to the surface. The OHP at negligible further distance from the surface has the same potential as the plane where the innersphere complexes reside. This is not the case for the Triple layer models of Yates et al. (1975) and Davis et al. (1978), for which cases the two planes do not coincide.

The various models clearly differ in complexity. Their input requirement may be large as it includes properties of different layers, such as charge, potential, and adsorption energy for the various ionic species. This may give rise to a large number of fitting parameters. To discriminate between different feasible descriptions is therefore possible only when a large data set is available, that has sufficient accuracy. An additional problem is the micro-scale heterogeneity of the solid surface that implies that measurements reveal only the average behavior.

3. Surface Heterogeneity

An extension of the above modeling approaches is based on heterogeneity of the surface at the microscopic scale. That such heterogeneity can be expected is

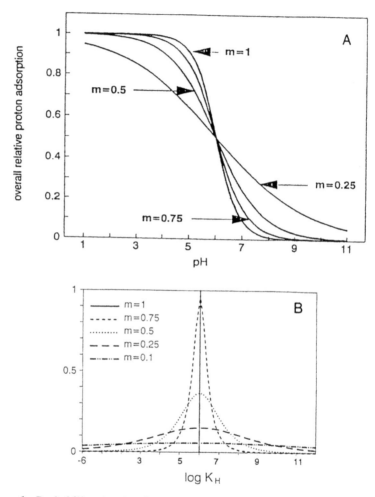

Figure 1. Probability density function (PDF) for $pK = -\log K$ with m the parameter controlling variability about the mean. Values of m shown at the curves (top). Proton adsorption isotherms for the PDFs showing adsorbed protons (relative to its maximum) Θ versus pH (bottom).

clear after consulting Dixon and Weed (1977) and Stevenson (1982). Different mineral faces and active groups should be distinguished for describing the microscopic scale phenomena. To deal with surface heterogeneity we may postulate the presence of a large number of sites with different equilibrium constants $\overline{K}$. This implies that we have a discrete or continuous probability density function (PDF) of $\overline{K}$ (or $p\overline{K} = -\log \overline{K}$) values. However, as adsorption manifests itself at the macroscopic level, we effectively measure a weighted average adsorption or fractional coverage. For constrained situations the averaging procedure can be done analytically assuming that Langmuir statistics for the placement at the microscopic level are valid (Van Riemsdijk et al., 1987; and references cited therein). For an almost Gaussian PDF (Figure 1A), with m being the parameter describing variability, Sips (1950) derived the Langmuir-Freundlich equation. Expressed as a fractional coverage or saturation ($\hat{\theta} = S/S_T$ with S_T the total number of sites and S the fraction of sites occupied by a solute with a concentration equal to c) the result is

$$\hat{\theta} = \frac{(\tilde{K}c)^m}{1 + (\tilde{K}c)^m} \tag{5}$$

for the species with concentration, c, and an apparent equilibrium coefficient, $\tilde{K}$, depending, among others, on the $\overline{K}$-values distribution. An example of the adsorption isotherms for protonation is also given in Figure 1, where the concentration of protons is given in terms of pH instead of c. The smaller m is, the steeper the shown isotherms, $\hat{\theta}$(pH), become. It is interesting that Equation (5) has the Freundlich (6a) and Langmuir (6b) equations as limiting cases, respectively

$$\hat{\theta} = K^* c^m \qquad ; \tilde{K}c \ll 1 \tag{6a}$$

$$\hat{\theta} = \frac{\tilde{K}c}{1 + \tilde{K}c} \qquad ; m = 1 \tag{6b}$$

where $K^* = \tilde{K}^m$). These equations are commonly given in terms of the adsorbed amount. For the Langmuir equation this yields

$$q = \frac{KQc}{1 + Kc} \tag{6c}$$

where q is the adsorbed amount at a particular equilibrium solution concentration (e.g., on a mass basis), Q is the adsorption maximum, and K is an apparent adsorption coefficient. For the Freundlich equation we obtain

$$q = kc^n \tag{6d}$$

where k and n are parameters. Although the Langmuir and Freundlich equations can be derived rigorously, they are mostly used as empirical models. Whereas the Langmuir (K) and Freundlich (k) coefficients are related with thermodynamic and statistical parameters, they also incorporate concentrations of, e.g., ions competing for the same adsorption sites. Consequently, they are applicable (constant) when the activities of other reactants are not varied.

4. Illustration of Some Effects

Provided that sufficiently accurate data are available, attempts can be made to describe the system with one of the models discussed. In practice this needs to be done numerically, as the number of algebraic equations to be solved is too large for an analytical treatise. Procedures were discussed by Westall (1980), and are similar to those needed for calculating chemical equilibria that include, e.g., precipitation, complexation in solution with MINEQL (Westall et al., 1976), and GEOCHEM (Sposito and Mattigod, 1980). Because precipitation/dissolution and complexation in solution can equally well be formulated as, e.g., the reactions (4), these processes require no new modeling approach. Depending on the chemical forms (species) that may be found, the total dissolved concentration of components may vary significantly. An example is given for cadmium. For two different sets of soil samples, Christensen (1980) and Chardon (1984) described Cd adsorption with the Freundlich equation.

The parameter k (Freundlich coefficient in Equation (6d)) in that case depends on all factors not taken explicitly into account, i.e., all factors but the solution concentration of Cd. As may be expected, a significant dependence of k on the calcium concentration and pH was found. Also, the effect of ionic strength and chloride concentration appeared to be large. Some of the experimental results are shown in Figure 2. For one soil the k-value could vary over several orders of magnitude. After correcting for pH and $[Ca^{2+}]$ a variation over one order of magnitude could still be found by varying ionic strength and chloride concentration. The latter effect can be described by Cd-chloride complexation. When in the Freundlich equation the Cd concentration is replaced by Cd activity and complexes with chloride are assumed to have a much smaller affinity for the surface, the curves for the different chloride concentration may coincide reasonably (Chardon, 1984). The Freundlich power (n) for Cd appeared to be less dependent on solution composition than k and was smaller than one. Assuming the ratio of Cd and other metals (Cu, Zn) in solution to be constant, smaller powers were found than in the case where Cd concentrations were varied while the competing heavy metal concentrations were kept constant (De Haan et al., 1987).

A different situation can be encountered for Zn adsorption onto Al-hydroxide (Micera et al., 1986). Ionic strength effects cannot explain the difference of Figure 3, as it is almost constant. Specific SO_4^{2-} adsorption leading to a more negative surface charge, which in turn favors Zn adsorption, could explain the effect of the Zn-salt on adsorption. Examples of pH effects have been reported for many components, such as Zn (Bolland et al., 1977) and phosphate (Bowden et al., 1980). As for the Cd example, complexation in solution may enhance the mobility. Uncharged ion pairs or complexes with lower charge than the constituting components are less susceptible to electrostatic adsorption than the components. Examples are found in saline soils, podsol formation, etc.

The main effect of the additional precipitation/dissolution phenomena is their control of upper and lower concentration or activity bounds for the components

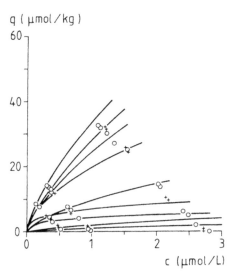

Figure 2. Cadmium adsorption under different conditions concerning ionic strength, chloride, and calcium concentrations. (From De Haan et al., 1987.)

involved. Precipitation as well as dissolution may be fast processes, depending on the equilibrium conditions and the solid phase under consideration. The formation of Ca and Mg carbonates and sulfates may be considered instantaneous in saline soils (even though their dissolution is not). Under reduced conditions, heavy metals may form sulfide fast. Likewise, dissolution of Ca/Mg-phosphate salts, as encountered in animal slurries, that are soluble at acid soil pH values may be a relatively fast and solubility controlled process (De Haan and Van Riemsdijk, 1986). Subsequently, the released phosphate may precipitate slowly (kinetically controlled) by reaction with Fe and Al compounds (van der Zee, 1988).

B. Kinetics

Whereas the approaches mentioned in the previous section describe the chemical equilibrium, we may often deal with non-equilibrium. Non-equilibrium conditions may arise due to slow kinetics of a chemical reaction, physical resistances to mass transfer, and slow (micro)biological processes. The last cause of non-equilibrium will not be considered in this chapter.

When we consider purely chemical kinetics and physical causes of non-equilibrium it is often extremely difficult to experimentally distinguish which is most important, at least in soil systems. One of the methods to distinguish between these causes is by assessing the temperature dependency of the rate process and comparing the activation energy with those of diffusion and

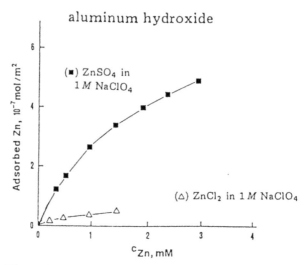

Figure 3. Zinc adsorption by aluminum hydroxide under varied conditions. (From Micera et al., 1986.)

chemical reaction processes. Despite sometimes conflicting evidence, processes such as electrostatic adsorption and complexation in solution often appear to be at chemical equilibrium (Bolt, 1982). The assumption of chemical equilibrium for chemical reactions involving the solid phase, e.g., surface complexation or precipitation/dissolution is debatable. As an example, we will discuss mineral weathering in more detail.

A definition of the acid neutralizing capacity as given by Van Breemen et al. (1983, 1984) in the acidification context may be useful, but the actually expected acidity depends on the acid consumption rate. Dissolution of, e.g., feldspars, (hydr)oxides, and carbonates appears to be a kinetic process. Reactant transfer through a product layer to and from the reacting surface was believed to be rate-limiting by Wollast (1967), Paces (1973), and Helgeson (1971). Because product layers were not found or observed reaction rates were too small, product-layer diffusion was discarded by Berner (1978), and Berner and Holdren (1979), and the chemical reaction rate was postulated as controlling the overall rate (Aagaard and Helgeson, 1982; Lasaga, 1983). The argument to discard of product layer transfer control for a homogeneous surface was weakened by observations of etch pit formation (Wilson, 1975; Berner and Holdren, 1979; Blum and Lasaga, 1987; Schott and Petit, 1987). For this local phenomenon both earlier arguments, based on a homogeneous surface, do not invalidate the description of Chou and Wollast (1984). They described dissolution peaks after sudden pH changes by locally diffusion limited etch pit formation. Surface morphology and energetics may affect dissolution by etch pit development and was studied by Blum and Lasaga (1987). Schott and Petit (1987) also cite several publications where dissolution is predominantly at high energy sites, e.g., edges, corners, cracks, and dislocations of the surface. This feldspar dissolution example reveals that non-equilibrium (kinetics) may be a result of physical diffusion and chemical reaction kinetics. The latter may be governed by complicated reaction orders involving different components or species. A thorough chemical analysis of oxide dissolution was given recently by Hiemstra and van Riemsdijk (1990). They developed a model that relates the dissolution rate with pH. All activated complexes are, in principle, able to dissolve, instead of only one as is conventional. The model of Hiemstra and van Riemsdijk (1990) predicts the pH and salt dependency of dissolution rates on the basis of thermodynamic quantities. No use of the dissolution data is needed provided the surface geometry and composition are known. More discussion on physical (diffusion affected) non-equilibrium is given in part C.

C. Effect of Chemical Behavior on Transport

In most cases transport through homogeneous porous media is described based on the convection dispersion equation (CDE)

$$\theta \frac{\partial c}{\partial t} = \theta D \frac{\partial^2 c}{\partial z^2} - \theta v \frac{\partial c}{\partial z} \qquad (7)$$

which combines the continuity and flux equations for non-reacting solutes in a one-dimensional homogeneous column. In Equation (7), θ is the volumetric water fraction, t is time, D is the diffusion/dispersion coefficient, v is the average pore water velocity, and z is distance. The version given by Equation (7) considers steady-state transport. Equation (7) has been used to describe breakthrough curves (i.e., c(L,t) at depth L) for different initial and boundary conditions, flow rates, and materials. Extensive experimentation for different conditions established that $D = D_m + \lambda v^a$ where parameter $a \approx 1$ and the dispersivity, λ, characterizes the homogeneous material, and has the dimension length (Pfannkuch, 1963; Perkins and Johnston, 1963). Obvious extensions of Equation (7) for field soils account for higher dimensionality and/or transient flow conditions, which will be considered in part III. In this part we discuss first the multicomponent approach. Later we consider some special cases in more detail.

1. Multicomponent Transport Modeling

In view of the scope of this section it is of main interest to extend Equation (7) with reaction terms. In the most general case we account for a solution that contains different species that react with each other as well as with the solid phase. Moreover, this solution moves through the porous medium. To describe the complicated displacement of different species two main approaches of modeling can be distinguished: 1. Integrated or Direct Multicomponent Approach, and 2. Modular or Two Step Approach.

<u>Integrated Approach</u>: The non-reactive transport equation or CDE is extended by reaction terms. For each species this yields a transport equation. As the different species are related with each other through the algebraic chemistry equations, this yields a (usually non-linear) set of differential equations. Examples of different chemical interactions and background were provided by Rubin and James (1973), Valocchi et al. (1981), Jennings et al. (1982), Miller and Benson (1983), Kirk and Nye (1985), and Singh and Nye (1986). In some of the cases the transport volume could also account for transfer into solid (mineral, salt) or gas phases. Among others, these examples differed with regard to the adsorption model chosen: Jennings et al. (1982) used a surface complexation model, Rubin and James (1973) and Valocchi et al. (1981) used mass action or Gapon models for exchange, and Nye and co-workers used several empirical models. The integrated approach of, e.g., Singh and Nye (1986), which is specific for a certain problem, lacks flexibility when different species are concerned. For each additional species the entire set of equations may have to be adapted. This approach is primarily used for relatively simple systems. Chemically simple systems are most commonly found under laboratory conditions (where the number of chemical parameters that is changed can be limited). Chemically simple systems under field conditions usually mean that the solutes with steep concentration gradients do not significantly affect chemical parameters that are assumed to be constant. In that case, the interactions have to be described for those constant environmental conditions using (semi-) empirical models. When formulating the integrated approach in a general way it becomes as flexible as the first (modular or decoupled) approach, but may require excessive computing times and computer storage (Pers. Comm. Hassanizadeh, Inst. Public Health & Environment, Netherlands, 1990, Yeh and Tripathi, 1989).

<u>Modular Approach</u>: As mentioned earlier, an alternative of the integrated modeling approach limits the transport volume such that all (bio)chemical interactions with the solid phase are excluded from the transport equation. The effect of different reactions is included via source/sink terms. To deal with solid/solution multicomponent chemistry properly, it is necessary to have a separate module that evaluates the transport volume sources and sinks. This approach was called the modular or two-step approach (Abriola, 1987; Herzer and Kinzelbach, 1989). It consists of a transport algorithm (Equation 7) for each component and, for instance, a chemical equilibrium module like MINEQL to evaluate the speciation for each component. This way of dealing with reactive transport may, for species subject to solid/solution equilibrium, lead to additional numerical dispersion that can be limited effectively by iterating between the two modules at each time step (Herzer and Kinzelbach, 1989). For cadmium transport with complexation by other solute species and adsorption described with the constant capacitance model, an example was given by Cederberg et al. (1985). The modular approach was also used by Kirkner et al. (1984, 1985) and

Jennings and Kirkner (1984). Of special interest is their consideration of non-equilibrium. For this case, the algebraic equations of the chemical module become a set of non-linear, ordinary differential equations.

Another application taking as many as 60 species and 20 solids into account was given by Walsch et al. (1984), where precipitation/dissolution and redox reactions were considered. To allow algebraic approximations that limit computation costs the concept of scalability was proposed, which is based on the chromatography phenomenon (front separation). In a subsequent paper, Bryant et al. (1986) added adsorption/exchange and complexation to the chemical module. They obtained a complicated interaction between different fronts when both precipitation/dissolution and adsorption occur. Their results suggest that analyses such as those given by Harmsen and Bolt (1982a,b) may be feasible only for relatively simple systems. The modular approach was used for problems related with saline soils by Van Beek and Pal (1978), Dutt et al. (1972), Tanji et al. (1972), Dance and Reardon (1983), and Grove and Wood (1979). From the scope of soil acidification examples were the models by Förster (1986), and by Van Grinsven et al. (1987). The latter accounts for exchange of monovalent and divalent cations with aluminum, $Al(OH)_3$ precipitation/dissolution, incongruent mineral dissolution (releasing cations), and a number of other (hydrologic, deposition and biocycle) processes. Even more comprehensive is the well-known ILWAS model (Goldstein et al., 1984) that does not account for Al exchange. To take this relevant process into consideration, the original Gapon-like exchange equations in ILWAS were replaced by the equations according to Gaines and Thomas (1953) and by Van Grinsven et al. (1989). Another versatile model is LEACHM developed by Hutson and Wagenet (1989), which in fact consists of four versions. These versions describe unsaturated flow and transport of nitrogen, pesticide, and inorganic ions, respectively. Because LEACHM is intended for field applications, the influences of plants on flow and solute transport were taken into account. Heat flow and volatilization were also included. Modular approach models such as ILWAS and LEACHM provide useful tools for the systematic analysis of transport in natural systems. The aims of such research tools can be extended when they are changed into management tools. This is the goal of the LEACHM model (Hutson and Wagenet, 1989).

So far the integrated approach has been used mostly for simple systems. This approach is used among others for analytical solutions, as given by Harmsen and Bolt (1982a,b). Numerical solutions are mostly obtained using the modular approach. A detailed discussion of the computational aspects was given by Yeh and Tripathi (1989). For this reason, these aspects are not taken into consideration in this chapter. Instead, some examples are discussed that illustrate the effect of chemistry on transport.

2. Examples of Multicomponent Transport

When an existing chemical equilibrium model is used, the chemical complexity that can be handled is significant. Details of chemical equilibrium packages were given by Sposito (1985). In practice, the potential of such packages has not yet been exploited. Although chemical packages have been developed to describe interactions according to electrochemical adsorption models, complexation in solution, precipitation, etc., most actually given examples are much more limited. An example of the modular approach using an electrochemical model was given by Cederberg et al. (1985). They used the constant capacitance model (as well as a mass action/exchange model). Their consideration of surface complexation in transport modeling is one of the few in its field. Mostly, the adsorption models used in transport modeling are based on the ion-exchange formulations as given in Table 1 or Equation (2) (Bryant et al., 1986; Dutt et al., 1972; Tanji et al., 1972; Grove and Wood, 1979; Dance and Reardon, 1983; Schultz and Reardon, 1983; Kirkner et al., 1984). Although the exchange models may give a good description, they are usually more constrained than the electrochemical models. For instance, a variable surface charge cannot be modeled well and ions of opposite charge cannot be treated simultaneously. That good predictions can be made of transport subject to ion exchange was shown by Jauzein et al. (1989) and Schulin et al. (1989). The experimental and predicted breakthrough curves of Figure 4 (Schulin et al., 1989) show a good agreement. The parameter values used for the predictions were obtained independently, i.e., the results were not fitted.

Most of the mentioned modular approach models correct for the activity in solution and take complexation in solution into account. Nice illustrations of the importance of complexation on transport were given by Cederberg et al. (1985). A solution containing chloride, bromide, and cadmium (in total, 13 species) was considered. Transport in a column was described for the three components, which limited the computation requirement compared to a description for each separate species. This reduction in the number of transport equations is possible because of the algebraic relations between the various species (see also Yeh and Tripathi, 1989). Due to the formation of Cd-chloride or Cd-bromide complexes, the distribution of Cd over the solution and adsorption phases depends on the background anion concentrations. The complexes were assumed to remain in solution. Hence, for instance $CdCl^+$ did not adsorb. Then the ratio sorbed Cd over dissolved Cd decreases as halide concentrations increase. The example of Figure 5 that shows fronts of the three components reveals that sorbed Cd ($\overline{Cd}$) is smallest for the case of larger chloride concentrations. The ratio $\overline{Cd}/Cd_{aq}$ affects front retardation. The smaller the ratio is (i.e., large (Cl_T), the farther downstream the fronts are situated. The trends found by Cederberg et al. (1985) were in agreement with experimental data of Lecky et al. (1980), Donner (1978), and Donner et al. (1982). Both pH and the total number of surface sites were assumed constant, and Cederberg et al. (1985) mentioned that a change in pH would lead to a change in the total number of sites. This is not a serious problem, as by a small reformulation of their model this complication can be avoided (van der Zee and van Riemsdijk, 1988). Additional chemical complications in multicomponent modeling of transport are due to precipitation/dissolution reactions, carbonate equilibria, and oxidation/reduction processes. Often

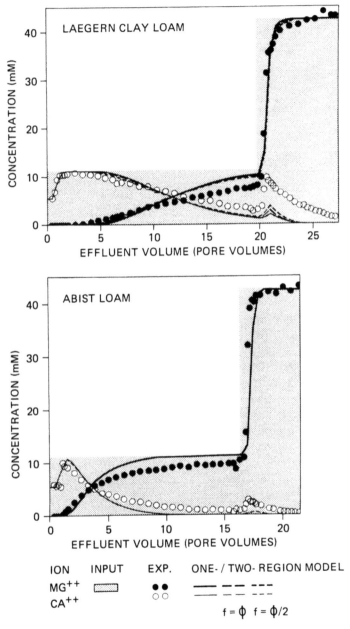

Figure 4. Predicted and measured breakthrough curves of Ca and Mg. (From Schulin et al., 1989.)

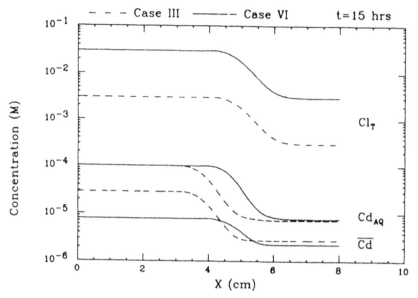

Figure 5. Fronts of dissolved and adsorbed Cd and of chloride for different feed solutions. (From Cederberg et al., 1985.)

these processes imply bounds to the concentration or activities of the various species. Especially when gaseous compounds (CO_2, O_2) and their transport and production need to be taken into account, the system becomes more complex. Different phases (gas, liquid, solid) need to be considered, and physical, chemical, and biological (decay, respiration, in relation with CO_2/O_2-status) processes become important. Besides these aspects, the assumption of local equilibrium (LEA) becomes debatable. Thus, LEA may be considered valid if the flow time scales are large compared to the interaction time scales. For precipitation/dissolution reactions and processes that involve biota (microorganisms, plant roots, etc.) LEA is often invalid, and the description of kinetics rather than of the equilibrium becomes of main importance. So far, few analyses of non-equilibrium multicomponent models are available (Kirkner et al., 1984, 1985; Jennings and Kirkner, 1984). Validity of LEA is not the only matter that has to be resolved. Many of the chemical coefficients that describe chemical interactions may not be known. Values as present in, e.g., data bases of chemical equilibrium packages may hold for pure minerals. In natural systems non-pure minerals may be present, for which more or less different values of equilibrium coefficients than those present in data bases are valid. In view of the logarithmical relationships between pK-values and concentrations, the effects of erroneous values may be large. The effects of such uncertainty may be large for the complicated system (20 solids and 60 dissolved species) of Walsch et al. (1984).

3. Monocomponent Transport Models

Two main approaches for multicomponent transport modeling were reviewed above. Partly, the effects of chemistry on transport may also be described using a monocomponent approach. In some cases even analytical solutions are feasible. Then, the description is always in agreement with the integrated approach as the chemistry equations are directly inserted in the differential equation that describes transport. The simplest monocomponent description is based on a linear relationship between sorbed and dissolved amounts.

A linear adsorption equation may be applicable when the solute of interest is present in a relatively small concentration. Then, the adsorption capacity is much larger than the actually adsorbed amounts and changes in the concentration hardly affect the equilibrium for other present species. Mostly, another condition that has to be met is that the changes in other species have an insignificant effect on the equilibrium of the solute under investigation, because such effects would be exhibited in variations of the adsorption coefficient (e.g., k in Equation (6d) when $n = 1$). When we assume that $n = 1$ in Equation (6d), a retardation factor (R) may be defined that is given by

$$R = 1 + \frac{\rho}{\theta}\frac{dq}{dc} = 1 + \frac{\rho}{\theta}k \qquad (8)$$

where ρ is the bulk density. With this definition the linear adsorption CDE given by

$$\rho\frac{\partial q}{\partial t} + \theta\frac{\partial c}{\partial t} = \theta D\frac{\partial^2 c}{\partial z^2} - \theta v\frac{\partial c}{\partial z} \qquad (9)$$

may be rewritten as

$$\frac{\partial c}{\partial t} = D^*\frac{\partial^2 c}{\partial z^2} - v^*\frac{\partial c}{\partial z} \qquad (10)$$

where the effective dispersion coefficient is $D^* = D/R$ and the effective solute velocity is $v^* = v/R$. We see that linear adsorption leads to the same transport equation as for the non-reactive case, except that the parameters are changed. For $k \geq 0$ we have $R \geq 1$, which implies that $D^* < D$ and $v^* < v$. Both convection and diffusion/dispersion are decreased by the same factor (R). Because the solute front moves with a velocity that is R times smaller than the carrier (water), R is called a retardation factor. However, both for a non-reactive and a reactive solute the front shapes are the same once they arrive at a particular mean depth. Analytical solutions were given by Van Genuchten and Alves (1982).

Another situation arises when adsorption is non-linear. In that case, the chemical interactions cause both a front retardation and a change in the front shape. Freundlich type of adsorption is found for, e.g., heavy metals (Christensen, 1980; Chardon, 1984; Lexmond, 1980). For non-linear adsorption, the retardation factor (8) becomes concentration dependent, i.e., different concentrations experience different apparent velocities $(v/R(c))$. For adsorption according to the Freundlich equation with the power n less than one and the initial resident concentration that is larger than the feed concentration, a rapidly spreading front develops. Large concentrations will leach quickly, whereas it may take a relatively long time before the concentration in the leachate becomes small. This is due to increasing $R(c)$-values with decreasing c because $R(c) = 1 + \rho n k c^{n-1}/\theta$. Quite the opposite occurs when the resident concentration c_i is smaller than the feed concentration, c_o (van der Zee, 1990; Van Duijn and Knabner, 1990). Then, a traveling wave type of displacement will develop where the front shape and its velocity become independent of time for a homogeneous column. In practice, such a traveling wave may be formed after a short time for a stepwise increase of the concentration. Its formation is due to the non-linear effects of $R(c)$ that oppose dispersional spreading if $c_i < c_o$. In particular, the apparent front velocity for that case does not depend on $R(c)$ as defined by (8), but depends instead on $R_f = 1 + (\rho\Delta q/\theta\Delta c) = 1 + \rho k[c_o^n - c_i^n] / \theta[c_o - c_i]$ Thus defined, R_f is the front retardation in case of traveling wave type of displacement (van der Zee, 1990).

Traveling waves may, in practice, form relatively soon, despite the fact that the analytical solutions as derived by van der Zee (1990) hold theoretically for infinite times. Typically, the fronts that develop are steeper than would be expected for a linearly adsorbing solute that has the same front retardation factor. For the interpretation of breakthrough data, it is important to be aware of the linearity or non-linearity of adsorption. In case of an erroneous assumption with regard to adsorption linearity, the assessment of the dispersive behavior of the soil and the front velocity for other feed concentrations are also in error. Hence, predictions for other conditions will be wrong.

Besides for the linear, Freundlich, and Langmuir adsorption equations, a monocomponent description of transport may also be based on an exchange formulation of adsorption. Whereas the earlier mentioned adsorption equations are common for trace compounds (pesticides, heavy metals), the exchange equations are often used for the main cations such as Na^+, Ca^{2+}, Mg^{2+}, K^+, etc. A monocomponent approach is feasible only when two ions are taken into consideration. Situations where this is the case are the saline/sodic soil problems. These problems may be dominated by the behavior of sodium and the divalent cations (Ca^{2+}, Mg^{2+}). The divalent cations may be considered to behave similarly in soil. Then, they are lumped in the divalent cation (X^{2+}), and we deal with Na^+/X^{2+}-exchange (Bresler et al., 1982).

Commonly, the exchange reaction between Na^+ and X^{2+} is described with the Gapon equation (Table 1). Because divalent cations are adsorbed preferentially, compared with sodium, again two situations may arise. For the sodication problem, where Na^+ exchanged with adsorbed divalent cations, we deal with unfavorable exchange. This situation resembles the case of Freundlich adsorption where $0 < n < 1$ and the feed concentration is smaller than the initial resident concentration. Consequently, Na^+ moves rapidly downstream and the Na/X-front extends over a major part of the soil profile. This causes a gradual amelioration of soil quality throughout the soil profile (Bolt, 1982).

The reverse situation is expected for sodic soil reclamation, where divalent cations exchange for adsorbed Na^+. The favorable displacement of Na^+ by X^{2+} leads again to traveling wave behavior, as was shown by Reiniger and Bolt (1972) and Bolt (1982). The resulting steep fronts imply that desodication does not occur gradually throughout the soil column. Instead, this process occurs more or less layer-wise (almost like piston flow). Assuming a constant total solute concentration in solution, a monocomponent approach is feasible, and analytical approximations can be obtained (Bolt, 1982). As Bond and Philips (1990) showed, the traveling wave solutions of Bolt (1982) can be of use even for transient flow conditions. Approximations are even possible for a complex dominated soil system when the total solute concentration is varied.

For the desodication of fine textured soils, an additional physical complication may be important. When a large fraction of the CEC is occupied by sodium this may lead to peptization and swelling, which decreases the hydraulic conductivity. This effect can be expected when sodic clayey soils are reclaimed by percolating good quality water. The impeded drainage at the relatively sharp Ca/Na-front may complicate reclamation by leaching considerably (Gardner et al., 1959; McNeal and Coleman, 1966; Russo and Bresler, 1977a,b). The risk of sodication due to the poor quality of irrigation water can be related to the SAR (sodium adsorption ratio) defined on the basis of a concentration ratio as in Equation (10), i.e.,

$$SAR = \frac{c_e(Na^+)}{[0.5c_e(Ca^{2+} + Mg^{2+})]^{0.5}} \tag{11}$$

with concentrations c_e in $mmol_e \, l^{-1}$ (Kamphorst and Bolt, 1976; Bresler et al., 1982). For a more detailed discussion of different aspects of salinization and sodication, see the latter reference.

An obvious extension of the above analytical models is to consider more than two exchanging species or precipitation/dissolution processes. Although such complications are usually dealt with numerically, analytical treatises were given by Harmsen and Bolt (1982a,b) and Harmsen (1987) who neglected pore scale dispersion. For multinary exchange the similarities with chromatography become most apparent. When an instantaneous change in the concentrations of several species is brought about (e.g., a step or pulse-like change), the different species will experience different retardation factors. This implies that a front or a pulse separation will occur. Once this has taken place, the different fronts and pulses can be considered separately, with simple theory. Whether such approximations are feasible and accurate depends on the system and its complexity, as can be inferred from cautioning remarks by Walsch et al. (1984) and Bryant et al. (1986).

A shortcoming of many currently available multicomponent transport models (with exception of Kirkner et al., 1985) is the assumption of local equilibrium. For the monocomponent approach, great attention has been given to non-equilibrium transport. Whereas chemical kinetics may give rise to complicated rate laws, simple rate order appears to be favored. In part, this is due to the applicability of a first-order rate law to physical (diffusion-limited) non-equilibrium.

The transport equation including first- and zero-order irreversible transformations reads

$$\rho\frac{\partial q}{\partial t} + \theta\frac{\partial c}{\partial t} = \theta D\frac{\partial^2 c}{\partial z^2} - \theta v\frac{\partial c}{\partial z} - \mu_1 c - \mu_0 \qquad (12)$$

where μ_i is the rate parameter for the process of order i. For linear adsorption, analytical solutions were summarized by Van Genuchten and Alves (1982). The first-order transformation is applicable for radionuclides (Rogers, 1978; Gureghian and Jansen, 1983), pesticides (Bromilow and Leistra, 1980, Boesten, 1986), and has also been used for nitrogen (Cho, 1971; Misra et al., 1974). McLaren (1975) suggested that zero-order kinetics may apply to denitrification when nitrate abounds. Further analytical approximations were given by Parlange et al. (1984) for Michaelis-Menten kinetics and by Barry et al. (1986) when the reaction order changes gradually from zero to first order. Assuming that the original solute and its decay members have different retardation factors, solutions were given by Van Genuchten (1985) and Lester et al. (1975). Of interest for, e.g., pesticides may be the behavior of non-linear adsorption and first-order degradation. In Figure 6 the front of a pesticide is shown with n = 0.65 (i.e., non-linear Freundlich adsorption) and $\mu_1 = 3.73$ (yr^{-1}). The analytical approximation given by van der Zee and Bosma (1990), which is based on an adaption of the traveling wave solution by van der Zee (1990), appears to adequately describe the downstream front shape and its deceleration with increasing time. The linearized case shows a much more dispersed front. Although the linearized fronts were forced to coincide with the mean downstream numerical fronts, they actually did not predict a deceleration of the front velocity.

A further complication in monocomponent transport modeling was to account for reversible non-equilibrium. Although part of the available sorption sites often had to be modeled as being at local equilibrium to provide a good description of experimental data (Van Genuchten and Cleary, 1982), a fraction of the sorption sites being at non-equilibrium has been postulated in many papers. Linear (Hornsby and Davidson, 1973; Cameron and Klute, 1977), Langmuir (Novak et al., 1975), and Freundlich (Rao and Jessup, 1982; Rao et al., 1979; Miller and Weber, 1988) non-equilibrium adsorption were considered. This gave rise to the two-site or dual porosity models that consider either two different sorption sites (of which one is at non-equilibrium) or two different pore regions. The analytical solutions of these models consider linear adsorption and first-

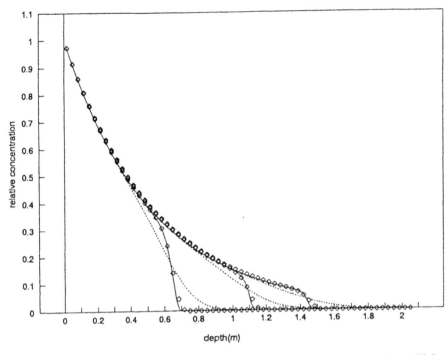

Figure 6. Fronts for three times for a solute with non-linear Freundlich adsorption (n=0.65) and first order decay. Solid: approximation by van der Zee and Bosma (1990); dashed: linearized solution; point: numerical results.

order rate expressions. The different solutions as given by Villermaux and Van Swaaij (1969), Bennet and Goodridge (1970), Lindstrom and co-workers (1973, 1975), Cameron and Klute (1977), and Van Genuchten and Wierenga (1976) can be cast in the same general form (De Smedt and Wierenga, 1979) and are mathematically identical (Van Genuchten, 1981; Nkedi-Kizza et al., 1984). This is the case both for linear adsorption (for which the solutions mentioned were developed) and for non-linear adsorption (van der Zee, 1990). The two site/dual porosity concept has been applied to such diverse cases as pesticide, tritium, chloride, cation, and heavy metal transport. Numerical examples for non-linear adsorption were provided by Schulin et al. (1989). A physical difference between the dual porosity and the two site models is that in the dual porosity model the two volume fractions of mobile and immobile regions add up to one. Hence, in the dual porosity model first breakthrough may occur long before one (total) pore volume is leached when immobile water is present. This is not the case for the two-site model. More sophisticated than the "overall" first-order rate expression used for the previously mentioned models is to describe the physical non-equilibrium by Fick's laws applied to, e.g., spherical, cylindrical, or slab-shaped immobile zone regions. Examples were provided by Rasmuson (1985a, 1986), Barker (1985), Van Genuchten and Dalton (1986), Valocchi

(1985), and Parker and Valocchi (1986), who gave analytical solutions. Numerical solutions were given by, e.g., Rao et al. (1980a) and Huyakorn et al. (1983) and applied to data by Miller and Weber (1986, 1988). Because the independent information with regard to the studied system often is limited, it is questionable whether this sophistication is justified for the heterogeneous, poorly defined soil system. Rather than forcing nature to behave like the simplified system of the model (which in some cases may be acceptable in view of arguments given by Van Genuchten and Dalton (1986) and Parker and Valocchi (1986)) one might simplify the model, until it is compatible with our knowledge of the system (van der Zee, 1991). In that case, the simplest model (with least adaptable parameters) is used when it describes experimental results acceptably. Usually this implies that additional experimentation is needed, e.g., by variation of the concentration of the solute under investigation besides varying pore water velocity, to discriminate between models (van der Zee, 1991). Furthermore, it is likely that to describe transport under field conditions, including temporal and spatial variability, only some broad features of local behavior are maintained. An example of this latter hypothesis by Rinaldo and Marani (1987) was given by van der Zee and Boesten (1991), who showed that many complications that are realistic for pesticide transport do not affect the overall first-order degradation and linear adsorption transport behavior adversely at the field scale. Although absolute values of estimated parameters may be affected by such complications (such as temperature variations, transientness of flow), the resulting leached pesticide fractions may be adequately described by an overall zero-dispersion, linear adsorption with first-order decay solution as advocated by Jury et al. (1987). This is due to the predominant effect of field scale heterogeneity addressed in the next section.

III. Soil Heterogeneity

The transport of pollutants from the soil surface into the groundwater is the result of a complex interaction between physical, chemical, and biological phenomena. Furthermore, natural fields exhibit a spatial variability in their properties that appears in many instances to be essentially random; this heterogeneity will generally have a significant effect on field-scale solute movement through the unsaturated zone. The effect of spatial variability in soil hydraulic properties has been investigated using either parametric models (Dagan and Bresler, 1979; Bresler and Dagan, 1981; Simmons, 1982; Amoozegar-Fard et al., 1982; Destouni and Cvetkovic, 1989), or the transfer function model (Jury et al., 1986; Butters and Jury, 1989). More recently, the effect of spatially variable chemical properties on mass transport of reactive solute has also been addressed (van der Zee and van Riemsdijk, 1986, 1987; Destouni and Cvetkovic, 1990, 1991).

Numerical solutions of three-dimensional solute transport under transient flow conditions in the unsaturated zone are complex and require excessive input information of both climatic and soil properties. Spatial heterogeneity in natural fields further complicates the solution and increases the need for input information. Thus, although water flow in the unsaturated zone generally is transient, and the path lines of particles moving in that flow field are three-dimensional, many investigators of field-scale solute transport have used the simplifying assumptions of steady and essentially vertical flow (e.g., Bresler and Dagan, 1981; Dagan and Bresler, 1979; Jury, 1982; van der Zee and van Riemsdijk, 1986; 1987; Butters and Jury 1989; Destouni and Cvetkovic, 1990, 1991).

In the following, we first discuss field-scale spreading of non-reactive solute due to spatial variability in physical properties. Because the steady-state and vertical flow assumptions are common and underlie a number of investigations on field-scale solute transport, we shall emphasize recent studies that scrutinize the applicability of these assumptions. Furthermore, we extend our discussion to consider the effect of spatial variability in sorption-desorption parameters on field-scale transport, of reactive solute in heterogeneous soils. In addition, we illustrate the effect of soil heterogeneity on transport of degradable solute with a nitrate case study.

A. Non-Reactive Solute

The number of field-scale experiments of solute transport conducted in the unsaturated zone is limited (Biggar and Nielsen, 1976; Wild and Babiker, 1976; Starr et al., 1978; Jury et al., 1982; Schulin et al., 1987; Butters et al., 1989). Nevertheless, the results indicate a considerable areal variability in pore water velocity, where the reported values of the velocity coefficient of variation (CV) range from 58% for mainly unsaturated conditions (Jury et al., 1982) to 194% for ponded conditions (Biggar and Nielsen, 1976). Furthermore, the spreading of non-reactive solute about its center of mass in field-scale experiments is much larger than could be expected as an effect of pore-scale dispersion, D, in Equation (7). Neglecting molecular diffusion, D is the coefficient of hydrodynamic dispersion, i.e., $D = \lambda v$, where v is the local mean pore water velocity and λ is defined as the pore-scale dispersivity. The values of λ for laboratory experiments, at a scale that is associated with pore-scale dispersion, are typically about 0.005 m or less, whereas λ values determined in field soils may be considerably larger and depend on the scale of the experiment (Nielsen et al., 1986).

A qualitative explanation of the field-scale transport process in aquifers referring to the spatially variable advective pattern has, for instance, been discussed by Dagan (1990). An analogous conceptualization may also be applicable for the unsaturated zone. Due to spatial heterogeneity, the stream tubes defining the local mean pore water velocity, v, are tortuous with varying cross-section, where v varies both along a given stream tube and between the different stream tubes. Neglecting pore-scale dispersion and considering a non-reactive solute, the advection variability results in a disintegration of the solute body into slugs carried at different velocities along the various stream tubes, implying an irregular field-scale solute spreading that is considerably larger than that associated with pore-scale dispersion. If the longitudinal pore-scale dispersion is accounted for, it results in an additional local spreading of the solute slug in each stream tube, while the transverse pore-scale dispersion causes transfer of solute among adjacent stream tubes. The latter mechanism is under usual circumstances assumed to affect the field-scale solute spreading only after large travel times (for aquifers see the discussion in, e.g., Dagan, 1987, 1990). In structured soil, however, the transverse pore-scale dispersion may play an important role by transferring solute between flow regions with different hydraulic and transport properties, i.e., intra-aggregate and inter-aggregate porosity (e.g., Jardine et al., 1990).

Because solute spreading at the field scale in general is highly irregular, local concentration values, c(x,t), are subject to large uncertainty. However, in many applications we are interested in more global quantities describing the field-scale behavior of solute transport. These global quantities can be predicted using stochastic methods and are generally less affected by uncertainty than the corresponding local quantities, depending on the initial size of the solute body and on the travel time (Dagan, 1990). Such a global quantity is the field-scale resident concentration, $\bar{c}(t)$, that quantifies the spatially averaged concentration per unit volume of fluid in the vicinity of a horizontal plane at a given depth below the soil surface. The spatially averaged mass flux through that horizontal plane $\bar{s}(t) = (\bar{c}\bar{Q})$, then defines the field-scale flux-averaged concentration, $\bar{s}(t)/\bar{Q}(t)$, where $Q(x,t)$ is the local volumetric flow rate. In fields with highly variable advection properties it is important to distinghuish between the two different field-scale concentrations (Cvetkovic and Destouni, 1989).

In a recent study, Russo (1991) simulated solute transport through a two-dimensional vertical cross-section of heterogeneous soil under unsaturated and transient flow conditions. The spreading of solute about its center of mass resulting from this numerical experiment was then compared with the predictions of the stochastic theory for solute transport in steady, and mainly horizontal groundwater flow (Dagan, 1982, 1984; Gelhar and Axness, 1983). The solute spread about its center of mass was quantified by the spatial covariance tensor, σ^2, with the components

$$\sigma_{xx}^2 = \frac{M_{20}}{M_{00}} - x_c \; ; \quad \sigma_{zz}^2 = \frac{M_{02}}{M_{00}} - z_c \qquad (13)$$

where x and z are the horizontal and vertical spatial coordinates, respectively, x_c and z_c denote the position of the center of mass in each direction, and M_{ij} is the ij-th moment of the concentration distribution in the xz vertical plane defined as:

$$M_{ij}(t) = \int_0^\infty \int_0^\infty \theta(x,z;t)c(x,z;t)x^{i}z^{j}dxdz \qquad (14)$$

with θ being the volumetric water content. The expressions for the components of σ^2 derived by Dagan (1984) and used by Russo (1991) for the comparison with Equation (13) are:

$$\sigma_{xx}^2 = \sigma_f^2\eta_z^2\{1n(\tau)-(3/2)+e-E_i(-\tau)-\exp(-\tau)(\tau^{-2}+\tau^{-1})-3\tau^{-2}\} \qquad (15a)$$

$$\sigma_{zz}^2(t) = \sigma_f^2\eta_z^2\{2\tau-3\ln(\tau)+(3/2)-3e+3E_i(-\tau)+3\exp(-\tau)(\tau^{-2}+\tau^{-1})-3\tau^{-2}\} \qquad (15b)$$

where $\tau \equiv z_c (t)/\eta_z$, η_z is the correlation length (or heterogeneity scale) in the vertical direction, e is Euler's constant (0.5772156649...), E_i is the exponential integral of τ, and σ_f^2 is the variance of $\ln(K_s)$, where K_s is the hydraulic conductivity at saturation. Figure 7 illustrates that the expressions for the components of σ^2 (15) given by Dagan (1984) were in good agreement with the general trend of the results of the single realization (13) considered by Russo (1991).

The time dependence of the longitudinal (i.e., vertical for the unsaturated zone) component of the spatial covariance tensor, $\sigma_{zz}^2(t)$, resulting from the numerical experiment of Russo (1991) was found to be proportional to t^2 during the first 100 h, which in the considered experiment is equivalent to a travel distance of about 1.6 m (Figure 7). The dependence of σ_{zz}^2 on t^2 is consistent with the behavior predicted by stochastic models considering transport of conservative solute in vertically homogeneous, noninteracting, parallel soil colums (Dagan and Bresler, 1979; Bresler and Dagan, 1981; Jury, 1982); the underlying assumptions for these models are that the fields have a horizontal extent much larger than the horizontal correlation length, and transport depths are small relative to the vertical correlation length. Furthermore, Figure 7 illustrates that for depths larger than 1.6 m, σ_{zz}^2 is no longer proportional to t^2 but exhibits temporal fluctuations about the linear, asymptotic behavior predicted by the stochastic two-dimensional analysis of Dagan (1984) for an infinite heterogeneous aquifer with a finite correlation length in the longitudinal direction. The horizontal extent of the flow domain considered by Russo (1991) was 15 m, covering about 15 horizontal correlation lengths, while the vertical extent was 10 m, covering more than 80 vertical correlation lengths. Therefore, it is not surprising that the results of solute spreading below 1.6 m were incompatible with the predictions of a parallel column model (Dagan and Bresler, 1979; Bresler and Dagan, 1981; Jury, 1982). In comparison with the results of Russo (1991), the experiment of Butters et al. (1989), that was conducted in a field of 0.64 ha, does not indicate asymptotic behavior for field-

S.E.A.T.M. van der Zee and G. Destouni

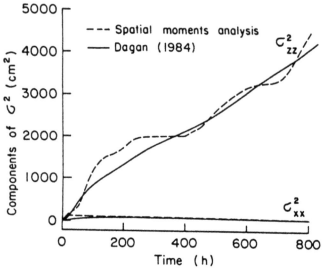

Figure 7. Components of the spatial covariance tensor, σ^2, as function of time, t. (From Russo, 1991.)

scale dispersivity even for travel depths greater than 10 m from the soil surface (Butters and Jury, 1989); this is illustrated in Figure 8 in terms of the estimated field-scale dispersivity values at different depths.

The field experiment of Butters et al. (1989) exhibited a decrease in the apparent field-scale dispersivity between 3 and 4.5 m from the soil surface, followed by a renewed growth below 4.5 m (Figure 8). This implies a compression-expansion behavior that was also observed by Ellsworth and Jury (1991) in the same field and is similar to the temporal fluctuations of σ_{zz}^2 found in the numerical experiment of Russo (1991) (Figure 7). The compression of the solute body may be a result of the low-conductivity zones experienced by the solute, e.g., finer-textured layers slowing down the longitudinal spreading of the pulse by damping out the short travel times (Butters and Jury, 1989; Russo, 1991; Ellsworth et al., 1991). Because the advective growth of the field-scale solute spreading resumes below a certain depth, the local velocity variations do not appear to be eliminated by increased lateral spreading; this indicates that solute transport in the unsaturated zone may also be essentially vertical in vertically heterogeneous soil. The results of Russo (1991) and Ellsworth et al. (1991) support this assumption by exhibiting limited solute spreading in the lateral direction. This is illustrated through the transversal component of the spatial covariance tensor, $\sigma_{xx}^2(t)$, in Figure 7. However, it may be noted that both flow and transport characteristics may be significantly different in particular situations, such as steep and layered hillslopes (Wilson et al., 1990).

We discuss the applicability of the steady flow assumption that has been applied in a number of investigations of solute transport in the unsaturated zone (Bresler and Dagan, 1981; Dagan and Bresler, 1979; Jury, 1982; van der Zee

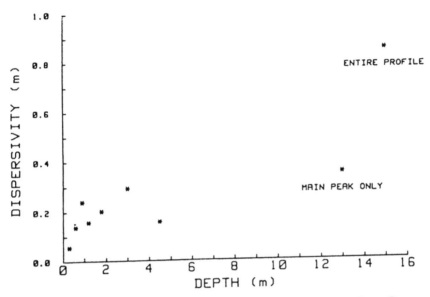

Figure 8. Field-scale dispersivity as a function of travel distance. (From Butters and Jury, 1989.)

and van Riemsdijk, 1986, 1987; Small and Mular, 1987; Butters and Jury, 1989; Destouni and Cvetkovic, 1990, 1991). A study by Wierenga (1977), for instance, indicates that prediction of solute transport based on a steady-state flow model is comparable to that based on a transient flow model if the breakthrough curve is presented as a function of cumulative drainage instead of the time variable. A study of field-scale solute transport by Bresler and Dagan (1983) suggests that simulated solute front propagation based on a transient flow model may be faster than that based on a steady flow model for relatively homogeneous fields. The compatibility between the flow models, however, appears to increase with increasing field heterogeneity (Bresler and Dagan, 1983). In contrast, a study by Russo et al. (1989) indicates that a steady-state flow model may considerably overestimate the effective solute velocity, i.e., underestimate solute travel time to a given depth.

The different indications with regard to the steady-state flow assumption resulting from the discussed papers (Wierenga, 1977, Bresler and Dagan, 1983, Russo et al., 1989) may be due to differences in the considered simulation conditions. Wierenga (1977) considered soil data for a laboratory column, uniformly filled with silty clay loam, whereas Russo et al. (1989) used data for a soil profile corresponding to a sandy loam, assuming vertical heterogeneity according to a random scaling factor. Bresler and Dagan (1983) considered statistical soil data for two different, areally heterogeneous fields, neglecting vertical variability. Wierenga (1977) and Russo et al. (1989) addressed transient flow due to periodic irrigation practices, while Bresler and Dagan (1983) investigated a single infiltration event followed by a redistribution phase.

Beese and Wierenga (1980) considered both root water uptake and equilibrium sorption-desorption in a comparison between solute transport in a homogeneous soil profile (clay loam) predicted by a transient and a constant-infiltration flow model. The transient flow was assumed to occur because of periodic irrigation practices and time-dependent root water uptake. Root water uptake was also included in the constant-infiltration model, which therefore was quasi steady-state with regard to flow. The study of Beese and Wierenga (1980) supports the results of Wierenga (1977) and indicates an increased compatibility between the two flow models when accounting for root uptake and equilibrium sorption processes. In addition, Jury and Gruber (1989) analyzed the effect of climatic variability and spatial heterogeneity on field-scale transport of pesticides using a stochastic representation of rainfall distribution and two hypothetical, vertically homogeneous soils. Their results indicate that a simple constant-infiltration model may be useful for assessing the probability of pesticide leaching through the biologically active soil zone provided that spatial variability is accounted for.

Natural precipitation and water uptake by roots have been considered in a recent study by Destouni (1991), where the applicability of the steady-state flow assumption for solute transport in different soil types was investigated for weather and vegetation conditions prevailing in the field. Specifically, soil data for five different profiles with detailed measurements of their hydraulic properties and their variation with depth were used. Furthermore, daily measurements of meteorological data were used as input parameters in the transient simulations that included snow and frost dynamics, interception of precipitation, and evapotranspiration. The results of Destouni (1991) indicate that a steady-state flow model may provide estimates of the mean solute advection in different soil profiles that are compatible with those based on a transient flow model. The constant rate of recharge in the steady-state flow model should then be interpreted as the average annual effective infiltration (i.e., infiltration minus actual evapotranspiration) in the considered area. This quantity reflects the prevailing climatic, soil, and vegetal conditions, and can be estimated from the average annual water balance equation (Eagleson, 1978). Furthermore, in order to account for vertical heterogeneity, the soil parameters of the steady-state flow model should be related to the measured soil properties (that determine the hydraulic conductivity curve) by a relevant averaging procedure over the depth of interest. The results of Destouni (1991) indicate that an arithmetic depth averaging of the relevant soil parameters may in some cases be sufficient for reflecting the effect of vertical heterogeneity in a given soil profile.

B. Sorptive Solute

The chemical processes in the unsaturated zone are generally complex. The detailed models of different chemical interactions discussed in previous sections are relevant for scrutinizing local solute behavior. However, in order to arrive at computable quantitative characterizations of field-scale solute transport in heterogeneous soil, significant simplifications may often be required. In that context, relatively simple equilibrium and nonequilibrium sorption-desorption models have provided suitable approximations for chemical interaction of the solute with the solid phase (Travis and Etnier, 1981; Nielsen et al., 1986). Moreover, the diffusive mass transfer resistances associated with immobile flow regions (such as intra-aggregate porosity) may often be significant and are manifested at the field scale as nonequilibrium sorption-desorption reactions, commonly referred to as physical nonequilibrium (Brusseau and Rao, 1989). In view of the large variability in soil properties of natural fields, a comparable degree of spatial variability in the sorption parameters, as in the hydraulic parameters, is plausible. van der Zee and van Riemsdijk (1987) studied the distribution of organic carbon content, OC, and of pH in 84 Dutch topsoils, and found OC to be lognormally distributed with a CV of 50% for ln(OC), whereas pH was normally distributed with a CV of 16%. van der Zee and van Riemsdijk (1987) related OC and pH to a retardation factor, R, accounting for equilibrium sorption-desorption of heavy metals. They analyzed the effect of spatially variable R, v, and solute input time T_c, on field-scale transport of reacting solute, using parameters relevant for cadmium and copper. Figure 9 illustrates that the effect of spatially varying R (curve 3) may be significant for the predicted solute penetration depth, especially if R is negatively correlated with v (curve 4).

Strongly structured soil and the existence of preferential flow paths are typical conditions in the unsaturated zone that may lead to nonequilibrium sorption-desorption at the field scale (Brusseau and Rao, 1989). Although experimental studies of the spatial variability in nonequilibrium sorption parameters are yet to be conducted, there are both theoretical and experimental indications that the rate of mass transfer in physical nonequilibrium depends on the soil structure, i.e., aggregate size and shape (Rao et al., 1980a,b; Nkedi-Kizza et al., 1983; Rasmuson, 1985a). Moreover, soil hydraulic properties may be influenced by the prevailing chemical processes. Some degree of correlation between the nonequilibrium sorption parameters and the hydraulic properties in heterogeneous soil is therefore plausible.

The effect of spatial variability in nonequilibrium sorption parameters and their correlation to the hydraulic conductivity on the field-scale mass flux of solute has been investigated by Destouni and Cvetkovic (1990, 1991). The obtained results indicate that spatially variable sorption rate coefficients that are uncorrelated or negatively correlated with the hydraulic conductivity may result in double peak behavior of the field-scale breakthrough curve with an earlier mass arrival than can be predicted by assuming constant sorption parameters. Figure 10 illustrates the double peak behavior of the field-scale breakthrough curve for sorption parameters that are uncorrelated with the hydraulic conductivity; the effect of negative correlation with the hydraulic conductivity is an even more pronounced early peak.

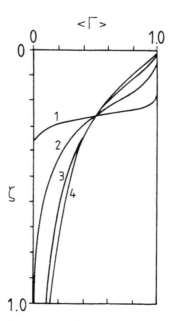

Figure 9. Profiles of field-averaged dimensionless sorbed solute, $<\Gamma>$, as function of dimensionless depth, ζ, where: (curve 1) R and T_c are constant, (curve 2) R is constant and T_c varies, (curve 3) R varies uncorrelated with v and T_c is constant, and (curve 4) R varies negatively correlated with v, and T_c is constant. (From van der Zee and van Riemsdijk, 1987.)

Because systematic field studies that are focused on the spatial distribution of sorption-desorption properties have not been conducted, the analyses discussed here (van der Zee and van Riemsdijk, 1987; Destouni and Cvetkovic, 1990, 1991) are primarily illustrative. Nevertheless, the results indicate that spatial variability in the sorption parameters, and their correlation to the hydraulic properties may be important when estimating field-scale transport of sorptive solute in heterogeneous soils.

C. Nitrate Case Study

To illustrate the effect of soil heterogeneity on transport of reactive solute we have chosen the nutrient ion nitrate. This compound received much attention in view of soil fertility as well as in connection with environmental issues. In North Western European countries nitrate applied to soil in excessive amounts leaches rapidly into the ground water. Although the EEC regulations for drinking water restrict nitrate concentrations to less than 50 mg/l, this standard is sometimes

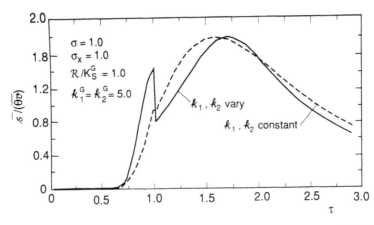

Figure 10. Dimensionless field-scale flux averaged concentration, $\overline{s}/(\overline{\theta\,v})$, as a function of dimensionless time, τ, for: (solid curve) spatially varying nonequilibrium sorption rate coefficients uncorrelated with hydraulic conductivity, and (dashed curve) constant nonequilibrium sorption rate coefficients. (Reproduced from Destouni and Cvetkovic, 1991.)

already exceeded in ground water used for drinking water at significant depths (e.g., 50 m below surface).

Nitrogen participates in a complicated cycle of transformations. Besides inputs at the soil surface, N may mineralize from soil organic matter, while on the other hand, it may be used by soil biota. In inorganic form, it may be present as the ammonia cation (NH_4^+) or the nitrate anion (NO_3^-). The former may adsorb, whereas the latter may be subject to denitrification (into N_2O or N_2-gases). A discussion of different aspects of nitrogen behavior has been given by Frissel and van Veen (1980) and Iskandar (1981). The work by Leffelaar (1987) considers N behavior at the microscopic level. It was shown that the denitrification may be a very localized phenomenon due to partial anaerobic soil conditions and may be very difficult to predict deterministically. Therefore, stochastic methods that account for soil heterogeneity may be appropriate for modeling transport of nitrate at the field scale.

In the following, we consider field-scale transport of nitrate in the unsaturated zone assuming a pulse injection due to fertilization, and a first-order loss of nitrate due to denitrification and biological consumption. Specifically, we illustrate the effect of spatial variability in the first-order rate parameter, μ, and its correlation with the hydraulic conductivity at saturation, K_s, following the stochastic methodology used by Destouni and Cvetkovic (1990, 1991).

A heterogeneous field extending in the horizontal x,y plane with the mean surface at $z = 0$ and z positive downwards is considered. Assuming, for simplicity, steady, vertical, gravitational water flow and homogeneous soil conditions in the vertical direction, we can relate the mean pore water velocity, v at x,y to the soil hydraulic properties and the recharge boundary conditions as (Dagan and Bresler, 1979; Bresler and Dagan, 1981):

$$v = \frac{K_s}{\theta_s} \qquad for \qquad R_e/K_s \geq 1 \qquad (16a)$$

$$v = \frac{K(\theta)}{\theta} = \frac{R_e}{\theta} \qquad for \qquad R_e/K_s < 1 \qquad (16b)$$

where R_e is the constant rate of recharge, and θ_s is the water content at saturation. The relationship for the hydraulic conductivity function, $K\ (\theta)$, adopted here is

$$K(\theta) = K_s \left[\frac{\theta - \theta_{ir}}{\theta_s - \theta_{ir}} \right]^{1/\beta} \qquad (17)$$

where θ_{ir} is the irreducible water content and β is a soil coefficient that can be related to the pore size distribution. In the following we shall assume $R_e = 0.3$ m/year (which is a reasonable value for the average annual effective infiltration in Northern Europe), $\beta = 1/7$, $\theta_s = 0.3$, and $\theta_{ir} = 0$.

For an instantaneous injection of nitrate at the surface of the field at the time $t = 0$ the mass balance equation at a given point x, y may be written as

$$\theta \frac{\partial c}{\partial t} + \frac{\partial s}{\partial z} = -\mu c \theta \qquad (18)$$

where $s(z, t)$ is the mass flux of solute per unit cross-sectional area. The mass flux in Equation (18) is composed of an advective and a dispersive part, where the dispersive part incorporates both local dispersion and molecular diffusion. In the present discussion, however, it is assumed that the solute spreading at the field scale is primarily due to advection variability and the effect of local dispersion may be neglected. For pure advection, the mass flux s is defined as $s = c\theta v$. Using the definition of s, the concentration is $c = s\ /\ (\theta v)$ and is substituted into Equation (18), which yields the mass balance equation as

$$\frac{\partial s}{\partial t} + v \frac{\partial s}{\partial z} = -\mu s \qquad (19)$$

For an instantaneous injection in a semi-infinite domain, the mass flux at a given depth, Z, is obtained by solving Equation (19):

$$s = \rho_A \exp(-\mu t)\delta(t-T) \tag{20}$$

where δ is the Dirac delta function, ρ_A is the surface density of the uniformly applied mass of nitrate, and $\tau = Z / v$ is the nitrate arrival time at Z.

In order to account for spatial variability, we regard K_s as a lognormally distributed random variable, i.e., $K_s = K_s^G \exp(Y)$, where K_s^G is the geometric mean of K_s, and Y is normally distributed with zero mean and variance σ^2. In addition, we regard μ as spatially variable, investigating the extreme cases of both negative and positive correlation with K_s, as well as the case where μ is independent of K_s. The rate coefficient, μ, is then defined as $\mu = \mu^G \exp(-Y)$ and $\mu = \mu^G \exp(Y)$ in the negative and positive correlation case, respectively, and as $\mu = \mu^G \exp(X)$ in the case where μ is independent of K_s. Here, μ^G is the geometric mean value of μ and X is a normally distributed random variable with zero mean and variance σ^2_X.

Assuming that the scale of the horizontal heterogeneity is small in comparison to the areal extent of the considered field, we may assume that the ergodic hypothesis is valid (Bresler and Dagan, 1981). Thus, the ensemble average of the mass flux is approximately equal to the mass flux over the field area. For the cases of correlated μ and K_s, the expression for the expected field-scale mass flux can be expressed as

$$\bar{s} = \int_{-\infty}^{\infty} s(t;T(Y),\mu(Y))p_Y(Y)dY \tag{21a}$$

and for the case of independent μ and K_s, as

$$\bar{s} = \int_{-\infty}^{\infty}\int_{-\infty}^{\infty} s(t;T(Y),\mu(X))p_Y(Y)p_X(X)dYdX \tag{21b}$$

where p_Y and p_X are the probability density functions for Y and X, respectively. In order to evaluate Equation (21), we have assumed $Z = 1$ m, and $\mu^G = 2$ year^{-1}, which is in agreement with field observations of nitrate leaching in the Netherlands (Kolenbrander, 1981). Furthermore, we have investigated different degrees of variability (i.e., σ and σ_X), where we chose the K_s^G values to arrive at the average pore water velocity of 1 m year^{-1}.

Figure 11 illustrates the field-scale cumulative mass arrival at Z in the cases of spatially variable μ with different correlation to K_s, in comparison to the case where μ is assumed to be constant over the entire field. The cumulative mass

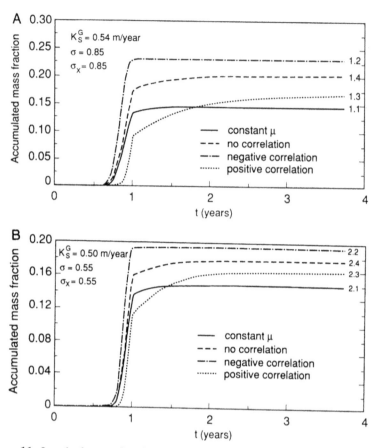

Figure 11. Leached mass fraction of nitrate as a function of time for random μ and K_s. Figure 11A shows cases 1.1 (μ constant), 1.2 (negative correlation of μ, K_s), 1.3 (pos. correlation) and 1.4 (uncorrelated). Figure 11B shows case 2.1 (μ constant), 2.2 (neg.correlation), 2.3 (pos. correlation) and 2.4 (uncorrelated). Coefficient of variation of v is 0.25 (Figure 11A) and 0.15 (Figure 11B).

arrival is determined as the time integral of the mass flux of Equation (21). A comparison between the different curves in Figure 11 shows that the assumption of constant μ may lead to a considerable underestimation of the leached nitrate fraction, especially if μ is negatively correlated with, or independent of K_s.

Besides the model based on Equation (21) we also evaluated the fraction that will be leached using a simple approximation developed by van der Zee and Boesten (1991) for pesticides, where the expected value of the leached fraction (m_F) is expressed as

$$m_F = \exp\left\{ -\frac{1}{2}Pe_f\left[\sqrt{1+\frac{4m_p}{Pe_f}}-1\right]\right\} \tag{22}$$

where m_p is the mean of the term $P = \mu L/v$, assuming this term is normally distributed. The apparent "field-Peclet number" (Pe_f) accounts for the variability of P and can be calculated from the variation coefficient (η_P) of P, by

$$\eta_p^2 = \frac{2}{Pe_f^2}\left[Pe_f - 1 + \exp(-Pe_f)\right] \tag{23}$$

Due to field heterogeneity, m_F differs from the leached fraction (F) in homogeneous fields (i.e., $F = exp\,(-\,\mu L/v)$ when pore scale dispersion is neglected). As can be seen from Table 2, the values of m_F calculated for the cases of Figure 11 are in good agreement with the values found using the model of Destouni and Cvetkovic (1990, 1991) (for long times),whereas this is not the case for F. Although Equation (22) apparently gives a useful approximation for the leached fraction of nitrate with an exception for the negatively correlated cases, its use is limited. The statistics of P (i.e., m_p and η_P) should be known, which in the present case required a numerical evaluation. Furthermore, the analysis leading to Equation (22) is based on a normal distribution of P, which is an incorrect assumption for, e.g., the negatively correlated cases of Figure 11.

Table 2. Leached fraction of nitrate after complete breakthrough for the cases of Figure 11A and 11B. The leached fraction for t → ∞ according to Equation (21) is given by M_F (num).

Case	m_p	η_p	M_F(num)	M_F(eq. 22)	F
1.1	2.22	0.59	0.15	0.176	0.110
1.2	5.12	3.76	0.23	0.311	0.006
1.3	2.48	0.77	0.17	0.181	0.084
1.4	3.19	1.35	0.20	0.212	0.041
2.1	1.99	0.26	0.15	0.154	0.137
2.2	2.65	1.22	0.19	0.236	0.071
2.3	2.12	0.47	0.17	0.168	0.120
2.4	2.32	0.66	0.18	0.178	0.100

The actual nitrate losses in a given field situation depend on the time of application during the year, atmospheric conditions, runoff, soil aeration and type of soil cover, as well as on the form and quantity of the applied fertilizer (Jarvis et al., 1987). These factors have not been considered in the present case study. Nevertheless, the results indicate that spatial variability in the conditions governing denitrification and biological consumption, and their correlation with hydraulic properties, may be important when estimating field-scale leaching of nitrate into the groundwater.

IV. Conclusions

In this work we have discussed effects of (bio)chemical interactions on solute transport and reviewed different approaches to model the processes that are involved. The (bio)chemical interactions may have profound effects on displacement, because they may control the concentration bounds, the solute front velocity (through retardation), and the front shape in a homogeneous flow domain. Although, in general, solute transport is a multicomponent and non-equilibrium process, adequate descriptions based on significant simplifications (such as considering only one solute or assuming local equilibrium conditions) may be feasible. Even with such simplifications the effects of (bio)chemical interactions on transport may be significant.

An additional complication when we are interested in solute behavior at the field scale is the spatial heterogeneity with regard to physical and (bio)chemical properties. At this larger scale the type of interactions and the correlation structure between different spatially variable parameters may control the displacement behavior. The relevance of such heterogeneity was illustrated here for transport of a degradable solute (nitrate), using a stochastic framework. The stochastic approaches reviewed in this chapter may be necessary for understanding transport of (reacting) solute in heterogeneous soils.

Computationally, it may be feasible in the near future to combine complicated (bio)chemical models with stochastic approaches. However, the data requirements may often be prohibitive for application to real field situations. It may also be questionable whether such detailed stochastic approaches will be necessary. Usually, microscopic phenomena are maintained only for some broad features at the macroscopic (field) scale. Heterogeneity may then be more important than details concerning flow, (bio)chemical transformations, etc. On the other hand, such transformations may control the rate of transport via retardation or may impose bounds on the concentration in solution. When the effects of the various interactions predominate they may overrule random heterogeneity. Hence, it seems likely that both aspects (i.e., the effect of both complicated solute interactions and heterogeneity) need to be accounted for and to be coupled. Most likely either heterogeneity or the interactions of solutes predominates and should be emphasized in modeling.

Acknowledgements
We thank V.D. Cvetkovic (Royal Inst. Techn., Stockholm, Sweden) and J.C.M. de Wit (Agric. Univ., Wageningen, The Netherlands) for the discussions regarding this work. The work was partly funded by the Dir. General Science, Research & Development (Commission Europ. Commun.) through the Europ. Commun. Environ. Res. Program STEP (S. van der Zee) and by the National Swedish Environ. Protection Board, SNV. (G. Destouni).

References

Aagaard, P. and H.C. Helgeson. 1982. Thermodynamic and kinetic constraints on reaction rates among minerals and aqueous solutions. I. Theoretical considerations. *Am. J. Sci.* 282:237-285.

Abriola, L.M. 1987. Modeling contaminant transport in the subsurface: an interdisciplinary challenge. *Reviews Geophysics.* 25:125-134.

Amoozegar-Fard, A., D.R. Nielsen, and A.W. Warrick. 1982. Soil solute concentration distributions for spatially varying pore water velocities and apparent diffusion coefficients. *Soil Sci. Soc. Am. J.* 46:3-8.

Barker, J.A. 1982. Laplace transform solutions for solute transport in fissured aquifers. *Adv. Water Resour.* 5:98-104.

Barker, J.A. 1985. Block-geometry functions characterizing transport in density fissured media. *J. Hydrol.* 77:263-279.

Barry, D.A., J.Y. Parlange, and J.L. Starr. 1986. Interpolation method for solving the transport equation in soil columns with irreversible kinetics. *Soil Sci.* 142:296-307.

Bennet, A. and F. Goodridge. 1970. Hydrodynamic and mass transfer studies in packed adsorption columns. 1. Axial liquid dispersion. *Trans. Inst. Chem. Eng.* 48:232-244.

Berner, R.A. 1978. Rate control of mineral dissolution under earth surface conditions. *Am. J. Sci.* 278:235-252.

Berner, R.A. and G.R. Holdren. 1979. Mechanism of feldspar weathering. II. Observations of feldspars from soils. *Geochim. Cosmochim. Acta.* 43:1173-1186.

Biggar, J.W. and D.R. Nielsen. 1976. Spatial variability of the leaching characteristics of a field soil. *Water Resour. Res.* 12:78-84.

Blum, A.E. and A.C. Lasaga. 1987. Monte Carlo simulations of surface reaction rate laws. p. 255-292. In: W. Stumm (ed.), Aquatic surface chemistry. J. Wiley & Sons, NY.

Boesten, J.J.T.I. 1986. Behaviour of herbicides in soil: simulation and experimental assessment. Ph.D thesis. Inst. Pest. Res. Wageningen/Agric. Univ. Wageningen, 263 pp.

Bolland, M.D.A., M. Posner, and J.P. Quirk. 1977. Zinc adsorption by goethite in the presence and absence of phosphate. *Austral. J. Soil Res.* 15,279-286.

Bolt, G.H., 1982. Movement of solutes in soil: Principles of adsorption/exchange chromatography. p. 285-348. In: G.H. Bolt (ed.), *Soil Chemistry, B.* Elsevier, Amsterdam.

Bolt, G.H. 1982. The ionic distribution in the diffuse double layers. In: Soil Chemistry, B. 2nd ed. (G.H. Bolt, ed)., Elsevier, Amsterdam, 1-26.

Bolt, G.H., and W.H. van Riemsdijk. 1991. The electrified interface of the soil solid phase, A. The electrochemical control system. p. 37-79. In: G.H. Bolt et al. (eds.), *Interactions at the soil colloid-soil solution interface*, Kluwer Ac. Publ., Dordrecht.

Bond, W.J. and I.R. Philips. 1990. Approximate solutions for cation transport during unsteady, unsaturated water flow. *Water Resour. Res.* 26:2195-2205.

Bowden, J.W., S. Nagarajah, N.J. Barrow, A.M. Posner, and J.P. Quirk. 1980. Describing the adsorption of phosphate, citrate, and selenite on a variable - charge surface. *Austral. J. Soil Res.* 18:49-60.

Bresler, E. and G. Dagan. 1981. Convective and pore scale dispersive solute transport in unsaturated heterogeneous fields. *Water Resour. Res.* 17:1683-1693.

Bresler, E. and G. Dagan. 1983. Unsaturated flow in spatially variable fields, 3. Solute transport models and their application to two fields. *Water Resour. Res.* 19:429-435.

Bresler, E., B.L. McNeal, and D.L. Carter. 1982. Saline and sodic soils. *Adv. Series Agric. Sci.* Springer, Berlin, 236 pp.

Bromilow, R.H. and M. Leistra. 1980. Measured and simulated behaviour of aldicarb and its oxidation products in fallow soils. *Pestic. Sci.* 11:389-395.

Bruggenwert, M.G.M. and A. Kamphorst. 1982. Survey of experimental information on cation exchange in soil systems. p. 141-203. In: G.H. Bolt, (ed.). *Soil Chemistry B.*, Elsevier, Amsterdam, 141-203.

Brusseau, M.L. and P.S.C. Rao. 1989. Sorption nonideality during organic contaminant transport in porous media. *CRC Crit. Rev. Environ. Control.* 19:33-99.

Bryant, S.L., R.S. Schechter, and L.W. Lake. 1986. Interactions of precipitation/dissolution waves and ion exchange in flow through permeable media. *Am. Inst. Chem. Eng.* 32:751-764.

Butters, G.L. and W.A. Jury. 1989. Field scale transport of bromide in an unsaturated soil, 2. Dispersion modeling. *Water Resour. Res.* 25:1583-1589.

Butters, G.L., W.A. Jury, and F.F. Ernst. 1989. Field scale transport of bromide in an unsaturated soil, 1. Experimental methodology and results. *Water Resour. Res.* 25:1575-1581.

Cameron, D.A. and A. Klute. 1977. Convective - dispersive solute transport with a combined equilibrium and kinetic adsorption model. *Water Resour. Res.* 13: 183-188.

Cederberg. G.A., R.L. Street, and J.O. Lecky. 1985. A ground water mass transport and equilibrium chemistry model for multi component systems. *Water Resour. Res.* 21:1095-1104.

Chardon, W.J. 1984. Mobility of cadmium in soil (in Dutch). *Serie Bodembescherming 36*. Staatsuitgeverij, Den Haag, 200 pp.

Cho, C.M. 1971. Convective transport of ammonium with nitrification in soil. *Can. J. Soil Sci.* 51:339-350.

Chou, L. and R. Wollast. 1984. Study of the weathering of Albite at room temperature and pressure with a fluidized bed reaction. *Geochim. Cosmochim. Acta.* 48:2205-2217.

Christensen, T.H. 1980. Cadmium sorption onto two mineral soils. Report, Dep. Sanit. Eng., Tech. Univ. Denmark, Lingby.

Cvetkovic, V.D. and G. Destouni. 1989. Comparison between resident and flux-averaged concentration models for field-scale solute transport in the unsaturated zone. p. 245-250. In: H.E. Kobus and W. Kinzelbach (eds.). *Contaminant Transport in Groundwater* A.A. Balkema, Rotterdam, Brookfield.

Dagan, G. 1982. Stochastic modeling of groundwater flow by unconditional and conditional probabilities, 2. The solute transport. *Water Resour. Res.* 18:835-848.

Dagan, G. 1984. Solute transport in heterogeneous porous formations. *J. Fluid Mech.* 145:151-177.

Dagan, G. 1987. Theory of solute transport by groundwater. *Ann. Rev. Fluid Mech.* 19:183-215.

Dagan, G. 1990. Transport in heterogeneous porous formations: spatial moments, ergodicity, and effective dispersion. *Water Resour. Res.* 26:1281-1290.

Dagan, G. and E. Bresler. 1979. Solute dispersion in unsaturated heterogeneous soil at a field scale. I. Theory. *Soil Sci. Soc. Am. J.* 43:461-467.

Dance, J.T. and E.J. Reardon. 1983. Migration of contaminants in ground water at a land fill: A case study, 5. Cation migration in a dispersion test. *J. Hydrol.* 63:109-130.

Davis, J.A., R.O. James, and J.O. Leckie. 1978. Surface ionization and complexation at the oxide/water surface I. *J. Colloid Interface Sci.* 63:480-499.

Davis, L.E. and J.M. Rible. 1950. Monolayers containing polyvalent ions. *J. Colloid Sci.* 5:81-83.

De Haan, F.A.M. 1964. The negative adsorption of anions (anion exclusion) in systems with interacting double layers. *J. Phys. Chem.* 68:2970-2976.

De Haan, F.A.M. and W.H. van Riemsdijk. 1986. Behaviour of inorganic contaminants in soil. p. 19-32. In: J.W. Assink and W.J. van den Brink (eds.) *Contaminated Soil*. Martinus Nijhoff, Dordrecht.

De Haan, F.A.M., S.E.A.T.M. van der Zee, and W.H. van Riemsdijk. 1987. The role of soil chemistry and soil physics in protecting soil quality: variability of sorption and transport of cadmium as an example. *Neth. J. Agric. Sci.* 35:347-359.

De Smedt, F. and P.J. Wierenga. 1979. A generalized solution for solute flow in soils with mobile and immobile water. *Water Resour. Res.* 15:1137-1141.

Destouni, G. 1991. Applicability of the steady-state flow assumption for solute advection in field soils. *Water Resour. Res.* 27:2129-2140.

Destouni, G. and V. Cvetkovic. 1989. The effect of heterogeneity on large scale solute transport in the unsaturated zone. *Nordic Hydrology.* 20:43-52.

Destouni, G. and V. Cvetkovic. 1990. Mass flux of sorptive solute in heterogeneous soils. p. 251-260. In: K. Roth, H. Flühler, W.A. Jury, and J.C. Parker. (eds), *Field-Scale Water and Solute Flux in Soils.* Birkhäuser Verlag, Basel.

Destouni, G. and V. Cvetkovic. 1991. Field-scale mass arrival of sorptive solute into the groundwater. *Water Resour. Res.* 27:1315-1325.

Dixon, J.B. and S.B. Weed. 1977. Minerals in Soil Environments. *Soil Sci. Soc. Am. Inc.* Madison, WI, 948 pp.

Donner, H.E. 1978. Chloride as a factor in mobilities of Ni (I), Cu (II) and Cd (II) in soil. *Soil Sci. Soc. Am. J.* 42:882-885.

Donner, H.E., A Pukite, and E. Yang. 1982. Mobility through soils of certain heavy metals in geothermal brine water. *J. Environ. Qual.* 11(3):339-394.

Dutt, G.R., R.W. Terkeltoub, and R.S. Rauschkolb. 1972. Prediction of gypsum and leaching requirements for sodium-affected soils. *Soil Sci.* 114:93-103.

Eagleson, P.S. 1978. Climate, soil, and vegetation, 1. Introduction to water balance dynamics. *Water Resour. Res.* 14:705-712.

Edwards, D.G., A.M. Posner, and J.P. Quirk. 1965. Repulsion of chloride ions by negatively charged clay surfaces. I-III, *Transact. Faraday Soc.* 61:2808-2823.

Edwards, D.G. and J.P. Quirk. 1962. Repulsion of chloride by montmorillonite. *J. Colloid Sci.* 17:872-882.

Ellsworth, T.R. and W.A. Jury. 1991. A three-dimensional field study of solute transport through unsaturated, layered, porous media, 2. Characterization of vertical dispersion. *Water Resour. Res.* 27:967-981.

Ellsworth, T.R., W.A. Jury, F.F. Ernst, and P.J. Shouse. 1991. A three-dimensional field study of solute transport through unsaturated, layered, porous media, 1. Methodology, mass recovery, and mean transport. *Water Resour. Res.* 27:951-965.

Förster, R. 1986. A multicomponent transport model. *Geoderma* 38:261-278.

Frissel, M.J. and J.A. van Veen (eds.). 1980. *Simulation of nitrogen behaviour of soil-plant systems.* 277pp. Centre Agric. Publ. Doc., Wageningen, Netherlands.

Gaines, G.L. and H.C. Thomas. 1953. Adsorption studies on clay minerals II A formulation of the thermodynamics of exchange adsorption. *J. Phys. Chem.* 21:714-718.

Gapon, E.N. 1933. On the theory of exchange adsorption in soil. *J. Gen. Chem.*, USSR. 3:144-152.

Gardner, W.R., M.S. Mayhugh, J.O. Goertzen, and C.A. Bower. 1959. Effect of electrolyte concentration and exchangeable sodium percentage on diffusivity of water in soils. *Soil Sci.* 88:270-274.

Gelhar, L.W. and C.L. Axness. 1983. Three-dimensional stochastic analysis of macrodispersion in aquifers. *Water Resour. Res.* 19:161-180.

Goldstein, R.A., S.A. Gherini, C.W. Chen, L. Mak, and R.J.M. Hudson. 1984. Integrated acidification study (ILWAS): a mechanistic ecosystem analysis. *Philos. Transact. R. Soc. London*, B. 305:409-425.

Grove, D.B. and W.W. Wood. 1979. Prediction and field verification of subsurface water quality changes during artificial recharge. Lubbock, TX, *Ground Water*. 17:250-257.

Gureghian, A.B. and G. Jansen. 1983. LAYFLO: A one dimensional semi-analytic model for the migration of a three member decay chain in a multilayered geologic medium. Rep. ONWI-466, Off. Nucl. West Isol. Batelle Memor. Inst. Columbus Ott. 83 pp.

Harmsen, K. 1987. Movement of zinc sulphate in a calcium-saturated soil: cation exchange and precipitation of gypsum. *Neth. J. Agric. Sci.* 35(3):296-314.

Harmsen, K. and G.H. Bolt. 1982a. Movement of ions in soil, I. Ion exchange and precipitation. *Geoderma*. 28:85-101.

Harmsen, K. and G.H. Bolt. 1982b. Movement of ions in soil, II. Ion exchange and dissolution. *Geoderma*. 28:10/3-116.

Helgeson, H.C. 1971. Kinetics of mass transfer among silicates and aqueous solutions. *Geochim. Cosmochim. Acta.* 35:421-469.

Herzer, J. and W. Kinzelbach. 1989. Coupling of transport and chemical processes in numerical transport models. *Geoderma*. 44:115-128.

Hiemstra, T. and W.H. van Riemsdijk. 1990. Multiple activated complex dissolution of metal (hydr)oxides: A thermodynamic approach applied to Quarz. *J. Colloid Interface Sci.* 136 (1):132-150.

Hiemstra, T., W.H. van Riemsdijk, and G.H. Bolt. 1989. Multisite proton adsorption modeling at the solid/solution interface of (hydr)oxides: A new approach I. Model description and evaluation of intrinsic reaction constants. *J. Colloid Interface Sci.* 133 (1):91-104.

Hiemstra, T., W.H. van Riemsdijk, and M.G.M. Bruggenwert. 1987. Proton adsorption mechanism at the gibbsite and aluminum oxide solid/solution interface. *Neth. J. Agric. Sci.* 35:281-293.

Hornsby, A.G. and J.M. Davidson. 1973. Solution and adsorbed fluometuron concentration distribution in a water-saturated soil - experimental and predicted evaluation. *Soil Sci. Soc. Am. Proc.* 37:823-828.

Huang, C.P. and W. Stumm. 1973. Specific adsorption of cations on hydrous γ-Al_2O_3. *J. Colloid Interface Sci.* 43:409-420.

Hutson, J.L. and R.J. Wagenet. 1989. LEACHM, Leaching estimation and chemistry model, user's guide. Continuum Water Resour. Institute 2, 148 pp, Cornell Univ., Ithaca, New York.

Huyakorn, P.S., B.H. Lester, and J.W. Mercer. 1983. An efficient finite element technique for modeling transport in fractured porous media, 1. Single species transport. *Water Resour. Res.* 20(8):841-854.

Iskandar, I.K., (eds.). 1981. *Modeling wastewater renovation and land treatment.* 802pp., John Wiley & Sons, New York.

Jardine, P.M., G.V. Wilson, and R.J. Luxmoore. 1990. Unsaturated solute transport through a forest soil during rain storm events. *Geoderma.* 46:103-118.

Jarvis, S.C., M. Sherwood, and J.H.A.M. Steenvoorden. 1987. Nitrogen losses from animal manures: from grazed pastures and from applied slurry. p. 195-212. In: H.G. van der Meer et al. (eds.). *Animal manure on grassland and fodder crops. Fertilizer or waste?.* Mart. Nijhoff Publ., Dordrecht.

Jauzein, M., C. André, R. Margita, M. Sardin, and D. Schweich. 1989. A flexible computer code for modeling transport in porous media: impact. *Geoderma* 44:2/3, 95-114.

Jennings, A.A. and J. Kirkner. 1984. Instantaneous equilibrium approximation analysis. *J. Hydraul. Eng.* 110(12):1700-1717.

Jennings, A.A., D.J. Kirkner, and T.L. Theis. 1982. Multicomponent equilibrium chemistry in groundwater quality models. *Water Resour. Res.* 18:1089-1096.

Jury, W.A. 1982. Simulation of solute transport with a transfer function model. *Water Resour. Res.* 18:363-368.

Jury, W.A. and J. Gruber. 1989. A stochastic analysis of the influence of soil and climatic variability on the estimate of pesticide groundwater pollution potential. *Water Resour. Res.* 25:2465-2474.

Jury, W.A., D.D. Focht, and W.J. Farmer. 1987. Evaluation of pesticide ground water pollution potential from standard indices of soil-chemical adsorption and bio-degradation. *J. Environ. Qual.* 16:422-428.

Jury, W.A., G. Sposito, and R.E. White. 1986. A transfer function model of solute transport through soil, 1. Fundamental concepts. *Water Resour. Res.* 22:243-247.

Jury, W.A., L.H. Stolzy, and P.H. Shouse. 1982. A field test of the transfer function model for predicting solute transport. *Water Resour. Res.* 18:359-375.

Kamphorst, A. and G.H. Bolt. 1976. Saline and sodic soils. p. 171-191. In: G.H. Bolt and M.G.M. Bruggenwert (eds.), *Soil Chemistry A.* Developm. Soil Sci. 5A, Elsevier, Amsterdam.

Kirk, G.J. and P.H. Nye. 1985. The dissolution and dispersion of dicalcium phosphate dihydrate in soils, I. A predictive model for a planar source. *J. Soil Sci.* 36:446-459.

Kirkner, D.J., A.A. Jennings, and T.L. Theis. 1985. Multisolute mass transport with chemical interaction kinetics. *J. Hydrol.* 76:107-117.

Kirkner, D.J., T.L. Theis, and A.A. Jennings. 1984. Multicomponent solute transport with sorption and soluble complexation. *Adv. Water Resour.* 7:120-125.

Kolenbrander, G.J. 1981. Leaching of nitrogen in agriculture. p. 199-217. In: J.C. Brogan (ed.), *Nitrogen losses and surface run-off from landspreading of manures.* Dev. Plant Soil Sci., Martinus Nijhoff/Junk Publ. 2.

Krishnamoorthy, C. and R. Overstreet. 1950. An experimental evaluation of ion exchange relationships. *Soil Sci.* 69:41-53.

Lasaga, A.C. 1983. Kinetics of silicate dissolution. *Proc. 4th Int. Conf. Water Rock Interaction.* 269-274.

Lecky, J.O., M.M. Benjamin, K. Hayer, G. Kaufman, and S. Altman. 1980. Adsorption coprecipitation of trace elements from water with iron oxyhydroxide. Fin. Rep. EPRI CS-1513, Electr. Power Res. Inst., Palo Alto, Ca.

Leffelaar, P.A. 1987. Dynamics of partial anaerobiosis, denitrification and water in soil: experiments and simulation. Ph.D thesis, Agric. Univ. Wageningen, The Netherlands, 117 pp.

Lester, D.H., G. Janssen, and H.C. Burkholder. 1975. Migration of radionuclide chains through an adsorbing medium. In: Adsorption and Ion exchange. *Am. Inst. Chem. Symp. Series* 71(152):202-213.

Lexmond, Th.M. 1980. The effect of soil pH on copper toxicity to forage maize grown under field conditions. *Neth. J. Agric. Sci.* 28:164-183.

Lindstrom, F.T. and I. Boersma. 1975. A theory of mass transport of previously distributed chemicals in a water-saturated sorbing porous medium, 4, distributions, *Soil Sci.* 119:411-420.

Lindstrom, F.T. and M.N.L. Narasimhan. 1973. Mathematical description of a kinetic model for dispersion of previously distributed chemicals in a sorbing porous medium. SIAM. *J. Appl. Math.* 24:496-510.

Lyklema, J. 1968. The structure of the electrical double layer on porous surfaces. *Electroanal. Chem.* 18:341-348.

McLaren, A.D. 1975. Comments on kinetics of nitrification and biomass of nitrifiers in a soil column. *Soil Sci. Soc. Am. Proc.* 35:597-598.

McNeal, B.L. and N.T. Coleman. 1966. Effect of solution composition on soil hydraulic conductivity. *Soil Sci. Soc. Am. Proc.* 30:308-312.

Micera, G., C. Gessna, P. Melis, A. Premboli, R. Dallocchio, and S. Deiana. 1986. Zinc (II) adsorption on aluminum hydroxide. *Coll. Surfaces.* 17:389-394.

Miller, C.T. and W.J. Weber. 1986. Sorption of hydrophobic organic pollutants in saturated soil systsems. *J. Cont. Hydrol.* 1:243-261.

Miller, C.T. and W.J. Weber. 1988. Modeling the sorption of hydrophobic contaminants by aquifer materials. II. Column reactor systems. *Water Res.* 22:465-474.

Miller, C.W. and L.V. Benson. 1983. Simulation of solute transport in a chemically reactive heterogeneous system. Model development and application. *Water Resour. Res.* 19:381-391.

Misra, C., D.R. Nielsen, and J.W. Biggar. 1974. Nitrogen transformations in soil during leaching. I. Theoretical considerations. *Soil Sci. Soc. Am. Proc.* 38:289-293.

Nederlof, M.M., W.H. van Riemsdijk, and L.K. Koopal. 1989a. Proc. "Heavy Metals in the Environment. Vol. II," 7th Int. Conf., Geneve, Sept. 12-15, CEP Cons. Ltd, Edinburgh, UK, 400 p.

Nederlof, M.M., W.H. van Riemsdijk, and L.K. Koopal. 1990. Determination of adsorption affinity distributions: a general frame work for methods related to local isotherm approximations. *J. Colloid Interface Sci.* 135:410-426.

Nielsen, D.R., M.Th. van Genuchten, and J.W. Biggar. 1986. Water and solute transport processes in the unsaturated zone. *Water Resour. Res.* 22:89S-108S.

Nkedi-Kizza, P., J.W. Biggar, H.M. Selim, M.Th. van Genuchten, P. Wierenga, J.M. Davidson, and D.R. Nielsen. 1984. On the equivalence of two conceptual models. *Water Resour. Res.* 20:1123-1130.

Nkedi-Kizza, P., J.W. Biggar, M.Th. van Genuchten, P.J. Wierenga, H.M. Selim, J.M. Davidson, and D.R. Nielsen. 1983. Modelling tritium and chloride 36 transport through an aggregated oxisol. *Water Resour. Res.* 19:691-700.

Novak, L.T., D.C. Adriano, G.A. Coulman, and D.B. Shah. 1975. Phosphorus movement in soils: Theoretical aspects. *J. Environ. Qual.* 4:93-99.

Paces, T. 1973. Steady state kinetics and equilibrium between ground water and granitic rock. *Geochim. Cosmochim. Acta.* 37:2641-2663.

Parker, J.C. and A.J. Valocchi. 1986. Constraints on the validity of equilibrium and first order kinetic transport models in structured soils. *Water Resour. Res.* 22:399-407.

Parlange, J.Y., J.L. Starr, D.A. Barry, and R.D. Braddock. 1984. Some approximate solutions of the transport equation with irreversible reactions. *Soil Sci.* 137:434-442.

Perkins, T.K. and O.C. Johnston. 1963. A review of diffusion and dispersion in porous media. *Soc. Pet. Eng. J.* 3:70-84.

Pfannkuch, A.O. 1963. Contribution à l' étude des desplacements de fluides miscibles dans un milieu poreux. *Rev. Inst. Fr. Petr.* 18:215-270.

Rao, P.S.C. and R.E. Jessup. 1982. Development and verification of simulation models for describing pesticide dynamics in soils. *Ecol. Modeling*, 16:67-75.

Rao, P.S.C., D.E. Rolston, R.E. Jessup, and J.M. Davidson. 1980a. Solute transport in aggregated porous media: Theoretical and experimental evaluation. *Soil Sci. Soc. Am. J.* 44:1139-1146.

Rao, P.S.C., R.E. Jessup, D.E. Rolston, J.M. Davidson, and P. Kilcrease. 1980b. Experimental and mathematical description of non-adsorbed solute transfer by diffusion in spherical aggregates. *Soil Sci. Soc. Am. J.* 44:684-688.

Rao, P.S.C., J.M. Davidson, and H.M. Selim. 1979. Evaluation of conceptual models for describing non-equilibrium adsorption-desorption of pesticides during steady flow in soils. *Soil Sci. Soc. Am. J.* 43:22-28.

Rasmuson, A. 1985. The effect of particles of variable size, shape and properties on the dynamics of fixed beds. *Chem. Eng. Sci.* 40:621-629.

Rasmuson, A. 1985a. The influence of particle shape on the dynamics of fixed beds. *Chem. Eng. Sci.* 40:1115-1122.

Rasmuson, A. 1986. Modeling of solute transport in aggregated/fractured media including diffusion into the bulk matrix. *Geoderma.* 38:41-60.

Reiniger, P. and G.H. Bolt. 1972. Theory of chromatography and its application to cation exchange in soils, *Neth. J. Agric. Sci.* 20:301-313.

Rinaldo, A. and A. Marani. 1987. Basin scale model of solute transport. *Water Resour. Res.* 23:2107-2118.

Rogers, V.C. 1978. Migration of radionuclide chains in ground water. *Nucl. Techn.* 40:315-320.

Rubin, J. and R.V. James. 1973. Dispersion-affected transport of reacting solutes in saturated porous media: Galerkin method applied to equilibrium-controlled exchange in unidirectional steady water flow. *Water Resour. Res.* 9:1332-1356.

Russo, D. 1991. Stochastic analysis of simulated vadose-zone solute transport in a vertical cross section of heterogeneous soil during nonsteady water flow. *Water Resour. Res.* 27:267-283.

Russo, D. and E. Bresler. 1977a. Effect of mixed Na/Ca solutions on the hydraulic properties of unsaturated soils. *Soil Sci. Soc. Am. J.* 41:713-717.

Russo, D. and E. Bresler. 1977b. Analysis of the saturated - unsaturated hydraulic conductivity in a mixed Na/Ca soil system. *Soil Sci. Soc. Am. J.* 41:706-710.

Russo, D., W.A. Jury, and G.L. Butters. 1989. Numerical analysis of solute transport during transient irrigation, 1. The effect of hysteresis and profile heterogeneity. *Water Resour. Res.* 25:2109-2118.

Schindler, P.W. and H. Gamsjäger. 1972. Acid-base reactions of the T_iO_2 (anatase)-water interface and the point of zero charge of T_iO_2-suspensions. *Kolloid Z. Polymere* 250:759-763.

Schott, J. and J.C. Petit. 1987. New evidence for the mechanisms of dissolution of silicate minerals. p. 293-317. In: W. Stumm (ed.) *Aquatic surface chemistry.* J. Wiley & Sons, NY.

Schulin, R., A. Papritz, H. Flühler, and H.M. Selim. 1989. Calcium and magnesium transport in aggregated soils at variable ionic strength. *Geoderma* 44:2/3, 129-142.

Schulin, R., M.Th. van Genuchten, H. Flühler, and P. Ferlin. 1987. An experimental study of solute transport in a stony field soil. *Water Resour. Res.* 23:1785-1794.

Schulz, H.C. and E.J. Reardon. 1983. A combined mixing cell/analytical model to describe two-dimensional reactive solute transport for undirectional ground water flow. *Water Resour. Res.* 19:493-502.

Simmons, C.S. 1982. A stochastic-convective transport representation of dispersion in one-dimensional porous media systems. *Water Resour. Res.* 18:1193-1214.

Singh, R. and P.H. Nye. 1986. A model for ammonia volatilization from applied urea, I. Development of the model. *J. Soil Sci.* 37:9-20.

Sips, R. 1950. On the structure of the catalyst surface, II. *J. Chem. Phys.* 18:1024-1026.

Small, M.J. and J.R. Mular. 1987. Long-term pollutant degradation in the unsaturated zone with stochastic rainfall infiltration. *Water Resour. Res.* 23:2246-2256.

Sposito, G. 1985. *CRC Crit. Rev. Environ. Control* 16:193-223.

Sposito, G. and S.V. Mattigod. 1980. GEOCHEM: a computer program for the calculation of chemical equilibria in soil solutions and other natural water systems. Kearney Found. *Soil Sci.*, Univ. Cal. Riverside, CA, USA.

Starr, J.L., H.C. DeRoo, C.R. Frink, and J.Y. Parlange. 1978. Leaching characteristics of a layered field soil. *Soil Sci. Soc. Am. J.* 42:386-391.

Stern, O. 1924. Zur Theorie der electrischen Doppelschicht. *Z. Electrochem.* 30:508-516.

Stevenson, F.J. 1982. *Humus chemistry: Genesis, Composition, and Reactions.* Wiley & Sons, New York.

Stumm, W., C.P. Huang, and S.R. Jenkins. 1970. Specific chemical interaction affecting the stability of dispersed systems. Croat. Chim. Acta 42:223-245.

Stumm, W., H. Hohl, and F. Dalang. 1976. Interaction of metal ions with hydrous oxide surfaces. *Croat. Chim. Acta.* 48:491-504.

Tanji, K.K., L.D. Doneen, G.V. Ferry, and R.S. Ayers. 1972. Computer simulation analysis on reclamation of salt affected soils in San Joaquin Valley, Ca. *Soil Sci. Soc. Am. Proc.* 36:127-133.

Travis, C.C. and E.L. Etnier. 1981. A survey of sorption relationships for reactive solute in soil. *J. Environ. Qual.* 10:8-17.

Valocchi, A.J. 1985. Validity of the local equilibrium assumption for modeling sorbing solute transport through homogeneous soils. *Water Resour. Res.* 21:808-820.

Valocchi, A.J., P.V. Roberts, G.A. Parks, and R.L. Street. 1981. Simulation of the transport of ion-exchanging solutes using laboratory-determined chemical parameter values. *Ground Water.* 19:600-607.

Valocchi, A.J., R.L. Street, and P.V. Roberts. 1981. Transport of ion-exchanging solutes in ground water: chromatographic theory and field simulation. *Water Resour. Res.* 17:1517-1527.

Van Beek, C.G.E.M. and R. Pal. 1978. The influence of cation exchange and gypsum solubility on the transport of sodium, calcium, and sulphate through soils. *J. Hydrol.* 36:133-142.

Van Breemen, N., C.T. Driscoll, and J. Mulder. 1984. Acidic deposition and internal proton sources in acidification of soils and waters. *Nature* 307:599-604.

Van Breemen, N., J. Mulder, and C.T. Driscoll, 1983, Acidification and alkalinization of soils. *Plant and Soil.* 75:283-308.

van der Zee, S.E.A.T.M. 1988. Transport of reactive contaminants in heterogeneous soil systems. Ph.D. dissertation, Agric. Univ. Wageningen, NL, 283 pp.

van der Zee, S.E.A.T.M. 1990. Analytical travelling wave solutions for transport with non-linear and non-equilibrium adsorption. *Water Resour. Res.* 26(10):2563-2578.

van der Zee, S.E.A.T.M. 1991. Reaction kinetics and transport in soil: Compatibility and differences between some simple models. *Transp. Porous Media* 6:703-738.

van der Zee, S.E.A.T.M. and J.J.T.I. Boesten. 1991. Effects of soil heterogeneity on pesticide leaching to ground water. *Water Resour. Res.* 27:3051-3063.

van der Zee, S.E.A.T.M. and W.H. van Riemsdijk. 1986. Transport of phosphate in a heterogeneous field. *Transp. Porous Media* 1:339-359.

van der Zee, S.E.A.T.M. and W.H. van Riemsdijk. 1987. Transport of reactive solute in spatially variable soil systems. *Water Resour. Res.* 23:2059-2069.

van der Zee, S.E.A.T.M. and W.H. van Riemsdijk. 1988. Model for long term phosphate reaction kinetics in soils. *J. Environ. Qual.* 17:35-41.

van der Zee, S.E.A.T.M. and W.H. van Riemsdijk. 1989. Transport of reacting solutes in soil. Report. Dept. *Soil Sci. & Plant Nutrition*, Agric. Univ. Wageningen 140 pp.

van der Zee, S.E.A.T.M. and W.J.P. Bosma. 1990. Analytical approximation for non linear absorbing solute transport and first order degradation. *Transp. Porous Media.* (in press).

Van Duijn, C.J. and P. Knabner. 1990. Travelling waves in the transport of reactive solutes through porous media: Adsorption and binary ion exchange, Part 2. Rep. 205. Univ. Augsburg, Inst. Mathematic.

Van Genuchten, M.Th. 1981. Analytical solutions for chemical transport with simultaneous adsorption, zero-order production, and first-order decay. *J. Hydrol.* 49:213-233.

Van Genuchten, M.Th. 1985. Convective-dispersive transport of solutes involved in sequential first-order decay reactions. *Comp. and Geosciences*, 11(2):129-147.

Van Genuchten, M.Th. and F.N. Dalton. 1986. Models for simulating salt movement in aggregated soils. *Geoderma* 38:165-184.

Van Genuchten, M.Th. and P.J. Wierenga. 1976. Mass transfer studies in sorbing porous media. I. Analytical solutions. *Soil Sci. Soc. Am. J.* 40:473-480.

Van Genuchten, M.Th. and R.W. Cleary. 1982. Movement of solutes in soil: Computer simulated and laboratory results. p. 349-386. In: G.H. Bolt (ed.), *Soil Chemistry B*, Elsevier, Amsterdam.

Van Genuchten, M.Th. and W.J. Alves. 1982. Analytical solutions of the one-dimensional convective-dispersive solute transport equation. USDA Tech. Bull. 1661, 151 pp, USDA, Washington, DC.

Van Genuchten, M.Th., D.E. Ralston, and P.F. Germann. (eds.). 1990. Transport of water and solutes in macropores. *Geoderma*. Special Issue. 46:1-3, 1-297.

Van Grinsven, J.J.M., J. Kros, N. van Breemen, W.H. van Riemsdijk, and E. van Eek. 1989. Simulated response of an acid forest soil to acid deposition and mitigation measures. *Neth. J. Agric. Sci.* 37:279-299.

Van Grinsven, J.J.M., N. van Breemen, W.H. van Riemsdijk, and J. Mulder. 1987. The sensitivity of acid forest soils to acid deposition. Proc. Int. Symp. Acidification and water pathways (Bolkesjo). *Norw. Nat. Comm. Hydrol.* 365-374.

van Riemsdijk, W.H., L.K. Koopal, and J.C.M. de Wit. 1987. Heterogeneity and electrolyte adsorption: Intrinsic and electrostatic effects. *Neth. J. Agric. Sci.* 35:241-257.

Vanselow, A.P. 1932. Equilibria of the base exchange reactions of bentonites. permutites, soil colloids, and zeolites. *Soil Sci.* 33:95-113.

Villermaux, J. and W.P.M. van Swaaij. 1969. Modele representatif de la distribution des temps de sejour dans un reacteur semi-infini a dispersion axiale avec zones stagnantes. *Chem. Eng. Sci.* 24:1097-1111.

Walsch, M.P., S.L. Bryant, R.S. Schechter, and L.W. Lake. 1984. Precipitation and dissolution of solids attending flow through porous media. *Am. Inst. Chem. Eng.* 30:317-328.

Westall, J. 1980. Chemical equilibrium including adsorption on changed surfaces. In: J. Lecky and M. Kavanaugh (eds.) *Adv. Chem. Ser.* 189 Am. Chem. Soc., Wash. D.C.

Westall, J., J.L. Zachary, and F. Morel. 1976. MINEQL-A computer program for the calculation of chemical equilibrium composition of aqueous systems. Tech. note 18, R.M. Parsons Lab., MIT, Cambridge, Mass.

Westall, J.C. and H. Hohl. 1980. A comparison of electrostatic models for the oxide/solutions interface. *Adv. Colloid Interface Sci.* 12:265-294.

Wever, W.R. and C.T. Miller. 1988. Modeling the sorption of hydrophobic contaminants by aquifer materials, I. Rates and equilibria. *Water Res.* 22:457-464.

Wierenga, P.J. 1977. Solute distribution profiles computed with steady-state and transient water movement models. *Soil Sci. Soc. Am. J.* 41:1050-1055.

Wild, A. and I.A. Babiker. 1976. The assymetric leaching pattern of nitrate and chloride in a loamy sand under field conditions. *J. Soil Sci.* 27:460-466.

Wilson, G.V., P.M. Jardine, R.J. Luxmoore, and J.R. Jones. 1990. Hydrology of a forested hillslope during storm events. *Geoderma*. 46:119-138.

Wilson, M.J. 1975. Chemical weathering of some primary rockforming minerals. *Soil Sci.* 119:349-354.

Wollast, R. 1967. Kinetics of the alteration of K-feldspar in buffered solutions at low temperature. *Geochim. Cosmochim. Acta.* 31:635-648.

Yates, D.E. 1975. The structure of the oxide-aqueous electrolyte interface. Ph.D. dissertation. Univ. Melbourne, Australia.

Yeh, G.T. and V.S. Tripathi. 1989. A critical evaluation of recent developments in hydrogeochemical transport models of reactive multichemical components. *Water Resour. Res.* 25(1):93-108.

Modeling Coupled Processes in Porous Media: Sorption, Transformation, and Transport of Organic Solutes

M.L. Brusseau, P.S.C. Rao, and C.A. Bellin

I.   Introduction 147
II.   Process-Specific Models 148
      A. Sorption 148
      B. Biotic Transformations 151
      C. Abiotic Transformations 156
      D. Transport 157
III. Coupled-Process Transport Models 158
      A. Complex Transport, Simple Sorption and Transformation . . . 158
      B. Complex Sorption, Simple Transport and Transformation . . . 160
      C. Complex Transformation, Simple Transport and Sorption . . . 162
      D. Complex Sorption and Transformation, Simple Transport . . . 167
      E. Complex Transport and Sorption, Simple Transformation . . . 167
      F. Complex Transport and Transformation, Simple Sorption . . . 172
IV.   Conclusions 176
References 176

I. Introduction

The transport and fate of organic chemicals in the subsurface is a topic of great interest. Mathematical models are used to simulate the transport and fate of dissolved chemicals through soils and aquifers for a variety of purposes. Research scientists use models to help gain insight into the processes that affect the fate of chemicals in the subsurface. Environmental consultants use models to assist in their design and evaluation of soil and aquifer remediation systems. Environmental regulators use models to evaluate the threat posed to groundwater by chemicals entering the subsurface. Industrial chemists use models to assess the fate of pre-production chemicals.

Given the nature of some organic chemicals, especially many of those that are most often found in groundwater, transport models will often have to account

ISBN 0-87371-889-5
© 1992 by Lewis Publishers

for sorption and transformation to be considered valid. The development of coupled chemical-biological process models for solute transport has expanded greatly during the past decade. Our purpose is not to provide a comprehensive review and discussion of the specific processes, but rather to review the manner in which these processes have been integrated into coupled-process models.

II. Process-Specific Models

A. Sorption

The distribution of organic chemicals between solution and solid phases has been and is being studied within several disciplines, including soil science, geology, environmental science and engineering, analytical chemistry, and chemical engineering. Characterizing the distribution at equilibrium has been the primary emphasis of this research. Several equations have been used to represent the equilibrium distribution of solute between solid and solution phases (i.e., the isotherm). Among these, the Freundlich isotherm is used extensively for sorption from aqueous solutions. This isotherm is:

$$S = K_f C^n \tag{1}$$

where C is the solution-phase solute concentration (M L^{-3}), S is the sorbed-phase concentration (M M^{-1}), and K_f (ML^{-3})$^{-n}$ and n are the parameters specifying the degree and nature of sorption. Often, the value of n is equal or close to unity, wherein Equation (1) reduces to the linear isotherm.

Primarily because of its simplicity, the assumption of equilibrium sorption has been used in the development of most solute transport models. The validity of this assumption has recently been questioned, as the results of theoretical and experimental research have shown that sorption of organic chemicals by natural sorbents can be significantly rate limited. Delineation of the mechanism(s) responsible for the non-instantaneous sorption of organic solutes has been the focus of recent research (cf., Karickhoff and Morris, 1985; Wu and Gschwend, 1986; Brusseau and Rao, 1989a, 1991; Brusseau et al., 1989a, 1989b, 1991a, 1991b; Nkedi-Kizza et al., 1989; Szecody and Bales, 1989; Pignatello, 1990; Ball and Roberts, 1991). For non-structured porous media, the rate-limiting mechanism appears to involve intrasorbent diffusion (i.e., diffusion within organic matter and/or mineral particles). For aggregated or macroporous porous media, nonequilibrium can be caused by rate-limited mass transfer between advective and non-advective pore-water domains, as well as by intrasorbent diffusion.

Early attempts to model the rate-limited uptake/release of organic chemicals by natural sorbents used a "one-site" model where all sorption was assumed to be rate-limited. This approach, however, has generally failed to adequately represent experimental data (Rao and Jessup, 1983; Boesten and Van Der Pas, 1988). Sorption data for organic chemicals obtained from batch experiments often exhibit a two-stage approach to equilibrium: a short initial phase of fast uptake or release, where roughly 30-50% of the total sorption occurs within minutes to hours, followed by an extended period of much slower uptake/release occurring over periods of days or months (Hamaker and Thompson, 1972; Karickhoff, 1980; Karickhoff and Morris, 1985; McCall and Agin, 1985; Oliver, 1985; Coates and Elzerman, 1986; Wu and Gschwend, 1986; Ball and Roberts, 1991). This type of behavior is readily simulated by a bicontinuum (or two-domain) conceptualization. Such a conceptualization can be effected by means of diffusion equations based on Fick's Law (e.g., Wu and Gschwend, 1986; Ball and Roberts, 1991) or with first-order mass transfer equations (e.g., Selim et al., 1976; Cameron and Klute, 1977; Karickhoff, 1980) based on the linear driving force approximation of Glueckauf and Coates (1947). The conceptual basis for, and the performance of, these two approaches has been reviewed extensively by Brusseau and Rao (1989b).

A major disadvantage associated with the diffusion-based models is that detailed information on the structure of the porous medium is required, whereas no such requirements exist for the mass-transfer models. In addition, the use of diffusion-based models assumes *a priori* knowledge of the nonequilibrium mechanism; that is, a commitment to a particular mechanism is required in designing/selecting the model to be used. For situations where the mechanism involved is not fully elucidated, the use of a model that is not mechanism-unique, such as the first-order mass transfer model, is preferable. The first-order bicontinuum model can represent each of the major processes potentially responsible for nonequilibrium and provides results similar to those obtained with diffusion-based models (see Brusseau and Rao (1989b) and references cited therein). The first-order mass transfer bicontinuum model has received widespread acceptance because of its relative simplicity and versatility.

Models formulated using the first-order bicontinuum conceptualization have been variously called "two-site", "two-compartment", or "two-box" models. In these models, nonequilibrium is assumed to be the result of a time-dependent interaction between the solute and the sorbent. Sorption may be represented by two "reactions", occurring either in series or in parallel. The former is appropriate for the case where sorption is limited by diffusive mass transfer, whereas the latter is used for the case of two types of sorption mechanisms. The sorbent is hypothesized as having two sorption domains, where sorption is essentially instantaneous for one and is rate-limited for the other. Sorption for the kinetically-controlled domain is described by a first-order rate equation, whereas the other domain is represented by an equilibrium isotherm equation. It should be noted that, as long as one of the domains is equilibrium-controlled, the series and parallel conceptualizations are mathematically indistinguishable (Karickhoff, 1980; Karickhoff and Morris, 1985). Also, the equilibrium-controlled domain can be replaced such that both domains are kinetically controlled (cf., Selim et al., 1976; Karickhoff, 1980). The suggestion that the existence of a second, rate-limited class of sorption sites may be responsible for nonequilibrium was first made by Giddings and Eyring (1955).

Although the first-order model has met with widespread success, there are conditions under which this model may fail. For example, a more continuous distribution, rather than the extreme bimodality of reaction rates associated with the bicontinuum models, may be a more accurate representation of true conditions. Accordingly, the discretization of the sorbent into two, parallel domains differing in reaction time may be extended to accommodate any number of domains, each with its own unique rate constant. For example, Boesten et al. (1989) presented a "three-site" model. The limiting case would be a continuous distribution of domains and associated rate constants. A model representing this case has been developed by Villermaux (1974), where the site population is represented by the transfer-time distribution (i.e., the rate-constant distribution). However, the number of parameters associated with such a model greatly exceeds our present capability to evaluate independently the processes represented by those parameters. Such models are, therefore, constrained to operation in a calibration mode.

Modeling the transport of organic chemicals influenced by rate-limited sorption has been a topic of interest for some time. Initial attempts incorporated the one-site sorption kinetics model into the advective dispersive transport equation (cf., Oddson et al., 1970). This approach was based on that taken by researchers in chemical engineering (i.e., Lapidus and Amundson, 1952). However, models using the one-site formulation have not been able to predict observed transport behavior (Davidson and McDougal, 1973; Hornsby and Davidson, 1973; Schwarzenbach and Westall, 1981; Rao and Jessup, 1983).

Transport models employing the bicontinuum sorption formulation, with one domain equilibrium-controlled, were presented by Selim et al. (1976) and Cameron and Klute (1977), while Selim et al. (1976) also presented a model where both domains were rate limited. The one-site model mentioned previously is a special case of the two-site model, where all sorption sites are assumed to be of the time-dependent class (Selim et al., 1976; van Genuchten, 1981). The bicontinuum-based model has generally been able to represent nonequilibrium data much better than has the one-site model.

It should be noted that the bicontinuum models presented by Selim et al. (1976) and Cameron and Klute (1977), which were developed to represent sorption nonequilibrium, are mathematically equivalent in nondimensional form, for the case of linear isotherms, to the first-order mass transfer model (two-region model) presented by van Genuchten and Wierenga (1976), which was developed to represent transport-related nonequilibrium. This equivalency is beneficial in that it lends a large degree of versatility to the first-order bicontinuum model. However, this equivalency also means that elucidation of nonequilibrium mechanisms is not possible using modeling-based analysis alone.

With the first-order model, sorption is conceptualized to occur in two domains:

$$S_1 \;=\; F\,K_p\,C \tag{2a}$$

$$dS_2/dt \;=\; k_1 S_1 \;-\; k_2 S_2 \tag{2b}$$

where S_1 is the sorbed-phase concentration in the instantaneous domain (M M^{-1}), S_2 is the sorbed-phase concentration in the mass-transfer constrained domain (M M^{-1}), K_p is the equilibrium sorption constant (L^3 M^{-1}), F is the fraction of sorbent for which sorption is instantaneous, t is time, and k_1 and k_2 are forward and reverse first-order rate constants (T^{-1}), respectively. At equilibrium, Equation (2b) reduces to $S_2 = (1-F)\,K_p\,C$.

B. Biotic Transformations

Developing and selecting mathematical models that describe biodegradation requires making several decisions regarding the nature of the system of interest. Factors to consider include:

(1) the disposition of biomass in the system;

(2) the forms of the mathematical equations describing the growth dynamics of biomass and the functional dependency of transformation rate on substrate concentration (S_b), electron-acceptor concentration (S_e), nutrient concentrations (S_n), and biomass (B_m);

(3) mass transfer constraints and bioavailability; and

(4) other factors (e.g., inhibition, acclimation).

Each of these will be briefly discussed in the following sections.

1. Disposition of Microbial Biomass

The disposition of biomass in the subsurface may be described using one of three approaches (Baveye and Valocchi, 1989): (1) biofilms; (2) microcolonies; and (3) macroscopically uniform distribution. For all three approaches, the microbes are attached to solid surfaces; the approaches differ in the manner in which the surface coverage is described. The first approach is based on the concept that the solid phase of porous media is covered by a uniform film of biomass (i.e., biofilm), wherein substrate utilization occurs. The biofilm concept has been used extensively for modeling biodegradation during wastewater treatment (cf., Rittman and McCarty, 1978). It has also been used to describe microbial processes in aquifers (cf., Bouwer and McCarty, 1984). The second approach considers bacteria to exist as small, discrete colonies ("microcolonies") attached to surfaces of solids, rather than as continuous films. Models based on this conceptualization were presented by Benefield and Molz (1983) for activated sludge and by Molz et al. (1986) for transport in groundwater. These two approaches are primarily for eutrophic systems (i.e., high bacterial concentrations, $> 10^8$ cfu/g) and may not be applicable to oligotrophic systems (bacterial concentrations $< 10^4$ cfu/g) (Wilson and McNabb, 1983).

Whereas the first two approaches are based on microscopic descriptions of the system, the third approach is based on a macroscopic representation. With this approach, a biomass phase is considered to be distributed homogeneously within an elemental volume of the porous medium. This approach has been the most widely used for incorporating biodegradation into solute transport models (cf., Sykes et al., 1982; Corapcioglu and Haridas, 1985; Borden and Bedient, 1986).

The spatial distribution of bacteria in porous media has been examined by few investigators. Based on the limited data available, it appears that bacteria exist primarily as microcolonies attached to surfaces of solid particles (Harvey et al., 1984; Balkwill and Ghiorse, 1985; Birnbaum et al., 1990). If this is true, the microcolony approach would seem to be the appropriate one for describing the disposition of biomass in the subsurface. A major problem exists, however, in applying this model, or the biofilm model, to solute transport. Namely, specifying the dimensions of the microcolony or biofilm is difficult. It is improbable that these microscale properties can be measured adequately, especially at the field scale. This is especially true when the heterogeneous nature of the physical, chemical, and biological properties of the subsurface are considered. With this in mind, the macroscopic approach may be the only viable way in which to describe disposition of biomass in the subsurface.

2. Mathematical Description of Substrate Utilization and Biomass Growth

There exists a multitude of models that have been used to describe biodegradation of substrates. A review of these models was recently presented by Alexander and Scow (1989). The power rate model (Hamaker, 1972):

$$dC/dt = - k_b C^n \qquad (3)$$

where k_b is the biodegradation rate constant (T^{-1}), is an example of an empirical model. When $n = 1$, Equation (3) is equivalent to the first-order kinetics model. This model has been used extensively in soil science and hydrology.

The most widely used model is that presented by Monod (1949):

$$\mu = \frac{\mu_{max} S_b}{K_s + S_b} \qquad (4)$$

where μ is the specific growth rate of the biomass (T^{-1}), μ_{max} is the maximum specific growth rate, and K_s is the substrate concentration constant (i.e., concentration when rate of growth is half the maximum rate). While a theoretical basis has been ascribed to Equation (4), it is important to note that Equation (4) was developed empirically. As stated by Monod (1949), Equation (4) was developed by analogy "to an adsorption isotherm or to the Michaelis equation".

Equation (4) describes the growth of biomass in response to the availability of a substrate. However, we are interested in the change in substrate concentration (i.e., $\partial S_b / \partial t$) as well as biomass growth. Equation (4) can be used, along with an equation relating biomass growth and substrate loss (i.e., yield; $Y = \partial B_m / \partial S_b$), to develop an equation of the Monod form that describes substrate loss:

$$\frac{\partial S_b}{\partial t} = \frac{B_m}{Y} \frac{\mu_{max} S_b}{K_s + S_b} \qquad (5)$$

The Monod equation was developed for systems comprising a single pure culture growing under controlled conditions (high concentrations of freely available substrate, no nutrient limitations). Considering the vast differences that exist between pure-culture systems and soils, the validity of using this model to describe biodegradation in the subsurface is questionable (Alexander and Scow, 1989; Baveye and Valocchi, 1989). For example, recent research has demonstrated the inadequacy of the Monod-type equations for describing biodegradation in mixed culture systems (Scow et al., 1986; Simkins et al., 1986).

The growth status of the biomass is a major factor controlling the form of transformation model that is appropriate. Models describing transformation dynamics may be grouped into two categories, reflecting the degradation of substrate that either does or does not support growth. The Monod equation and various adaptations have been used for the case of degradation of substrates that support microbial growth; see Simkins and Alexander (1984) for a presentation of various models. Models for the biodegradation of organic chemicals not supporting growth are presented by Schmidt et al. (1985). These models are of special interest to the transport and fate of contaminants in the subsurface. Biodegradation may not support growth for three reasons (Alexander and Scow, 1989): (1) the concentration of substrate may be too low to support growth; (2) an essential nutrient is limiting; and (3) degradation occurs by cometabolism.

The biodegradation of substrate by microorganisms that do not grow (i.e., fixed population density) is often modeled using the Michaelis-Menten equation developed for enzyme reactions:

$$v = \frac{V_{max}S_b}{K_m + S_b} \tag{6}$$

where v is reaction rate, V_{max} is maximum reaction rate, and K_m is the Michaelis constant. Equation (6) is identical in form to Equation (4), with the essential difference that μ represents the rate of growth of biomass (i.e., growth), whereas v represents the rate of reaction or biodegradation of substrate (i.e., no growth). When initial substrate concentrations are low ($S_0 \ll K_m$), Equation (6) reduces to a first-order reaction equation. Thus, although first-order kinetics has usually been used on empirical grounds, it does have a theoretical basis. Considering that the concentrations of organic contaminants found in the subsurface are often relatively low, it is possible that the first-order model may be appropriate for some solute transport problems.

3. Mass Transfer Constraints and Bioavailability

Several factors relating to the availability of the substrate may affect the rate of biodegradation. Mass transfer of substrate from solution into the biomass may be rate limited. This is the concept upon which the biofilm model is based. Availability of substrate has also been considered to be constrained by mass transfer across a fluid boundary layer (Benefield and Molz, 1983; Molz et al., 1986). Biodegradation of slightly soluble compounds may be constrained by low concentrations and limited by the rate of dissolution when pure compound is present (Thomas et al., 1986; Stucki and Alexander, 1987; Miller and Bartha, 1989).

Association of substrate with the solid phase of porous media may influence appreciably, and in some instances mediate, the rate of biodegradation. The effect of solid surfaces on biodegradation of organic chemicals has been under investigation for quite some time. Numerous observations have been reported suggesting that organic chemicals are not, or are only slowly, degraded while associated with solid phases. However, other reports have suggested that surfaces may enhance degradation. This apparent contradiction results, in part, from the qualitative nature of the research. In addition, several issues have complicated data interpretation, including failure to separate abiotic and biotic transformation pathways, intracellular versus extracellular biodegradation, and the effect of solid phases on substrate concentration and on microbial activity. Unfortunately, there have been relatively few reports of experiments designed specifically to investigate and quantify this phenomenon.

Currently, most researchers believe that organic chemicals are not (or only slowly) degraded via microbial pathways while associated with solid phases.

Several studies have shown reduced rates of biodegradation of various organic chemicals in the presence of mineral surfaces and activated carbon, including diquat (Weber and Coble, 1968), atrazine and linuron (Moyer et al., 1972), alkylamines (Wszolek and Alexander, 1979), and picloram (Bondarev et al., 1985). Experiments reported by Robinson et al. (1990) showed that sorbed-phase substrate was not degraded and that long-term biodegradation was limited by the slowly-desorbing fraction of substrate. Fluorescence spectroscopy was used by Ainsworth et al. (1990) to evaluate the effect of sorption on the biodegradation of quinoline. They showed that quinoline sorbed to clay surfaces was not degraded and that degradation was limited by desorption.

Several investigations, based on fitting mathematical models, coupling sorption, and biodegradation, to data obtained from batch experiments have suggested that substrate associated with solid phases is not biodegraded. Steen et al. (1980) observed slower rates of biodegradation of chlorpropham and di-n-butyl phthalate in the presence of sediments and simulated the data using a second-order rate equation modified to account for equilibrium sorption. The model was developed assuming degradation in only the aqueous phase. Three sorption-degradation models were tested against data collected for the degradation of 2,4-D in three soil systems (Ogram et al., 1985). The only successful model was the one developed assuming sorbed substrate was not degraded and that sorbed- and solution-phase bacteria degraded solution-phase substrate. Sorption was assumed to be instantaneous and degradation was modeled as a first-order process. The degradation of naphthalene in soil-water suspensions under denitrification conditions was simulated using a model wherein sorption was represented by retarded intraparticle diffusion and degradation by Michaelis-Menten kinetics (Mihelcic and Luthy, 1991). Degradation was assumed to occur in only the solution-phase of the macropore (interaggregate) regions. Mass transfer appeared to be rapid in comparison to the rate of biodegradation. Models accounting for nonequilibrium sorption, described using both first-order and retarded intraparticle diffusion approaches, were used by Rijnaarts et al. (1990) to analyze the biodegradation of α-hexachlorocyclohexane (lindane) in a calcareous soil. However, the models did not contain separate terms for biodegradation. The results of experiments showed that the rate of biodegradation was controlled by mass transfer after 3 days.

The location, as well as the rate, of biodegradation may be influenced by physical properties of the porous medium. For example, bacteria may be excluded from the microporous domain of structured (e.g., aggregated, fractured) porous media, since most bacteria range in size from 0.5 to 0.8 μm (Casida, 1971). If such exclusion were to occur, biodegradation would not take place in the microporous domain and could, therefore, be limited by diffusional mass transfer of solute to the macroporous domain. Scow and Alexander (1991) investigated the effect of synthetic clay aggregates that excluded bacteria on the rates of biodegradation of phenol and glutamate. The rates of mineralization in the presence of the aggregates was observed to be lower than the rates obtained in the absence of aggregates.

The results of the above experiments, which suggest that substrate situated in micropores and sorbed-phase substrate is protected from biodegradation, are supported by the observation of "residues" in agricultural fields long after pesticide application. For example, residues of EDB (1,2-dibromoethane) were found to persist in fields for as long as 19 years after application, even though EDB is volatile and readily degraded (Steinberg et al., 1987). It was suggested that the EDB comprising the residue was "trapped" in regions of the soil that rendered it unavailable to biological transformation. Similar behavior was reported for DBCP (1,2-dibromo-3-chloropropane) by Buxton and Green (1987).

4. Other Factors

At high concentrations, many compounds are toxic to the microorganisms that use them as carbon sources. This inhibition results in a reduction in growth rate when the concentration of substrate exceeds the critical concentration. Substrate inhibition can be represented by the Haldane modification of the Monod equation (Alexander and Scow, 1989):

$$\mu = \frac{\mu_{max} \, S_b}{K_s + S_b + \dfrac{S_b^2}{K_I}} \tag{7}$$

where K_I is an inhibition constant. Substrate inhibition has rarely been incorporated into solute transport models; one example is provided by the model of Kindred and Celia (1989).

Use of the Monod and Michaelis-Menten equations is based on the presence of bacteria that are acclimated to the substrate. Many organic compounds, however, undergo an acclimation period before the onset of degradation (Alexander and Scow, 1989). Hence, application of the Monod and Michaelis-Menten equations to solute transport may be inappropriate in some cases. Acclimation has yet to be incorporated into solute transport models.

Another factor that has rarely been addressed is the effect of consortia on biodegradation and transport. The microbial communities that exist in soils and aquifers can consist of many populations of different species. It is possible that more than one species can degrade a particular substrate. The effect of such occurrences on modeling biodegradation, especially for solute transport, have not been investigated.

C. Abiotic Transformations

Several abiotic processes can affect the disposition of organic chemicals in the subsurface. Radioactive decay, a process of interest primarily for inorganic solutes, is described using a first-order equation. Hydrolysis is a significant abiotic-degradation process for many organic compounds of interest (Mabey and Mill, 1978), including triazine pesticides (cf., Armstrong et al., 1967) and halogenated-aliphatic solvents (cf., Vogel et al., 1987). In general, hydrolysis can be represented as a first-order process for neutral hydrolysis reactions, and for base- and acid-catalyzed reactions under conditions of fixed pH (Macalady et al., 1989; Wolfe et al., 1989). Hence, transport models with first-order degradation terms may be adequate to simulate the behavior of solute undergoing hydrolysis.

Sorption of solutes by solid phases can affect the rate of hydrolysis. Sorption appears to increase the rate of acid-catalyzed hydrolysis (Armstrong et al., 1967; Macalady and Wolfe, 1984; Hoag and Mill, 1988) and to decrease the rate of base-catalyzed hydrolysis (Macalady and Wolfe, 1985). Conversely, the rate of neutral hydrolysis appears to be unaffected by the presence of surfaces (Macalady and Wolfe, 1985; Hoag and Mill, 1988). As for the case of biotic degradation, the effect of sorption on hydrolysis can be simulated by using a coupled-process model.

D. Transport

Several approaches have been used to develop models that simulate the transport of fluids and solute in porous media. These approaches have included micro-scopic- and macroscopic-based descriptions of transport phenomena. The continuum approach is by far the most widely used method for developing transport models and, as such, models based on this approach will be the focus of this paper.

The general equation describing advective-dispersive transport of a solute in porous media, developed by performing a mass balance on the solute mass within an elemental volume, is given by:

$$\frac{\partial(\theta C)}{\partial t} = (\nabla \cdot D\theta \ \nabla C) - (\nabla \cdot qC) + \sum G_i \qquad (8)$$

where D is the hydrodynamic dispersion coefficient (L^2T^{-1}), θ is the fractional volumetric water content, and G_i represents various chemical and biological reactions. We will focus on steady-state water flow, since the vast majority of coupled sorption-transformation models have been developed for this case. Assuming spatially-uniform θ, and steady-state water flow conditions, Equation (8) reduces to:

$$\frac{\partial C}{\partial t} = (\nabla \cdot D \ \nabla C) - (\nabla \cdot vC) + \frac{1}{\theta}\sum G_i \qquad (9)$$

where $v = q/\theta$. By assuming that v and D are spatially invariant and that sorption is essentially instantaneous, the following widely used, one-dimensional transport equation is obtained:

$$R \frac{\partial C}{\partial t} = D \frac{\partial^2 C}{\partial x^2} - v \frac{\partial C}{\partial x} \qquad (10)$$

where the retardation factor $R = 1 + (\rho/\theta) K_p$ and ρ is the bulk density of the porous medium. Equation (10) serves as the point of departure for the coupled process models to be discussed in the next section.

As previously mentioned, Equation (10) is widely used to simulate the transport of solutes in soils and aquifers. It is widely recognized, however, that the assumption of homogeneous soil/aquifer properties is invalid. Soils are often structured with aggregates and macropores, as well as being composed of horizons that have different properties. Aquifers are heterogeneous, with macroscopic and megascopic variations in hydraulic conductivity and sorption capacity. Whereas the effect of heterogeneity on water flow and the transport of conservative solutes has been under investigation for the past two decades, the effect of heterogeneity on the transport of organic solutes is just beginning to be investigated.

III. Coupled-Process Transport Models

To develop a coupled-process model for solute transport, the form of equations representing water flow, solute transport, sorption, and transformation must be selected. As discussed, we will focus on steady-state water flow. The other major specification for flow and transport is the spatial character of hydraulic, chemical, biological properties; i.e., homogeneous or heterogeneous. The major specification for sorption is the description of dynamics; i.e., local equilibrium or rate limited. As discussed above, the factors to consider for transformation include the form of the equation and the disposition of biomass.

The simplest transport models incorporating coupled sorption-transformation are those specifying homogeneous porous media, instantaneous sorption, zero- or first-order transformation, and steady-state biomass. Many investigators have developed and used such models. We will focus on models that provide a more complex description of one or more of the three processes of interest (transport, sorption, transformation). The various models reported in the literature will be classified and discussed on the basis of the level of complexity used to describe the three processes. The classification matrix we will use to facilitate discussion is presented in Table 1.

A. Complex Transport, Simple Sorption and Transformation

Helfferich (1962) presented a model to simulate the catalysis of solute influenced by intra-particle diffusion, where catalysis was represented by a first-order irreversible surface reaction. Application of this model to soil systems is limited by the specification of no loss in the solution phase. Apparently, the first coupled-process transport model developed by soil scientists/hydrologists that considered soil heterogeneity was presented by Mironenko and Pachepsky (1984). They used the mobile-immobile approach (cf., Coats and Smith, 1964; van Genuchten and Wierenga, 1976) to describe solute transport in porous media under conditions of nonuniform pore-water velocity. With this approach, mass transfer of solute between domains wherein advective-dispersive transport occurs and domains where it does not is described by a first-order, linear driving force equation. Degradation was described using a first-order equation and was assumed to occur in only the solution phase. The authors provided an analytical solution to the model. The model was applied to column experiments involving the transport and reduction of nitrate. The optimized simulations matched the data well.

Table 1. Matrix describing organization of coupled-process models

Category	Transport	Sorption	Transformation
1	complex	simple	simple
2	simple	complex	simple
3	simple	simple	complex
4	simple	complex	complex
5	complex	complex	simple
6	complex	simple	complex

Lassey (1988) presented an analytical solution for a transport model that included a generalized description of rate-limited mass transfer and first-order loss. He showed how this model could be reduced to the two-region (mobile-immobile) model with degradation. An analytical solution for a model essentially identical to that of Mironenko and Pachepsky (1984), except that degradation could occur in both solution and sorbed phases, was presented by van Genuchten and Wagenet (1989). The model may be written in the following nondimensional form:

$$\beta R \frac{\partial C}{\partial T}m - + (1-\beta)R \frac{\partial C}{\partial T}im - = \frac{1}{P}\frac{\partial^2 C}{\partial X^2}m - - \frac{\partial C}{\partial X}m - - \xi C_m - \eta C_{im} \quad (11)$$

$$(1-\beta)R \frac{\partial C}{\partial T}im - = \omega(C_m - C_{im}) - \eta C_{im} \quad (12)$$

by defining the following dimensionless parameters:

$$
\begin{aligned}
C_m^* &= C_m/C_0 \\
C_{im}^* &= C_{im}/C_0 \\
R &= 1 + (\rho/\theta)\, K_p \\
\omega &= \alpha L/v \\
\beta &= [\theta_m + f\rho K_p]/[\theta + \rho K_p] \\
\xi &= \theta_m\, \mu_{lm}\, L/q + f\rho K_p\, \mu_{sm}\, L/q \\
\eta &= \theta_{im}\, \mu_{lim}\, L/q + (1\text{-}f)\rho K_p\, \mu_{sim}\, L/q \\
T &= tv/L \\
P &= v_m L/D_m
\end{aligned}
$$

where θ_m is the volumetric water content of the mobile domain, f is the fraction of sorbent associated with the mobile domain, α is the first-order mass transfer constant, and μ_i is the first-order degradation rate constant for the mobile solution (lm), sorbed-phase in mobile domain (sm), immobile solution (lim), and sorbed-phase in the immobile domain (sim), respectively.

This model was applied by Gamerdinger et al. (1990) to a column experiment involving the transport of 2,4,5-trichlorophenoxy acetic acid herbicide in an aggregated soil. The optimized simulation that included degradation fit the data better than did the simulation that excluded degradation (see Figure 1).

Sorption was assumed to be linear in the models described above. A model incorporating nonlinear sorption as well as mobile-immobile mass transfer and first-order degradation was presented by Jessup et al. (1989). The model was solved using a finite-difference numerical scheme.

B. Complex Sorption, Simple Transport and Transformation

Several models have been developed to describe the transport of radioactive solutes. Since first-order reaction is used to describe decay, these models could be used to simulate the transport of organic solutes influenced by first-order loss (biodegradation, hydrolysis). Eldor and Dagan (1972) presented an approximate analytical solution for a two-dimensional transport model that included rate-limited sorption and first-order decay. Rate-limited sorption was described using the one-site approach and loss could occur in both the solution and solid phases. Lindstrom (1976) presented an analytical solution for a one-dimensional transport model that included one-site rate-limited sorption and first-order loss from solution and sorbed phases. Carnahan and Remer (1984) presented an analytical solution for a three-dimensional transport model that included one-site rate-limited sorption and first-order loss from solution and sorbed phases.

Hoffman and Rolston (1980), and Mironenko and Pachepsky (1984) presented one-dimensional transport models that included rate-limited sorption and first-order loss from solution. The two-domain approach (Selim et al., 1976; Cameron and Klute, 1977) was used to describe rate-limited sorption. The

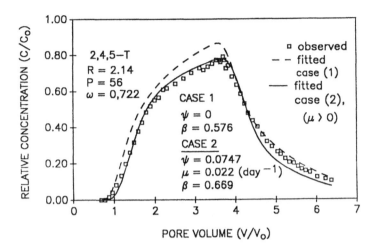

Figure 1. Measured and simulated breakthrough curves for the pesticide 2,4,5-T (pore water velocity, v = 8.73 cm d^{-1}) developed with the two-region model. (From Gamerdinger et al., 1990.)

generalized model of Lassey (1988) described in the section above can also be reduced to the two-domain model with degradation. An analytical solution for a one-dimensional transport model that included two-domain rate-limited sorption and first-order loss from solution and sorbed phases was presented by van Genuchten and Wagenet (1989). This model was applied to experiments involving transport of atrazine in packed soil columns by Gamerdinger et al. (1990). The optimized simulations compared relatively well to the data.

The influence of nonequilibrium sorption and biodegradation on the transport of several alkylbenzenes in an aquifer material was investigated by Angley et al. (1992). The breakthrough curves obtained from the two experiments where biodegradation occurred exhibited steady-state behavior exemplified by plateaus in effluent concentration that were lower than the influent concentration. Conversely, the breakthrough curves for the experiment performed under sterile conditions reached relative concentrations of one (see Figure 2A). These data were used by Angley et al. (1992) to test the performance of the two-domain rate-limited sorption, first-order degradation model of van Genuchten and Wagenet (1989). This was done by attempting to predict the breakthrough curves obtained from the fast-velocity biodegradation experiment, with values for all parameters of the transport model being obtained independent of the data being 6predicted. The predicted simulations matched the data well (see Figure 2B).

A model incorporating nonlinear sorption as well as two-domain rate-limited sorption and first-order degradation from solution and sorbed phases was presented by Jessup et al. (1989). The model may be written in the following nondimensional form:

$$\frac{\partial C^*}{\partial T} + (\beta R_n - 1)nC^{*-1}\frac{\partial C^*}{\partial T} + (1-\beta)R_n\frac{\partial S^*}{\partial T} = \frac{1}{P}\frac{\partial^2 C^*}{\partial X^2} - \frac{\partial C^*}{\partial X^-}\xi C^* - \eta S^* \quad (13)$$

$$(1-\beta)R_n\frac{\partial S^*}{\partial T} = \omega(C^{*n} - S^*) - \eta S^* \quad (14)$$

by defining the following dimensionless parameters:

$$
\begin{aligned}
S^* &= S_2 / (1\text{-}F) K_f C_0^n \\
R_n &= 1 + (\rho/\theta) K_f C_0^{n-1} \\
\omega &= k_2 (1\text{-}\beta) RL/v \\
\beta &= [1 + (\rho/\theta) F K_f C_0^{n-1}] / R_n \\
\xi &= \mu_l L/v + (\beta R_n\text{-}1) \mu_{s1} L/v \\
\eta &= (1\text{-}\beta) R_n \mu_{s2} L/v \\
P &= vL/D
\end{aligned}
$$

where k_2 is the first-order reverse sorption rate constant $(1/T)$, n is the Freundlich nonlinear sorption isotherm coefficient, and μ_i is the first-order degradation rate constant for the solution (l), equilibrium-sorbed (s1) and rate-limited-sorbed (s2) phases, respectively. This model reduces to that of van Genuchten and Wagenet when n is unity.

C. Complex Transformation, Simple Transport and Sorption

The models in this section will be divided between those developed assuming steady-state biomass and those describing growth.

1. Steady-State Biomass

A one-dimensional model developed to simulate transport and oxidation of nitrite was presented by Ardakani et al. (1973). Sorption of substrate was not included and transformation was described by the Monod equation. The model was applied to column experiments and provided fair approximations of the data. Models somewhat similar to that of Ardakani et al. (1973) were presented by Kam (1974) and Parlange et al. (1984), a major difference being the incorporation of dispersion.

 A two-dimensional model developed to simulate biodegradation of solute that is rate-limited by oxygen (electron acceptor) was presented by Borden and Bedient (1986). They assumed that substrate and oxygen in the solution phase are influenced by advective-dispersive transport, that only solution-phase substrate is subject to biodegradation, and that sorption of the substrate is governed by local equilibrium. Degradation of solute and consumption of oxygen

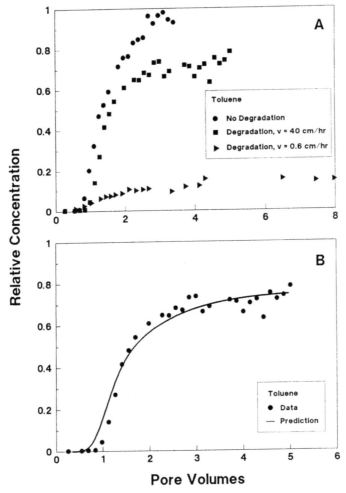

Figure 2. The influence of nonequilibrium sorption and biodegradation on the transport of toluene in a column packed with an aquifer material (A). (From Angley and Brusseau, 1991). Comparison of the breakthrough curve, obtained from the fast-velocity biodegradation experiment, and predicted simulation obtained with the two-domain nonequilibrum sorption, first-order transformation, transport model (B).

was approximated as an instantaneous reaction between oxygen and substrate. This model, therefore, can be used to simulate conditions wherein biodegradation is controlled by availability of oxygen. The model was solved using a finite-difference approach. Borden et al. (1986) presented a simplified solution technique, using the principle of superposition, for this model; however, sorption had to be ignored to apply the simplified approach. This model was applied to an organic contaminant plume emanating from an abandoned creosoting site (Borden et al., 1986). Transport parameters (e.g., longitudinal and transverse dispersivities) were obtained by calibrating the model to a chloride plume present at the site. The optimized simulations provided relatively good approximations of measured chloride, oxygen, and hydrocarbon concentrations.

Rifai et al. (1988) extended the simplified approach of Borden et al. (1986) to the case of sorbing solutes. The extended model was applied to a hydrocarbon plume located below a Coast Guard station. The simulations produced with the model, which was calibrated to the field data, matched the data fairly well. A model similar to that of Borden and Bedient (1986), but in three-dimensions, was presented by Baehr and Corapcioglu (1984). They also simulated biodegradation as a function of oxygen availability. In addition, however, they included sorption and incorporated provisions for a non-aqueous liquid phase.

A one-dimensional model that incorporates both aerobic and anaerobic degradation was presented by Srinivasan and Mercer (1988). Substrate degradation was described with the Monod-type equation for aerobic and anaerobic conditions. A provision was included for first-order kinetics when S_b is less than 0.25 K_m. Both substrate and oxygen are affected by advective-dispersive transport and equilibrium-governed sorption. This model was applied to a column experiment involving the transport and anaerobic degradation of trichloroethene. The optimized simulation closely matched the data.

A one-dimensional model designed to simulate the transport of chlorinated hydrocarbons under methanogenic conditions was presented by Corapcioglu et al. (1991). The model allows simulation of the advective-dispersive transport and anaerobic transformation of four species (i.e., tetrachloroethene, trichloroethene, dichloroethene, and vinyl chloride). The Michaelis-Menten equation is used to describe mass loss, and sorption of substrates is assumed to be instantaneous. The model was used to simulate a column experiment involving anaerobic biotransformation of tetrachloroethene and trichloroethene; in this application a first-order equation was used in place of the Michaelis-Menten equation. The model simulations provided fair approximations of the data.

A two-dimensional transport model that uses the Monod-type equation to describe biodegradation of solute, wherein biodegradation is limited by oxygen, was presented by Borden et al. (1984). In this model, both substrate and oxygen are influenced by advective-dispersive transport, sorption is ignored, and only solution-phase substrate is subject to biodegradation. A model incorporating a description of growth of biomass was used to show that output was insensitive to growth dynamics, primarily because of the high rates of microbial growth relative to typical rates of groundwater flow. On this basis, the authors eliminated growth dynamics and used an assumption of steady-state biomass. This finding is significant in that it simplifies the transport equation. The validity of the steady-state biomass assumption needs to be evaluated at the higher velocities typical of pump-and-treat remediation systems.

All of the models discussed in the preceding paragraphs were developed using the macroscopic approach for the disposition of biomass; in addition, no mass-transfer constraints for substrate were included. A one-dimensional transport model using the biofilm approach to describe disposition of biomass was presented by Bouwer and Cobb (1987). They considered advective-dispersive transport of substrate and ignored sorption. Substrate utilization in the biofilm was simulated using the Monod-type equation, and mass transfer was constrained by diffusion across a boundary layer as well as diffusion within the biomass.

2. Non-Steady-State Biomass

A one-dimensional transport model that uses the Monod-type equation (the authors use Michaelis-Menten terminology) to describe biodegradation and growth of biomass was presented by Sykes et al. (1982). In this model, sorption is ignored and only solution-phase substrate is subject to degradation. They demonstrate how the model reduces to the first-order degradation model when biomass is at steady-state and the concentration of the substrate is much lower than K_s. Corapcioglu and Haridas (1985) presented one- and two-dimensional models that used the Monod-type equation to describe growth of biomass and a first-order equation for biodegradation of substrate. Sorption was governed by local equilibrium and only solution-phase substrate was subject to biodegradation. A one-dimensional transport model that uses the Monod-type equation to describe biodegradation of substrate and growth of biomass was presented by Bosma et al. (1988). Sorption was governed by local equilibrium and only solution-phase substrate was subject to biodegradation. A term was included to account for the effect of chemotaxis (movement of bacteria in response to concentration gradient of substrate).

A two-dimensional transport model that uses the Monod-type equation to describe biodegradation of solute and growth of biomass, wherein biodegradation is limited by oxygen, was presented by Borden et al. (1984). In this model, both substrate and oxygen are influenced by advective-dispersive transport, sorption is ignored, and only solution-phase substrate is subject to biodegradation. A one-dimensional form of this model was presented by Borden and Bedient (1986) for the case of sorbing solute. One- and two-dimensional models that use the Monod-type equation to describe biodegradation of solute and growth of biomass, wherein biodegradation is limited by oxygen, were presented by MacQuarrie et al. (1990). In this model, both substrate and oxygen are influenced by advective-dispersive transport, sorption is governed by local equilibrium, and only solution-phase substrate is subject to biodegradation.

A one-dimensional model similar to those of Borden and Bedient (1986) and MacQuarrie et al. (1990) was presented by Semprini and McCarty (1991). They used the model to simulate the results of a series of controlled field experiments at the Moffet Naval Air Station, where the growth of an indigenous population of methane-utilizing bacteria was stimulated by the addition of dissolved methane

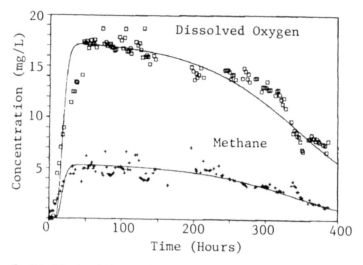

Figure 3. Model simulation and observed methane and dissolved oxygen response at the S2 observation well in the Biostim1 experiment. (From Semprini and McCarty, 1991.)

and oxygen into a semiconfined aquifer. The model, which was calibrated to the data, provided good simulations of the data (see Figure 3).

A one-dimensional model that uses forms of the Monod-type equation (the authors use Michaelis-Menten terminology) to describe loss of substrate and electron acceptor and growth of biomass was presented by Kindred and Celia (1989). The model includes provisions for aerobic metabolism, anaerobic nitrate-reducing metabolism, fermentation, and cometabolism of a non-growth supporting substrate. It also allows for uptake inhibition due to substrate inhibition and limitation of substrate transfer through biomass. The latter was simulated using a simple inhibition-factor approach, rather than a diffusion-based approach. Sorption of substrate is governed by local equilibrium and only solution-phase substrate is subject to biodegradation.

All of the models discussed in the preceding three paragraphs were developed using the macroscopic approach for the disposition of biomass; in addition, no mass-transfer constraints for substrate were included (except for the Kindred and Celia model). A one-dimensional transport model that uses the microcolony approach and wherein biodegradation is limited by oxygen was presented by Molz et al. (1986). In this model, both substrate and oxygen are influenced by advective-dispersive transport, sorption is governed by local equilibrium, and the Monod-type equation is used to simulate loss of substrate and oxygen as well as growth of biomass. Biodegradation occurs within the microcolonies; however, mass transfer of substrate and oxygen into the microcolonies occurs only from solution. This mass transfer is constrained by diffusion across a boundary layer. The model of Molz et al. (1986) was extended by Widdowson et al. (1988a) to account for biodegradation occurring by nitrate-based respiration and limited by a mineral nutrient (i.e., NH_4^+). This model was extended to the case of multiple substrates by Widdowson et al. (1988b).

D. Complex Sorption and Transformation, Simple Transport

The models in this category were developed for wastewater treatment applications. One-dimensional transport models that use the Monod equation to describe growth of biomass and a dual-resistance model (film mass transfer and solid-phase intraparticle diffusion) for sorption dynamics were presented by Ying and Weber (1979) and Speitel et al. (1987).

For both models, degradation is limited only by substrate, loss occurs only in the solid phase, and the distribution of biomass is described with the macroscopic approach. The difference between the two models is that Ying and Weber include hydrodynamic dispersion in their model, whereas Speitel et al. do not. These models are not directly applicable to most soil/aquifer systems because degradation is not allowed to occur in the solution phase.

E. Complex Transport and Sorption, Simple Transformation

A one-dimensional transport model that included film mass transfer, intra-particle diffusion, one-site rate-limited sorption, and first-order loss from sorbed phase was presented by Rasmusson (1982). This model was developed for applications in chemical engineering and is not useful for soil/aquifer systems because of the constraint of no degradation in solution. A one-dimensional multi-process nonequilibrium model developed by Brusseau et al. (1989a), which included porous-media heterogeneity and rate-limited sorption, was extended by Brusseau et al. (1992) to include transformation in solution and sorbed phases. The two-region approach was used to represent heterogeneity, and the two-domain approach was used to describe rate-limited sorption. Degradation was described as a first-order process. The equations were solved by means of a finite-difference numerical scheme.

The four dimensionless governing equations for the multi-process nonequilibrium with transformation (MPNET) model are:

$$R_{ml}\frac{\partial C_m^*}{\partial p} + k_m^0(C_m^* - S_m^*) + \omega(C_m^* - C_{im}^*) = \frac{1}{P}\frac{\partial^2 C_m^*}{\partial X^2} - \frac{\partial C_m^*}{\partial X} - \xi_{ml}C_m^* \quad (15)$$

M.L. Brusseau, P.S.C. Rao, and C.A. Bellin

$$R_{im1} \frac{\partial C_{im}^*}{\partial p} + k_{im}^0 (C_{im}^* - S_{im}^*) = \omega (C_m^* - C_{im}^*) - \xi_{im1} C_{im}^* \qquad (16)$$

$$R_{m2} \frac{\partial S_m^*}{\partial p} = k_m^0 (C_m^* - S_m^*) - \xi_{m2} S_m^* \qquad (17)$$

$$R_{im2} \frac{\partial S_{im}^*}{\partial p} = k_m^0 (C_{im}^* - S_{im}^*) - \xi_{im2} S_{im}^* \qquad (18)$$

by defining the following dimensionless parameters:

$$X = \frac{x}{L} \qquad (19a)$$

$$p = \frac{qt}{\theta L} \qquad (19b)$$

$$\phi = \frac{\theta m}{\theta} \qquad (19c)$$

$$P = \frac{qL}{\theta_m D} \qquad (19d)$$

$$C_m^* = \frac{C_m}{C_o} \qquad (19e)$$

$$C_{im}^* = \frac{C_{im}}{C_o} \qquad (19f)$$

$$\omega = \frac{\alpha L}{q} \tag{19g}$$

$$k_m^0 = \frac{k_{m2} L \theta R_{m2}}{q} \tag{19h}$$

$$k_{im}^0 = \frac{k_{im2} L \theta R_{im2}}{q} \tag{19i}$$

$$S_m^* = \frac{S_{m2}}{(1-F_m)k_m C_o} \tag{19j}$$

$$S_{im}^* = \frac{S_{im2}}{(1-F_{im})K_{im} C_o} \tag{19k}$$

$$\xi_{m1} = \mu_m \phi L/v + (\rho/\theta)\mu_{m1} f F_m K_m L/v \tag{19l}$$

$$\xi_{im1} = \mu_{im}(1-\phi)L/v + (\rho/\theta)\mu - im1(1-f)F_{im}K_{im}L/v \tag{19m}$$

$$\xi_{m2} = R_{m2}\mu_{m2} L/v \tag{19n}$$

$$\xi_{im2} = R_{im2}\mu_{im2} L/v \tag{19o}$$

$$R_{m1} = \phi + \frac{f\rho}{\theta} F_m K_m \tag{19p}$$

M.L. Brusseau, P.S.C. Rao, and C.A. Bellin

$$R_{m2} = \frac{f\rho}{\theta} (1-F_m)K_m \qquad (19q)$$

$$R_{im1} = (1-\phi) + \frac{(1-f)\rho}{\theta} F_{im}K_{im} \qquad (19r)$$

$$R_{im2} = \frac{(1-f)\rho}{\theta} (1-F_{im})K_{im} \qquad (19s)$$

Brusseau et al. (1992) investigated the effect of biodegradation on breakthrough curves for a hypothetical system in which nonequilibrium was caused by transport-related and sorption-related mechanisms. The base simulation for this system (case 1), where no transformation occurs, is shown in Figure 4. A case (case 2) where transformation (e.g., biodegradation) occurs in all domains is

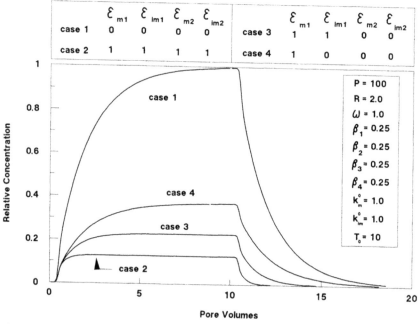

Figure 4. Simulated breakthrough curves, obtained with the MPNET model, illustrating the effect on solute transport of constraints on the location where biodegradation takes place. (From Brusseau et al., 1992.)

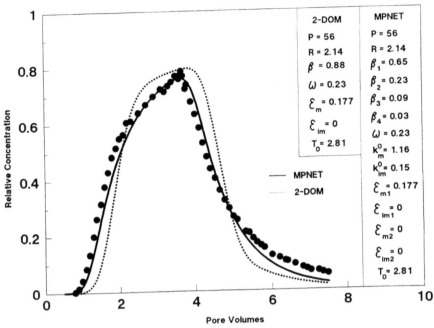

Figure 5. Evaluation of the performance of the MPNET model. (From Brusseau et al., 1992.) Predicted simulations obtained with the MPNET and two-domain models are compared to a breakthrough curve reported by van Genuchten (1974) for the transport of 2,4,5-T through a column packed with an aggregated soil.

also shown in Figure 4; the unity values assigned to the transformation parameters were chosen for simplicity. The comparison between the cases of degradation and no degradation illustrates the impact of transformation reactions on the concentration profile of a reactive solute. As discussed, recent research has suggested that sorbed-phase organic solute is resistant to biodegradation. To simulate this, degradation was allowed only in the solution phases (i.e., mobile and immobile pore water). This simulation (see case 3, Figure 4) reveals that rate-limited desorption can have a significant impact on biodegradation. A simulation for the case where biodegradation does not take place in the immobile domain (e.g., because of exclusion of bacteria), and where sorbed-phase contaminant is resistant, is provided in Figure 4 (case 4). The complex system examined above could not be simulated with a bicontinuum transformation model.

The performance of the MPNET model was tested by attempting to predict the breakthrough curve reported by van Genuchten (1974) for the transport of 2,4,5-trichlorophenoxy acetic acid in an aggregated soil (the same data set used by Gamerdinger et al. (1990) to test the two-region model and shown in Figure 1). Values for all parameters of the model were obtained independent of the

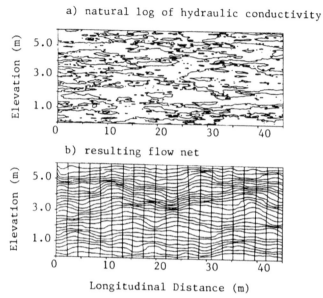

a) natural log of hydraulic conductivity

b) resulting flow net

Longitudinal Distance (m)

Figure 6. Representation of a random flow field. (From MacQuarrie and Sudicky, 1990.)

2,4,5-T breakthrough curve. The predicted simulation obtained with the MPNET model matched the data well, much better than did the prediction obtained with the two-region model (see Figure 5).

F. Complex Transport and Transformation, Simple Sorption

The influence of vertical aquifer heterogeneity on the transport of biodegrading solutes was investigated by Widdowson et al. (1987) and Molz and Widdowson (1988). This was accomplished by incorporating vertical variations in pore-water velocity into a vertical two-dimensional flow domain. The model did not have the capability to simulate non-uniform velocity fields; they were created deterministically by specification of nodal values. Biodegradation takes place in the biomass phase (microcolonies); however, transfer of substrate into the biomass occurs only from solution and is constrained by diffusion across a boundary layer. The researchers demonstrated that vertical heterogeneity can have a major impact on the distribution of oxygen and substrate and, thus, on microbial activity. Hence, two-dimensional vertically averaged models will not provide accurate simulations of transport of biodegradable solutes in heterogeneous aquifers.

The two-dimensional transport model presented by MacQuarrie et al. (1990) was applied to random flow fields by MacQuarrie and Sudicky (1990).

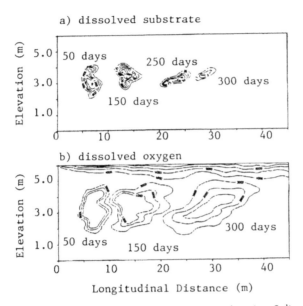

Figure 7. Simulated substrate and oxygen concentration (μg L^{-1}) contours for transport in a random flow field. Dissolved oxygen distribution near the water table is for 300 days at conditions of steady state. (From MacQuarrie and Sudicky, 1990.)

Hydraulic-conductivity heterogeneity was represented stochastically, and sorption was assumed to be linear and governed by local equilibrium. The Monod-type equation was used to simulate loss of substrate and of the electron acceptor (O_2), and to simulate growth of biomass. The macroscopic approach was used for the disposition of biomass (i.e., no microscopic description of biomass distribution). Biodegradation was assumed to occur only in solution and was not constrained by mass transfer of substrate (e.g., diffusion of substrate across boundary layer). Acclimation of bacterial population and substrate inhibition were not included in the model. The transport equation for substrate is:

$$\frac{\partial S_b}{\partial t} = \frac{\partial}{\partial x}(\frac{D_{xx}}{R}\frac{\partial S_b}{\partial x}) + \frac{\partial}{\partial z}(\frac{D_{zz}}{R}\frac{\partial S_b}{\partial z}) - \frac{v_x}{R}\frac{\partial S_b}{\partial x} - \frac{M_b\mu_{max}}{R}(\frac{S_b}{K_s+S_b})(\frac{S_e}{K_e+S_e})$$

$$(20)$$

MacQuarrie and Sudicky (1990) used the turning bands method of Mantoglou and Wilson (1982) to generate a statistical realization of the random flow field, which is shown in Figure 6a along with the corresponding flow net in Figure 6b. The statistical parameters used to generate the flow field were based on experimental data obtained at the Borden aquifer site. The "aquifer" is

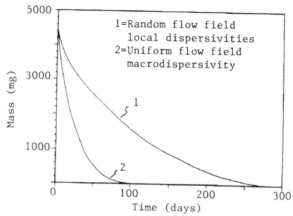

Figure 8. Comparison of substrate mass loss in random and uniform flow fields. (From MacQuarrie and Sudicky, 1990.)

characterized by horizontal, discontinuous lenses resulting in a flow domain with regions of tortuous flow paths. Note that high-velocity flow regions are depicted by areas of closely-spaced flow lines in Figure 6b.

Three consequences of the aquifer heterogeneity are of specific interest here: (1) the emergence of the plume shapes as influenced only by the physical heterogeneity of the aquifer; (2) the development of the substrate plume shape as affected by oxygen plume shape and position; and (3) the dynamics of overall mass loss of the substrate in a heterogeneous aquifer.

Simulations of the plume shapes for the substrate and oxygen during their transport through the heterogeneous flow field are shown in Figure 7. The plumes are strikingly irregular compared to the symmetric, elliptical plumes expected for solute transport in homogeneous flow fields. It is evident that the substrate plume shapes at different times are in direct correspondence with the local variations in the velocity fields (Figure 6). Higher concentrations and larger volumes associated with the substrate plume at earlier times result in zones of substantial depletion of oxygen, which then limits the rate of mass loss. However, at longer times the supply of oxygen via diffusion into the interior regions of the substrate plume is sufficient to meet the biological demands because of the diminished substrate plume size and concentration. These two factors contribute to much smaller rates of substrate loss in heterogeneous flow fields in contrast to losses in an equivalent uniform flow field (Figure 8). Another significant feature of the oxygen concentration contours depicted in Figure 7b should be noted. In the absence of significant recharge from the vadose zone, diffusive transport of oxygen into the saturated zone is likely to enhance biodegradation only if the contaminant plume were confined to a depth no greater than about a meter below the water table.

For practical reasons, it is often assumed that heterogeneous flow fields can be represented as being macroscopically homogeneous, and the solute dispersion

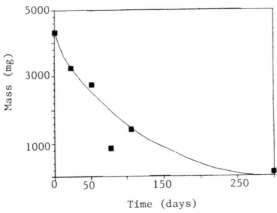

Figure 9. Model (curve) and field (symbol) plume mass loss comparison for benzene. (From MacQuarrie and Sudicky, 1990.)

can be approximated using a macro-dispersion term based on a conservative solute. However, this approach leads to far greater mixing of the substrate and oxygen plumes; as a consequence, the rate of substrate loss is over-predicted. The simulations presented by MacQuarrie and Sudicky (1990) clearly demonstrate the importance of coupling the local-scale physical heterogeneity of the porous medium (i.e., transport properties) and the macro-scale observations of biodegradation.

To evaluate the performance of their model, MacQuarrie and Sudicky (1990) attempted to predict the behavior of a benzene plume in a shallow sandy aquifer located at the Borden Air Base in Ontario, Canada. The mass loss curve predicted by the model closely matched the data (see Figure 9). This was especially encouraging because all input parameters were determined independently.

A three-dimensional model based on the approach of MacQuarrie and Sudicky (1990) was presented by Frind et al. (1989). They found that, similarly for the use of 1-D models for a 2-D problem, the use of a 2-D model for a 3-D problem can lead to errors in simulating plume dynamics. A three-dimensional transport model for substrate and oxygen, that includes heterogeneous hydraulic conductivity and biodegradation, was presented by Miralles-Wilhelm et al. (1990). Hydraulic conductivity was represented stochastically and sorption is assumed to be instantaneous. Growth of biomass was described with the Monod equation.

IV. Conclusions

Monod-type equations are widely used to describe substrate loss via biodegradation for solute transport models. The validity of these equations under conditions typical to contaminant transport in porous media requires investigation. Most of the models were developed with the assumption that only solution-phase substrate was subject to biodegradation. While the results of several experiments support this, the absence of sorbed-phase degradation may deleteriously affect the performance of transport models under some conditions.

Our approach in this discussion of models for coupled processes emphasized two perspectives -- one of process-scale models of varying levels of complexity, and the other of spatial scales of interest. It is important to recognize the connection between the two. The coupling of transport models can be argued from a local-scale perspective on the basis of physical-chemical-biological considerations. However, the significance of a given process and, thus, the degree of complexity required to describe the process, may depend entirely on the spatial scale of application. For example, factors or processes considered at a local scale (as in column experiments) may not have an observed effect on solute transport at the field scale. Hence, accurate and complete descriptions of processes should be balanced with a consideration of the importance of the processes as governed by scale effects and other factors.

A striking but understandable phenomenon that is apparent from this work is that the number of models developed and reported in the literature far exceeds the number of models that have been tested and validated against results of experiments. The dearth of well designed and controlled experiments, especially at the field scale, must be addressed. Some of the models discussed above have been used to provide insights into the transport of sorbing, degrading solute that we could not have otherwise obtained. However, we must reserve final judgment on the accuracy of these insights until the models have been validated to an extent much greater than they have so far.

References

Ainsworth, C.C., J.M. Zachara, and S.C. Smith. 1990. The effect of sorption on the degradation of quinoline and salicylate. *Trans. Am. Geophys. Union*, Proc. of Fall Natl. Meetings, San Francisco, CA, pp.1323.

Alexander, M. and K. Scow. 1989. Kinetics of biodegradation in soil. Chap. 10. In: B.L. Sawhney and K. Brown (eds.); *Reactions and Movement of Organic Chemicals in Soils,* Soil Sci. Soc. Am., Madison, WI. SSSA Spec. Pub. No. 22.

Angley, J.T., M.L. Brusseau, W.L. Miller, and J.J. Delfino. 1992. Nonequilibrium sorption and aerobic biodegradation of dissolved alkylbenzenes during transport in aquifer material: column experiments and evaluation of a coupled-process model. *Environ. Sci. Technol.*, (in press).

Ardakani, M.S., J.T. Rehbock, and A.D. McLaren. 1973. Oxidation of nitrite to nitrate in a soil column. *Soil Sci. Soc. Am. Proc.* 37:53-56.

Armstrong, D.E., G. Chesters, and R.F. Harris. 1967. Atrazine hydrolysis in soil. *Soil Sci. Soc. Am. Proc.*, 31:61-67.

Baehr, A. and M.Y. Corapcioglu. 1984. A predictive model for pollution from gasoline in soils and groundwater. In: *Petroleum Hydrocarbons and Organic Chemicals in Groundwater*. Houston, TX, Nov. 5-7, 1984. Nat. Water Well Assoc., Dublin, OH.

Balkwill, D.L. and W.G. Ghiorse. 1985. Characterization of subsurface bacteria associated with two shallow aquifers in Oklahoma. *App. Environ. Micro.* 50:580-588.

Ball, W.P. and P.V. Roberts. 1991. Long-term sorption of halogenated organic chemicals by aquifer material. Part 2. Intraparticle diffusion. *Environ. Sci. Technol.* 25:1237-1249.

Baveye, P. and A. Valocchi. 1989. An evaluation of mathematical models of the transport of biologically reacting solutes in saturated soils and aquifers. *Water Resour. Res.* 25:1413-1421.

Benefield, L.D. and F.J. Molz. 1983. A kinetic model for the activated sludge process which considers diffusion and reaction in the microbial floc. *Biotechnol. Bioeng.* 25:2591-2615.

Birnbaum, S.J., D.R. Olmos, P.E. Long, and S.G. McKinley. 1990. Direct observation of attached microbes in rock matrices using scanning electron microscopy techniques. *Trans. Amer. Geophys. Union*, Proc. of Fall Natl. Meetings, San Francisco, CA, pp.1318.

Boesten, J.J.T.I. and L.J.T. Van Der Pas. 1988. Modeling adsorption/desorption kinetics of pesticides in a soil suspension. *Soil Science* 146:221.

Boesten, J.J.T.I., L.J.T. Van Der Pas and J.H. Smelt. 1989. Field test of a mathematical model for non-equilibrium transport of pesticides in soil, *Pestic. Sci.*, 25:187.

Bondarev, V.S., Y.Y. Spiridonov, and V.G. Shestakov. 1985. Effect of sorption on rate of degradation of picloram in soil. Translated from *Pochvovedeniye.* 7:54-60.

Borden, R.C. and P.B. Bedient. 1986. Transport of dissolved hydrocarbons influenced by oxygen-limited biodegradation. 1. Theoretical development. *Water Resour. Res.* 22:1973-1982.

Borden, R.C., M.D. Lee, J.T. Wilson, C.H. Ward, and P.B. Bedient. 1984. In: *Petroleum Hydrocarbons and Organic Chemicals in Groundwater*. Houston, TX, Nov. 5-7, 1984. Nat. Water Well Assoc., Dublin, OH.

178 M.L. Brusseau, P.S.C. Rao, and C.A. Bellin

Borden, R.C., P.B. Bedient, M.D. Lee, C.H. Ward, and J.T. Wilson. 1986. Transport of dissolved hydrocarbons influenced by oxygen-limited biodegradation 2. Field application. *Water Resour. Res.* 22:1983-1990.

Bosma, T.N.P., J.L. Schnoor, G. Schraa, and A.J.B. Zehnder. 1988. Simulation model for biotransformation of xenobiotics and chemotaxis in soil columns. *J. Contam. Hydrol.* 2:225-236.

Bouwer, E.J. and P.L. McCarty. 1984. Modeling of trace organics biotransformation in the subsurface. *Groundwater* 22:433-440.

Bouwer, E.J. and G.D. Cobb. 1987. Modeling of biological processes in the subsurface. *Water Sci. Tech.* 19:769-779.

Brusseau, M. L. and P.S.C. Rao. 1989a. The influence of sorbate-organic matter interactions on sorption nonequilibrium. *Chemosphere* 18:1691-1706.

Brusseau, M. L. and P.S.C. Rao. 1989b. Sorption nonideality during organic contaminant transport in porous media. *CRC Critical Reviews in Environ. Control* 19:33-99.

Brusseau, M. L. and P.S.C. Rao. 1991. The influence of sorbate structure on nonequilibrium sorption of organic compounds. *Environ. Sci. Technol.* 25:1501-1506.

Brusseau, M. L., R.E. Jessup, and P.S.C. Rao. 1989a. Modeling the transport of solutes influenced by multi-process nonequilibrium. *Water Resour. Res.* 25:1971-1988.

Brusseau, M. L., P.S.C. Rao, R.E. Jessup, and J.M. Davidson. 1989b. Flow interruption: a method for investigating sorption nonequilibrium. *J. Contam. Hydrol.* 4:223-240.

Brusseau, M. L., R.E. Jessup, and P.S.C. Rao. 1991a. Nonequilibrium sorption of organic chemicals: Elucidation of rate-limiting processes. *Environ. Sci. Technol.* 25:134-142.

Brusseau, M. L., A.L. Wood, and P.S.C. Rao. 1991b. The influence of organic cosolvents on the sorption kinetics of hydrophobic organic chemicals. *Environ. Sci. Technol.* 25:903-910.

Brusseau, M. L., R.E. Jessup, and P.S.C. Rao. 1992. Modeling solute transport influenced by multi-process nonequilibrium and transformation reactions. *Water Resour. Res.* 28:175-182.

Buxton, D.S. and R.E. Green. 1987. In: *Agronomy Abst.* Am. Soc. Agron., Madison, WI.

Cameron, D. R. and A. Klute. 1977. Convective-dispersive solute transport with a combined equilibrium and kinetic adsorption model. *Water Resour. Res.* 13:183.

Carnahan, C.L. and J.S. Remer. 1984. Nonequilibrium and equilibrium sorption with a linear sorption isotherm during mass transport through an infinite porous medium: some analytical solutions. *J. Hydrol.* 73:227-258.

Casida, L.E. 1971. Microorganisms in unamended soil as observed by various forms of microscopy and staining. *App. Environ. Microbiol.* 21:1040-1045.

Coates, J.T., and A.W. Elzerman. 1986. Desorption kinetics for selected PCB congeners from river sediments. *J. Contam. Hydrol.* 1:191.

Coats, K.H. and B.D. Smith. 1964. Dead-end pore volume and dispersion in porous media. *J. Soc. Pet. Eng.* 4:73-80.

Corapcioglu, M.Y. and A. Haridas. 1985. Microbial transport in soils and groundwater: A numerical model. *Adv. Water Resour.* 8:188-200.

Corapcioglu, M.Y., M.A. Hossain, and M.A. Hossain. 1991. Methanogenic biotransformation of chlorinated hydrocarbons in groundwater. *J. Environ. Engin.* 117:47-65.

Davidson, J.M. and J.R. McDougal. 1973. Experimental and predicted movement of three herbicides in water-saturated soil. *J. Environ. Qual.* 2:428.

Eldor, M. and G. Dagan. 1972. Solutions of hydrodynamic dispersion in porous media. *Water Resour. Res.* 8:1316-1331.

Frind, E.O., E.A. Sudicky, and J.W. Molson. 1989. Three-dimensional simulation of organic transport with aerobic biodegradation. p. 89-96 In: L.M. Abriola (ed.) *Groundwater Contamination* IAHS Public. No. 185.

Gamerdinger, A.P., R.J. Wagenet, and M.Th. van Genuchten. 1990. Application of two-site/two-region models for studying simultaneous nonequilibrium transport and degradation of pesticides. *Soil Sci. Soc. Am. J.* 54:957-963.

Giddings, J.C. and H. Eyring. 1955. A molecular dynamic theory of chromatography. *J. Phys. Chem.* 59:416.

Glueckauf, E. and J.I. Coates. 1947.Theory of chromatography. IV. The influence of incomplete equilibrium on the front boundary of chromatograms and on the effectiveness of separation. *J. Chem. Soc.* p.1357.

Hamaker, J.W. 1972. Decomposition: quantitative aspects. p. 253-240 In: C.A.I. Goring and W.J. Hamaker (eds.) *Organic Chemicals in the Soil Environment.* Marcel Dekker, NY.

Hamaker, J. W. and J.M. Thompson. 1972. p. 49-143. In: C.A.I. Goring and J.W. Hamaker (eds.) *Organic Chemicals in the Soil Environment.* Marcel Dekker, NY.

Harvey, R.W., R.L. Smith, and L. George. 1984. Effect of organic contamination upon microbial distributions and heterotrophic uptake in a Cape Cod, Mass. aquifer. *App. Environ. Micro.* 48:1197-1202.

Helfferich, F. 1962. Ion Exchange. McGraw-Hill, New York, NY.

Hoag, W.R. and T. Mill. 1988. Effects of a subsurface sediment on hydrolysis of haloalkanes and epoxides. *Environ. Sci. Technol.* 22:658-663.

Hoffman, D.L., and D.E. Rolston. 1980. Transport of organic phosphate in soil as affected by soil type. *Soil Sci. Soc. Am. J.* 44:46-52.

Hornsby, A.G. and J.M. Davidson. 1973. Solution and adsorbed fluometu-ron concentration distribution in a water-saturated soil: experimental and predicted evaluation. *Soil Sci. Soc. Am. Proc.* 37:823-828.

Jessup, R.E., M.L. Brusseau, and P.S.C. Rao. 1989. Modeling one-dimensional solute transport, Florida Agric. Exper. Station Report, Univ. Florida.

Kam, T.K. 1974. Analytical and numerical solutions of convective diffusive equations of solutes subject to linear and Michaelis-Menten types of kinetics in soils. Unpublished Ph.D. dissertation, University of California, Berkeley.

Karickhoff, S. W. 1980. Sorption kinetics of hydrophobic pollutants in natural sediments, Chap. 11. In: R.A. Baker (ed.) *Contaminants and Sediments.* Vol. 2. Ann Arbor Press, Ann Arbor, MI.

Karickhoff, S.W. and K.R. Morris. 1985. Sorption dynamics of hydrophobic pollutants in sediment suspensions. *Env. Tox. Chem.* 4:469.

Kindred, J.S. and M.A. Celia. 1989. Contaminant transport and biodegradation. 2. Conceptual model and test simulations. *Water Resour. Res.* 25:1149-1159.

Lapidus, L. and N.R. Amundson. 1952. Mathematics of adsorption in beds. VI. The effects of longitudinal diffusion in ion exchange and chromatographic columns. *J. Phys. Chem.* 56:984.

Lassey, K.R. 1988. Unidimensional solute transport incorporating equilibrium and rate-limited isotherms with first-order loss. 1. Model conceptualizations and analytic solutions. *Water Resour. Res.* 24:343-350.

Lindstrom, F.T. 1976. Pulsed dispersion of trace chemical concentrations in a saturated sorbing porous medium. *Water Resour. Res.* 12:229-238.

Mabey, W. and T. Mill. 1978. Critical review of hydrolysis of organic compounds in water under environmental conditions. *J. Phys. Chem. Ref. Data.* 7:383-415.

Macalady, D.L. and N.L. Wolfe. 1984. Abiotic hydrolysis of sorbed pesticides. Chap. 14. In: R.F. Krueger and J.N. Seiber (eds.) *Treatment and Disposal of Pesticide Wastes.* Am. Chem. Soc., Wash. D.C., ACS Symp. Ser. 259.

Macalady, D.L. and N.L. Wolfe. 1985. Effects of sediment sorption and abiotic hydrolysis: 1. Organophosphorothioate esters. *J. Agric. Food Chem.* 33:167-173.

Macalady, D.L., P.G. Tratnyek, and N.L. Wolfe. 1989. Influences of natural organic matter on the abiotic hydrolysis of organic contaminants in aqueous systems. Chap. 21 In: I.H. Suffet and P. MacCarty (eds.) *Aquatic Humic Substances.* Am. Chem. Soc., Wash. D.C.

MacQuarrie, K.T.B. and E.A. Sudicky. 1990. Simulation of biodegradable organic contaminants in groundwater. 2. Plume behavior in uniform and random flow fields. *Water Resour. Res.* 26:223-240.

MacQuarrie, K.T.B., E.A. Sudicky, and E.O. Frind. 1990. Simulation of biodegradable organic contaminants in groundwater. 1. Numerical formulation in principle directions. *Water Resour. Res.* 26:207-223.

Mantoglou, A. and J.L. Wilson. 1982. The turning bands method for simulation of random fields using line generation by a spectral method. *Water Resour. Res.* 18:1379-1394.

McCall, P. J. and G.L. Agin. 1985. Desorption kinetics of picloram as affected by residence time in the soil. *Env. Tox. Chem.* 4:37.

Mihelcic, J.R. and R.G. Luthy. 1991. Sorption and microbial degradation of naphthalene in soil-water suspensions under denitrification conditions. *Environ. Sci. Technol.* 25:169-177.

Miller, R.M. and R. Bartha. 1989. Evidence from liposome encapsulation for transport-limited microbial metabolism of solid alkanes. *App. Environ. Micro.* 55:269-274.

Miralles-Wilhelm, F., L.W. Gelhar, and V. Kapoor. 1990. Effects of heterogeneities on field-scale biodegradation in groundwater. *Trans. Am. Geophys. Union*, Proc. of Fall Natl. Meetings, San Francisco, CA, pp.1304.

Mironenko, E.V. and Y.A. Pachepsky. 1984. Analytical solution for chemical transport with nonequilibrium mass transfer, adsorption and biological transformation. *J. Hydrol.* 70:167-175.

Molz, F.J. and M.A. Widdowson. 1988. Internal inconsistencies in dispersion-dominated models that incorporate chemicals and microbial kinetics. *Water Resour. Res.* 24:615-619.

Molz, F.J., M.A. Widdowson, and L.D. Benefield. 1986. Simulation of microbial growth dynamics coupled to nutrient and oxygen transport in porous media. *Water Resour. Res.* 22:1207-1216.

Monod, J. 1949. The growth of bacterial cultures. *Ann. Rev. Microbiol.* 3:371-394.

Moyer, J.R., R.J. Hance, and C.E. McKone. 1972. The effect of adsorbents on the rate of degradation of herbicides incubated with soil. *Soil Biol. Biochem.* 4:307-311.

Nkedi-Kizza, P., M.L. Brusseau, P.S.C. Rao, and A.G. Hornsby. 1989. Nonequilibrium sorption during displacement of hydrophobic organic contaminants and Ca^{45} through soil columns with aqueous and mixed solvents. *Environ. Sci. Technol.* 23:814-820.

Oddson, J.K., J. Letey, and L.V. Weeks. 1970. Predicted distribution of organic chemicals in solution and adsorbed as a function of position and time for various chemical and soil properties. *Soil Sci. Soc. Am. Proc.* 34:412.

Ogram, A.V., R.E. Jessup, L.T. Ou, and P.S.C. Rao. 1985. Effects of sorption on biological degradation rates of (2,4-Dichlorophenoxy)acetic acid in soils. *App. Environ. Micro.* 49:582-587.

Oliver, B.G. 1985. Desorption of chlorinated hydrocarbons from spiked and anthropogenically contaminated sediments. *Chemosphere* 14:1087.

Parlange, J.Y., J.L. Starr, D.A. Barry, and R.D. Braddock. 1984. Some approximate solutions of the transport equation with irreversible reactions. *Soil Science* 137:434-442.

Pignatello, J.J. 1990. Slowly reversible sorption of aliphatic halocarbons in soils. II. Mechanistic aspects. *Environ. Toxic. Chem.* 9:1117-1126.

Rao, P.S.C. and R.E. Jessup. 1983. Sorption and movement of pesticides and other toxic organic substances in soils. p.183-201. In: *Chemical Mobility and Reactivity in Soil Systems.* ASA, SSSA, Madison, WI.

Rasmusson, A. 1982. Transport processes and conversion in an isothermal fixed-bed catalytic reactor. *Chem. Engin. Sci.* 37:411-415.

Rifai, H.S., P.B. Bedient, J.T. Wilson, K.M. Miller, and J.M. Armstrong. 1988. Biodegradation modeling at aviation fuel spill site. *J. Environ. Engin.* 114:1007-1029.

Rijnaarts, H.H.M., A. Bachmann, J.C. Jumelet, and A.J.B. Zehnder. 1990. Effect of desorption and intraparticle mass transfer on the aerobic biomineralization of α-hexachlorocyclohexane in a contaminated calcareous soil. *Environ. Sci. Technol.* 24:1349-1354.

Rittman, B.E. and P.L. McCarty. 1978. Variable-order model of bacterial-film kinetics. *J. Environ. Engin.* 104:889-900.

Robinson, K.G., W.S. Farmer, and J.T. Novak. 1990. Availability of sorbed toluene in soils for biodegradation by acclimated bacteria. *Water Res.* 24:345-350.

Schmidt, S.K., S. Simkins, and M. Alexander. 1985. Models for the kinetics of biodegradation of organic compounds not supporting growth. *App. Environ. Micro.* 50:323-331.

Schwarzenbach, R.P. and J. Westall. 1981. Transport of nonpolar organic compounds from surface water to groundwater. Laboratory sorption studies, *Environ. Sci. Technol.* 15:1360.

Scow, K.M. and M. Alexander. 1991. Effect of diffusion on the kinetics of biodegradation: Experimental results with synthetic aggregates. *Soil Sci. Soc. Am. J.* (in review).

Scow, K.M., W. Simkins, and M. Alexander. 1986. Kinetics of mineralization of organic compounds at low concentrations in soil. *App. Environ. Micro.* 51:1028-1035.

Selim, H.M., J.M. Davidson, and R.S. Mansell. 1976. Evaluation of a two-site adsorption-desorption model for describing solute transport in soils. In: Proc. Summer Computer Simulation Conf., Washington, D.C.

Semprini, L. and P.L. McCarty. 1991. Comparison between model simulations and field results for in-situ biorestoration of chlorinated aliphatics. Part 1. Biostimulation of methanotrophic bacteria. *Groundwater* 29:365-374.

Simkins, S. and M. Alexander. 1984. Models for mineralization kinetics with the variables of substrate concentration and population density. *App. Environ. Micro.* 47:1299-1306.

Simkins, S., R. Mukherjee, and M. Alexander. 1986. Two approaches to modeling kinetics of biodegradation by growing cells and application of a two-compartment model for mineralization kinetics in sewage. *App. Environ. Micro.* 51:1153-1160.

Speitel, G.E., K. Dovantzis, and F.A. DiGiano. 1987. Mathematical modeling of biogeneration in GAC columns. *J. Environ. Engin.* 113:32-48.

Srinivasan, P. and J.W. Mercer. 1988. Simulation of biodegradation and sorption processes in groundwater. *Groundwater* 26:475-487.

Steen, W.C., D.F. Paris, and G.L. Baughman. 1980. Effects of sediment sorption on microbial degradation of toxic substances. Chap. 23 In: R.A. Baker (ed.) *Contaminants and Sediments*, Vol. 1. Ann Arbor Sci., Ann Arbor, MI.

Steinberg, S.M., J.J. Pignatello, and B.L. Sawhney. 1987. Persistence of 1,2-dibromomethane in soils: entrapment in intraparticle micro pores. *Environ. Sci. Technol.* 21:1201.

Stucki, G. and M. Alexander. 1987. Role of dissolution rate and solubility in biodegradation of aromatic compounds. *App. Environ. Micro.* 53:292-297.

Sykes, J.F., S. Soyupak, and G.J. Farquhar. 1982. Modeling of leachate organic migration and attenuation in groundwaters below sanitary landfills. *Water Resour. Res.* 18:135-145.

Szecody, J.E. and R.C. Bales. 1989. Sorption kinetics of low-molecular-weight hydrophobic organic compounds on surface-modified silica. *J. Contam. Hydrol.* 4:181-203.

Thomas, J.M., J.R. Yordy, J.A. Amador, and M. Alexander. 1986. Rates of dissolution and biodegradation of water-insoluble organic compounds. *App. Environ. Micro.* 52:290-296.

van Genuchten, M. Th. 1974. Ph.D. dissertation, New Mexico State University, Las Cruces, NM.

van Genuchten, M. Th. 1981. Non-equilibrium transport parameters from miscible displacement experiments. USDA, U.S. Salinity Lab., Research Report No. 119.

van Genuchten, M. Th. and P.J. Wierenga. 1976. Mass transfer studies in sorbing porous media: Analytical solutions. *Soil Sci. Soc. Am. J.* 40:473.

van Genuchten, M.Th. and R.J. Wagenet. 1989. Two-site/two-region models for pesticide transport and degradation: theoretical development and analytical solutions. *Soil Sci. Soc. Am. J.* 53:1303-1310.

Villermaux, J. 1974. Deformation of chromatographic peaks under the influence of mass transfer phenomena. *J. Chromat. Sci.* 12:822.

Vogel, T.M., C.S. Criddle, and P.L. McCarty. 1987. Transformations of halogenated aliphatic compounds. *Environ. Sci. Technol.* 21:722-736.

Weber, J.B. and H.D. Coble. 1968. Microbial decomposition of diquat adsorbed on montmorillonite and kaolinite clays. *J. Agric. Food Chem.* 16:475-478.

Widdowson, M.A., F.J. Molz, and L.D. Benefield. 1987. Development and application of a model for simulating microbial growth dynamics coupled to nutrient and oxygen transport in porous media. In: Proc. of NWWA/IGWMC Conf. on Solving Ground Water Problems with Models. Denver, CO, Feb. 10-12, 1987. Nat. Water Well Assoc., Dublin, OH.

Widdowson, M.A., F.J. Molz, and L.D. Benefield. 1988a. A numerical transport model for oxygen- and nitrate-based respiration linked to substrate and nutrient availability in porous media. *Water Resour. Res.* 24:1553-1565.

Widdowson, M.A., F.J. Molz, and L.D. Benefield. 1988b. Modeling multiple organic contaminant transport and biotransformations under aerobic and anaerobic (denitrifying) conditions in the subsurface. In: *Petroleum Hydrocarbons and Organic Chemicals in Groundwater.* Houston, TX, Nov. 9-11, 1988. Nat. Water Well Assoc., Dublin, OH.

Wilson, J.T. and J.F. McNabb. 1983. Biological transformation of organic pollutants in groundwater. *Trans. Am. Geophys. Union* 64:505.

Wolfe, N.L., M.E. Metwally, and A.E. Moftah. 1989. Hydrolytic Transformations of Organic Chemicals in the Environment; Chap. 9. In: B.L. Sawhney and K. Brown (eds.) *Reactions and Movement of Organic Chemicals in Soils.* Soil Sci. Soc. Am., Madison, WI. SSSA Spec. Pub. No. 22.

Wszolek, P.C. and M. Alexander. 1979. Effect of desorption rate on the biodegradation of n-alkyamines bound to clay. *J. Agric. Food. Chem.* 27:410-414.

Wu, S., and P.M. Gschwend. 1986. Sorption kinetics of hydrophobic organic compounds to natural sediments and soils. *Environ. Sci. Technol.* 20:725.

Ying, W. and W.J. Weber. 1979. Bio-physiochemical adsorption model systems for wastewater treatment. *J. Water Poll. Cont. Fed.* 51:2661-2677.

Microbial Distributions, Activities, and Movement in the Terrestrial Subsurface: Experimental and Theoretical Studies

R.W. Harvey and M.A. Widdowson

I.   Introduction 185
II. Distributions and Community Structure 186
     A. Distributions 187
     B. Community Structure 189
III. Activities 192
     A. Growth 192
     B. Degradation of Organic Carbon 200
IV. Movement 208
V.   Conclusions 212
References 212

I. Introduction

Although the numbers, types, and activities of microorganisms inhabiting the terrestrial subsurface below the root zone have historically been subject to debate, there is now considerable evidence that active and diverse microbial communities populate vast regions of the subsurface (Ghiorse and Wilson, 1988). Much of the more definitive evidence has been compiled over the last ten years as a result of rigorous attempts to minimize contamination by non-indigenous microbes during recovery and handling of whole aquifer material in groundwater microbiology studies. Motivation for a number of innovations in subsurface sampling and sample handling technology has been the direct result of increased interest in the role of subsurface microbial communities in controlling the ultimate fate of many organic and inorganic contaminants and in mediating groundwater geochemistry. There has also been a growing interest in the movement of microorganisms through the subsurface. The transport behavior

ISBN 0-87371-889-5

of microbial pathogens in an aquifer has long been a public health concern in the U.S., since biological contamination of drinking water aquifers substantively contributes to the number of waterborne disease outbreaks (Keswick, 1984). More recent interest in the transport behavior of bacteria through the subsurface is a consequence of the planned use of contaminant-adapted (both indigenous and non-indigenous) and genetically altered ("engineered") bacteria in *in situ* aquifer restoration of polluted water-supply aquifers.

Unlike soil microbiology, groundwater microbiology may still be considered a relatively "new" science, inasmuch as most of the reliable information involving the abundance, distribution, diversity, activity, and transport of microorganisms in aquifers has been published within the last ten years. Because modeling efforts to describe geochemical changes in groundwater and the fate and transport of subsurface contaminants and pathogens depend to varying degrees upon information on the subsurface microbial communities, it is important to periodically review and summarize what is and is not known about this rapidly growing field. The purpose of this chapter is twofold, i.e.: (i) to review important and recent findings involving the distribution, activities, and transport of subsurface microorganisms, particularly in contaminated zones and (ii) to describe and evaluate conceptual and mathematical models being used to describe subsurface microbial processes (including growth, contaminant degradation, and movement). The emphasis here is on the more recent microbiological reports, particularly those published after the last major review involving the microbiology of the terrestrial subsurface (Ghiorse and Wilson, 1988). All but a few of these studies have focused specifically on the saturated zone. For this reason, the emphasis here involves the microbiology of groundwater habitats. Although beyond the scope of this chapter, it should be recognized that microbial communities in the region between the root zone and the water table may significantly affect the overall fate of biological and chemical contaminants in the subsurface and may substantively contribute to subsurface geochemical changes.

II. Distributions and Community Structure

Accurate delineation of the distribution and community structure of microorganisms have historically been hampered by lack of funding, suitable aseptic-sampling technology, and methodology for detecting and characterizing sparse microbial populations in the presence of aquifer solids. Investigations on the distribution of microbes in the terrestrial subsurface began in the 1920s, when Soviet geomicrobiologists working with deep oil-bearing formations reported that certain physiological groups of microorganisms, including hydrogen-oxidizing and sulfate-reducing bacteria, were present to depths of at least 2000 m. This early work is summarized by Kuznetsov et al. (1963). In general, appreciation for the importance of subsurface microbial studies has been slow to develop.

However, funding for groundwater microbiology research has increased dramatically in a number of industrialized countries over the last decade, largely because of increasing awareness and detections of contaminants in water-supply aquifers. In the United States, the Department of Energy (USDOE), the Geological Survey (USGS), and the Environmental Protection Agency (USEPA) all have major research efforts involving microbial populations and processes in the subsurface and presently fund a variety of groundwater microbiology research at a number of universities.

Delineation of numbers and types of microbes in the groundwater environment has been greatly facilitated by the development of aseptic procedures for obtaining cores of aquifer material. This is because the integrity of well-water samples, which have traditionally served as source material for groundwater microbial studies, are often compromised by contamination from the surface environment. Also, a majority of groundwater microorganisms in both contaminated and uncontaminated zones are attached to solid surfaces (Harvey et al., 1984) and would not be detected using well-water samples alone. Contamination by non-indigenous microorganisms, which are abundant in drilling fluids commonly used in many drilling techniques, and technical problems associated with acquisition of uncompromised aquifer material have impeded characterization of aquifer microbial communities. Initial attempts at aseptic sampling at USEPA greatly facilitated delineation of microbial distribution in a shallow, pristine water-table aquifer in an Oklahoma flood plain (Balkwill and Ghiorse, 1985; Wilson et al., 1983; Webster et al, 1985; Beloin et al., 1988). Use of a wireline piston sampler developed at the University of Waterloo (Zapico et al., 1987) and modified by the USEPA, the Kansas Geological Survey, and the USGS has further facilitated delineation of microbiological investigations involving shallow, unconsolidated sandy aquifers (Leach, 1990; Smith and Harvey, 1990). Investigations of microbial distributions in deep subsurface (hundreds to thousands of meters below land surface) have been hampered by the complexity and high costs of aseptic drilling (Ghiorse and Wobber, 1989), but have been facilitated by recent advances in methodology for recovery, handling, and sample integrity monitoring of deep aquifer sediments (Phelps et al., 1989a).

A. Distributions

Microbial abundances for the saturated and unsaturated terrestrial subsurface, reported over a ten-year period from 1977 to 1987, are summarized by Ghiorse and Wilson (1988). Total bacterial abundance in pristine (uncontaminated) zones for depths between 1 and 1752 m below land surface (bls) range from 10^3 to 10^7/mL for groundwater samples and from 10^4 to 10^8/g for sediment samples. Bacterial abundance in contaminated zones for depths between 2 and 146 m bls range from 10^5 and 10^7/mL for groundwater samples and from 10^5 to 10^8/g for sediment samples. In most of the cited investigations, bacterial abundance varied between 10^5 and 10^8 cells/g sediment or mL of porewater. Reported bacterial abundances from more recent microbiological investigations of contaminated and uncontaminated zones of the subsurface have generally been within this range (Sinclair et al., 1990; Harvey and Barber, 1992; Hirsch and Rades-Rohkohl, 1988; Marxsen, 1988; Beloin et al., 1988; Balkwill et al., 1988; Pedersen and Ekendahl, 1990; Sinclair and Ghiorse, 1989; Phelps et al., 1988). Reported microbial abundances found in subsurface sediments are generally several orders of magnitude lower than those reported for surface-water sediments. For example, bacterial abundances reported for estuarine sediments are typically 10^9 - 10^{10}/g (Novitsky, 1983a, 1983b). However, bacterial abundances observed in well samples fall within the range of those reported for the water column of marine habitats, which range from 10^4/mL reported for Antarctic water under the Ross Ice Shelf (Azam et al., 1979) and at the 1 km depth in the Sargasso Sea (Liebezeit et al., 1980) to 10^7/mL reported for a number of estuaries (Goulder, 1977; Saltzman, 1980).

Since bacterial populations in the subsurface are often limited by the availability of organic carbon transported from the surface environment (Smith and Duff, 1988), a steady decrease in abundance with depth might be a reasonable assumption. However, recent investigations of bacterial distributions along vertical transects through subsurface sediments to depths of 210 to 260 m bls at three locations in a coastal plain aquifer in South Carolina suggest that this is not the case (Sinclair and Ghiorse, 1989). In the latter study, bacterial abundance generally varied between 10^6 and 10^7/g, apparently unrelated to depth. However, there appeared to be a strong effect of sediment transmissivity; highest bacterial counts were associated with sandy samples, whereas the lowest bacterial counts were associated with clay layers. In general, population densities correlated positively with sand content and pH and negatively with metal concentrations and clay content.

In addition to the hydrogeology, bacterial distribution also appears to be strongly affected by the presence of organic contaminants. In a sandy aquifer in Cape Cod, Massachusetts, abundance of unattached (free-living) bacteria within the dissolved oxygen-depleted core of a plume of secondarily treated sewage effluent decreased steadily from 4.5 x 10^6/mL at 0.25 km from the source of contamination to 7 x 10^4/mL in the most distal section of the plume 4 km downgradient (Harvey and Barber, 1992). A decrease in abundance of unattached groundwater bacteria with increasing distance from a contaminant source was also reported in other investigations at the Cape Cod site (Harvey et al., 1984; Harvey and George, 1987) and for another sandy, contaminated aquifer in Fulda, Germany (Marxsen, 1981). Decreases in population densities of groundwater bacteria with increasing distance downgradient from a point source of contamination would generally be expected, since the more readily degraded compounds in complex contaminant mixtures have been shown to be preferentially degraded by the microbiota during transport through the aquifer (Godsy et al., 1992).

It has generally been observed that the vast majority (>90%) of the bacteria in groundwater environments are attached to surfaces (Marxsen, 1981; Harvey et al., 1984). Studies involving other low-nutrient habitats suggest that the starvation response of some bacteria, including many copiotrophic bacteria, involves an increase in cell hydrophobicity and tendency for irreversible binding to solid surfaces (Kjelleberg, 1984). Since bacteria under severe carbon limitation often grow faster at solid surfaces (Jannasch and Pritchard, 1972), attached bacteria may have a distinct advantage over free-living bacteria in uncontaminated groundwater habitats. However, the partitioning of aquifer bacteria between the aqueous and solid phase appears to vary considerably in the presence of organic contaminants (Bengtsson, 1989). In several recent investigations, it has been observed that bacterial partitioning to the aqueous phase decreases with decreasing distance to the contaminant source (Harvey and Barber, 1992; Godsy et al., 1992). This suggests that at least some unattached bacteria may have a selective advantage over particle-bound bacteria in groundwater contaminated with readily degraded organic compounds. It is not known what concentrations of readily degraded dissolved organic compounds are required to offset the advantage offered by solid surfaces in promoting growth. This may vary among bacterial populations. Hence, the partitioning response of bacterial populations to dissolved organic contaminants in the aquifer is undoubtedly complex. However, a decreased bacterial partitioning to the solid phase may generally be expected in aquifers highly contaminated with organic compounds (Harvey and Barber, 1992).

B. Community Structure

A number of studies involving subsurface bacterial communities have attempted to differentiate between total numbers and so-called "viable" counts. The term "viable" has often been used to describe the numbers of bacteria that form colonies on solid nutrient media within the time frame of the incubation. This subpopulation is also referred to as "colony forming units" (cfu) or plate count bacteria. For most aquatic habitats, the number of cfu is a small fraction of the total population. In a number of recent subsurface investigations, reported plate counts were within experimental error of the total direct counts for at least one of several core samples (Sinclair et al., 1990; Sinclair and Ghiorse, 1989; Balkwill and Ghiorse, 1985; Bone and Balkwill, 1988). However, the ratio of plate count to total counts generally varied by several factors of ten among samples. In several other studies, plate counts generally represented a very small fraction (<0.1%) of the total bacterial number (Hirsch and Rades-Rohkohl, 1988; Ghiorse and Balkwill, 1983). In general, the largest plate counts for subsurface samples are observed using very dilute media and long incubation times. However, large numbers of "unculturable" subsurface bacteria hamper efforts to characterize the bacterial community using collections of cultured isolates.

Characterizations of subsurface microbial community structure in shallow aquifers, published before 1988, are presented by Ghiorse and Wilson (1988).

Much of what is now known about the structure of bacterial communities in the subsurface has resulted from detailed characterizations of isolates collected from deep aquifers. Recent characterizations of bacterial isolates obtained from deep (up to 265 m) subsurface sediments obtained from a coastal plain aquifer at the USDOE deep drilling site at the Savannah River Site, South Carolina suggest a surprisingly large morphologic and physiologic diversity of the culturable population (Balkwill, 1989; Balkwill et al., 1989; Fredrickson et al., 1991). In the Balkwill et al. (1989) study, there were 626 distinct physiologic types (unique response patterns to 21 tests) among 1112 isolates. However, the culturable bacterial populations obtained from different strata (geological formations) appeared to be physiologically distinct as did the culturable populations collected from the same geologic formation, but from different boreholes. The physiologic dissimilarities among the culturable bacterial populations from different geological formations at the Savannah River Site were further corroborated by the Fredrickson et al. (1991) and Jimenez (1990) studies. In the former study, distinct response patterns were registered using 108 different physiological measurements for the culturable populations from two vertically separated formations. A total of 198 morphologically distinct colony types were observed. In the latter study, the guanine plus cytosine $(G+C)$ content of the DNA varied between the culturable populations obtained from deep and from shallow formations. These studies suggest substantial vertical and longitudinal heterogeneity in the makeup of the bacterial population.

Bacterial populations in the subsurface appear to be well represented by genera commonly found in soils, such as *Pseudomonas* spp. Subsurface isolates from the Savannah River Site that had a $G+C$ content similar to those of Pseudomonas and Acinetobacter spp. were 60% and 12%, respectively, of the tested population (Jimenez, 1990). Identification to the species level of 1252 strains isolated from groundwater samples taken 5 to 150 m bls from 59 different bore holes in Denmark revealed that 56% of the cultured population were *Pseudomonas* spp., 11% were corneforms, 11% were *Aeromonas* spp., 7% were coliforms, 5% were Acinetobacter, and 5% were Cytophaga/Flavobacterium. However, unlike bacterial populations in soils, the vast majority of bacteria in aquifers appear to be Gram-negative (Bonde, 1987). Gram-negative species comprised about 95% of the cultured population in the Denmark study and over 80% in the Savannah River Site study (Balkwill, 1989). In a study of 2700 different isolates from a Pleistocene sandy aquifer near Bolcholt, Germany, 72% of the cultured population consisted of Gram-negative organisms and 52% consisted of Gram-negative rods (Kolbel-Boelke et al., 1988). Hirsch and Rades-Rohkohl (1988) reported that 86.8% of the cultured bacteria isolated from a shallow aquifer in Germany were Gram-negative, but that the ratio of Gram-negative to Gram-positive bacteria changed significantly within 24 hours of sample storage.

The nature of the differences in community structure between surface-attached (adherent) and unattached bacterial populations in aquifer sediments is unclear and likely depends, in part, upon the nutritional status of the groundwater. In a study involving 2700 strains collected from seven boreholes (up to 49 m deep) in a Pleistocene sandy and gravel aquifer in Germany, no higher morphological and/or physiological similarities existed among adherent or among unattached populations than were detected between adherent and unattached populations (Kolbel-Boelke, 1990). However, in a study comparing bacteria from deep (up to 550 m) subsurface sediments in South Carolina and adjacent groundwater, Hazen et al. (1990) report that in oligotrophic aquifers attached bacterial populations bear little resemblance to the unattached bacteria found in the same strata, which seem to have lower metabolic potentials. It was also observed that unattached and attached bacterial populations at the South Carolina site differed significantly in group physiological characteristics, but that the differences were not consistent from one aquifer to another (Krzanowski et al., 1990). However, differences between attached and unattached bacterial populations in shallow, organically contaminated aquifers, such as the one described by Harvey et al. (1984) and Harvey and Barber (1992), may be distinct. In at least one shallow hydrocarbon-contaminated aquifer, most of the degradation potential appeared to be associated with the unattached population (Aamand et al., 1989). It is clear that more research is needed on the role of unattached bacteria in the aquifer and the effect of organic contaminants upon that role.

Sensitive biochemical assays for characterizing subsurface microbial communities that do not require *a priori* intact recovery or culturing of the microbes from the recovered aquifer core material promise to yield important information about subsurface biota. Methods that involve signature biomarkers based on phospholipid ester-linked fatty acid (PLFA) pattern analysis have provided valuable insight on the total viable populations in several subsurface habitats. These techniques are particularly valuable where only a small fraction of the total microbial population in subsurface sediments is culturable. Using PLFA pattern analysis on recovered aquifer sediments from the USDOE groundwater study site in South Carolina, White et al. (1990) showed similarities in community structure among the more conductive layers. However, distinct dissimilarities were observed between communities from the more permeable zones and those in recovered drilling muds, clayey aquitard material, and surface soils. Permeability was found to be a more important determinant of PLFA patterns than were proportions of clay, sand, and silt. More insight on whole indigenous microbial populations in subsurface environments will also develop as a result of modifications of DNA probing/fingerprinting and ribosomal RNA sequencing techniques, which are presently being used to characterize bacterial isolates from subsurface samples (Jimenez, 1990; Reeves et al., 1990; Stim et al., 1990).

Results from a number of earlier microbiological investigations involving sandy aquifer sediments suggested an absence of eukaryotes from such aquifers (Harvey et al., 1984; Ghiorse and Balkwill, 1983; Wilson et al., 1983). However, among 90 morphotypically distinct microbes isolated from 10 m bls in a shallow aquifer in Northern Germany, 10 were represented by protozoa and 8 by fungi. More recent investigations with subsurface sediments, in which the

Singh dilution plate/most probable number technique was employed (Sinclair and Ghiorse, 1989; Beloin et al., 1988; Sinclair et al., 1990; Sinclair and Ghiorse, 1987), indicate that protozoa are widely distributed in aquifers, although generally in low numbers ($< 100/g$). Kinner et al. (1990, 1991), using direct counting and modified culturing procedures, found large numbers of protozoa (10^3 to $10^5/g$) in organically contaminated and uncontaminated zones of the Cape Cod aquifer. At this level, protozoa may be an important control of bacterial abundance as is the case in many surface water habitats. Fungi have been consistently isolated at depths up to 250 m bls at the Savannah River Site, but generally less than 100 cfu/g (Sinclair and Ghiorse, 1989). These reports suggest that eukaryotes may be important components of subsurface microbial communities and that an ecosystem approach may need to be taken in ascertaining the biological response to subsurface contamination.

III. Activities

A. Growth

Determination of *in situ* growth rates for aquifer bacteria can be problematic, due to difficulties in obtaining uncontaminated (uncompromised) samples, in the methods typically used to estimate bacterial productivity (e.g., frequency of dividing cells, tritiated thymidine uptake, uptake of labeled carbon sources, and closed bottle incubations). Also, some inherent assumptions in the techniques may not be applicable to groundwater (Harvey and George, 1987). Consequently, information on the growth rates of bacteria in groundwater is scarce and may be subject to some inaccuracy. For obligate pathogens and for many "displaced" bacteria from surfacial habitats, growth in the subsurface is thought to be quite slow or even negligible. This is because copiotrophic bacteria typically shift their metabolic capabilities from growth to starvation survival in response to a dearth of nutrients (Kurath and Morita, 1983). Not surprisingly, reported bacterial productivity in aquifer sediments collected in Wisconsin, Oklahoma, and Michigan (Thorn and Ventullo, 1988) were up to three orders of magnitude lower than those estimated for surface sediments (Fallon et al., 1983). Generation times calculated from Thorn and Ventullo's data ranged from 4.5 days (average of four samples) for the Oklahoma site to 100 days (average of two samples) for the Michigan site. Although currently available methodology for estimating *in situ* growth of aquifer bacterial populations are problematic and imprecise (Harvey and George, 1987), relatively long generation times (up to several months) would appear to be consistent with observations of slow mineralization of simple organic substrates, such as glucose and acetate, by aquifer bacteria. For example, turnover times of up to 52 days were observed for glucose that was amended to well-water samples collected from several locations in Ontario and Ohio (Ventullo and Larson, 1985).

The ability of bacteria in aquifers, which are among the most oligotrophic environments on earth, to grow on organic contaminants under *in situ* conditions has a number of important implications in the fate and transport of organic pollutants in the terrestrial subsurface. In many aquifers, indigenous bacterial populations may be ill-equipped to utilize substantial short-term increases in dissolved organic carbon and may even be inhibited to some degree. Stetzenbach et al. (1986) report that the growth rates of selected well-water isolates in filtered groundwater were significantly lower when amended with modest concentrations (0.1 and 1 mg C/L) of glucose, acetate, pyruvate, and succinate. However, bacterial populations in aquifers contaminated over long periods of time with organic wastes appear to be able to adapt. Investigations involving aquifer sediment microcosms suggest highly diverse and variable patterns of microbial community response to the introduction of various organic chemicals (Aelion et al., 1987). For some organic contaminants, adaptation may take many months or longer. Estimated growth rates (using frequency of dividing cells) for free-living bacterial populations sampled in contaminated zones of a sandy aquifer (Cape Cod, MA) that had been subjected to a 50-year influx of treated sewage were high (up to 0.04 h^{-1}) and comparable to at least some productive surface waters (Harvey and George, 1987). However, the adaptation time required for aquifer bacteria to growth on even the least recalcitrant of organic contaminants and the contributions of less-oligotrophic bacteria that are cotransported with the contaminants is not well understood.

Very little is known about the bacterial growth process in the aquifer. Models involving transport of bacteria or biodegradable contaminants typically employ Monod kinetics to describe bacterial growth dynamics and/or utilization of organic carbon (Corapcioglu and Haridas, 1984; Molz et al., 1986). One reason that Monod kinetics are invoked is that they generally work well for bacteria populations having low saturation constants for organic substrates, as should be the case in most subsurface habitats. However, the Monod relationship was developed using experimental data deriving from laboratory batch systems that involved well-defined substrates and pure cultures. Although Monod kinetics appear to describe organic substrate utilization and bacterial growth in at least one aquifer-derived laboratory microcosm (Godsy et al., 1990), its general applicability to the subsurface has not been demonstrated. Difficulties arise in the identification of growth substrates and the misapplication of Monod-type growth kinetics to non-growth-supporting organic compounds present at trace concentrations. Another possible complication in using the Monod approach to describe bacterial growth in the subsurface is that the kinetic "constants" in the Monod equation can be subject to change in response to temperature changes, the nature of the substrate, and other factors (Gaudy and Gaudy, 1980). However, fluctuations in physical and chemical factors are greatly moderated in the subsurface relative to surface environments, and Monod kinetics form the theoretical basis for much of the modeling of bacterial growth in the subsurface discussed below. Nevertheless, caution should be used in the application of Monod growth kinetics to subsurface bacterial populations, particularly those in contaminated groundwater where there are temporal and spatial changes in nutrient and physical conditions or in the bacterial populations themselves.

Much attention has been given to the application of bacterial growth models to the environments of surface soils. However, reported estimates of microbial growth rates for soils vary from days (e.g., Hissett and Gray, 1976) to nearly a year (Jenkinson and Ladd, 1981). Unfortunately, differences in assumptions regarding microbial abundance, degree of variation in substrate input, fraction of active (metabolizing) microbes, energy of maintenance, yield (production rate of new biomass per unit of substrate removed), and the importance of cryptic growth (growth on recycled organic matter locked up in dead microbial cells and released upon lysis) make such variations in growth rate difficult to interpret. There are a number of assumptions in growth models applied to subsurface microbial populations that may not be accurate. For example, maintenance energy estimates, often determined from indirect measurements using a number of questionable assumptions, can be over an order of magnitude too high (Smith, 1989). High energy of maintenance values that leave almost no energy for growth (e.g., Lynch and Panting, 1980) may be incorrect if there is significant recycling of organic carbon due to cryptic growth.

The concept of cryptic growth, introduced by Ryan (1955), has often been ignored in bacterial growth models, but may lead to improved estimates of subsurface bacterial growth. Although early bacterial growth studies suggest cryptic growth to be insignificant (Postgate and Hunter, 1962), later investigations involving more realistic and strictly controlled conditions indicate that cryptic growth can be quite important. Mason and Hamer (1987) observed cryptic growth yields ranging from 0.42 to 0.52 for *Klebsiella pneumoniae* grown in batch from its own soluble lysis products produced in a steady-state chemostat. In the latter study, it was demonstrated that the substantive deviation (depression) of theoretical (calculated) relative to observed maximum yields can be explained in terms of cryptic growth without invoking the use of mathematical constants which are undefined physiologically. Microbial-growth models that incorporate cryptic growth have been applied to soil environments (Chapman and Gray, 1986; Smith, 1989) and are likely to be germane to bacterial populations in aquifers, where the input rate of organic substrate is severely limiting.

Literature pertaining to modeling microbial activity in the general areas of soil science, environmental engineering, and groundwater microbiology has tended to focus on rates of degradation for organic compounds, particularly harmful contaminants, and less on the dynamics of growth and decay. However, many theoretical expressions for time-dependent growth are based on well developed theories for metabolic activities and the utilization of nutrients for biomass production and energy requirements. In particular, works by Herbert (1958) and Pirt (1975) dealing with carbon assimilation and oxidation by aerobic heterotrophic microorganisms form the basis for mathematical descriptions of the life cycle of individual and colonized bacteria. Although these theories are often viewed as competing in the scientific literature (Herbert proposed that the energy of maintenance is satisfied through endogenous respiration while Pirt assumed the need for an external nutrient source), mathematical expressions for net growth based on either theory are similar in form. For example, an equation for rate of net growth for a viable population of cells (y_v) in batch culture is proposed by Pirt (1975) as:

$$dy_v/dt = (\mu - b)y_v \tag{1}$$

where t is time, and μ and b are specific growth and death rates, respectively. Similarly, an equation for biomass (x) net growth based on Herbert's theory yields similar expression:

$$dx/dt = (\mu - a)x \tag{2}$$

where a is called the specific maintenance rate (p. 68 of Pirt, 1975). Because these equations are similar in form, it appears that identical results are produced from significantly different theories. It is the mathematical expression for the specific growth rate of an active population that varies greatly in the literature and is addressed herein.

Equations (1) and (2) are by no means the only general equations for microbial growth dynamics. A second class of growth model is identified in the literature for cases in which substrates at low concentrations do not support microbial growth (Schmidt et al., 1985). Four equations for growth are offered in which time-dependent population density is calculated based on empirical constants. Growth rates are independent of a primary (growth) substrate concentration and a non-growth supporting substrate (trace contaminant). These simplified growth models seem justifiable only on the basis of the ease of mathematical solution. However, recent research is evolving to more comprehensive models involving primary substrate availability (Monod-based equations) and secondary substrate utilization (non-Monod) which rely on numerical solution techniques for realistic predictive simulation. These models can require a considerable number of kinetic parameters which are best derived from laboratory data (e.g., microcosms). While realistic and accurate mathematical models are desired, the need for simple, numerically efficient solutions continues.

Theoretical expressions for growth rates reflect nutrient levels, electron acceptor availability, and many environmental conditions that affect microbial parameters for biomass growth and decay. Harris and Hansford (1976) discussed two forms of the Monod equation for the growth rate of aerobic heterotrophs; one in which the population is subject to single substrate limitation (standard Monod equation):

$$\mu = \mu_m \left[\frac{S}{K_s + S} \right] \tag{3}$$

and the other in which the population is subject to dual limitation (modified
Monod, e.g., by carbon substrate and terminal electron acceptor availability):

$$\mu = \mu_m \left[\frac{S}{K_s+S}\right]\left[\frac{O}{K_o+O}\right] \qquad (4)$$

where μ_m is the maximum specific growth rate [t^{-1}]; S and O are organic carbon
substrate and oxygen concentration in the biomass phase [M/L^3], respectively;
and K_S and K_o are substrate and oxygen half-saturation constants [M/L^3],
respectively. Equation (4) suggests the two reactants are simultaneously limiting
but predicts a zero rate of growth if either substrate or oxygen is absent. An
alternative expression to a dual reactant-limited growth rate (Roels, 1983; Baltzis
and Fredrickson, 1988) is given by:

$$m = \min\left\{\mu_{ms}\left[\frac{S}{K_s+S}\right]; \quad \mu_{mo}\left[\frac{O}{K_o+O}\right]\right\} \qquad (5)$$

where μ_{ms} and μ_{mo} are maximum specific growth rates applying to substrate and
oxygen, respectively.

   Growth rates predicted by Equations (3) through (5) can differ significantly.
Calculated growth rates are highest using Equation (3) and lowest for Equation
(4). Because oxygen is often cited as the limiting factor in subsurface biodegradation of organic compounds (Wilson et al., 1986), it follows that double Monod
kinetic expressions (Equations (4) and (5)) are superior to single substrate
models. It is emphasized, however, that while these kinetic models for growth
provide seemingly more realistic results than others, the above equations do not
adequately describe the rate of growth for all environmental conditions, such as
the case of inhibition of substrate oxidation.

   The Haldane equation defines a growth rate parameter which is, in part,
dependent on the inhibitory (toxic) nature of the substrate (Rozich et al., 1983;
Grady et al., 1989):

$$\mu = \mu_m\left[\frac{S}{K_s+S+S^2/K_I}\right] \qquad (6)$$

in which K_I is an inhibition coefficient [M/L^3]. For the case where there is
competitive inhibition from two substrates, an equation for the growth rate by
a single population is expressed as:

$$\mu = \mu_{m1} \left[\frac{S_1}{K_{s1} + S_1 + \alpha_2 S_2} \right] + \mu_{m2} \left[\frac{S_2}{K_{s2} + S_2 + \alpha_1 S_1} \right] \qquad (7)$$

where μ_{m1} and μ_{m2} are maximum specific growth rates, S_1 and S_2 are concentrations, α_1 and α_2 are empirical coefficients, and K_{s1} and K_{s2} are half-saturation constants, for substrates 1 and 2, respectively (Yoon et al., 1977).

The approach of modified Monod kinetics has been extended to include multiple respiratory processes in which oxygen acts as both the primary electron acceptor and inhibitor to denitrification. Kissel et al. (1984) proposed the transition of aerobic respiration to anaerobic, nitrate-based respiration to occur at a specific oxygen concentration (1% of K_O) for facultative heterotrophic bacteria utilizing oxygen, nitrate, and nitrite as electron acceptors. Grady et al. (1986) presented the overall growth rate of a heterotrophic population as consisting of an aerobic contribution to growth similar to Equation (2) and a rate for nitrate-based growth (μ_n):

$$\mu_n = \mu_{mn} \left[\frac{S}{K_s + S} \right] \left[\frac{N}{K_n + N} \right] \eta f_s \qquad (8)$$

in which η is a growth rate conversion factor [-]; N is a biomass phase nitrate concentration [M/L^3]; and K_N is the nitrate half-saturation constant. The transition from oxygen to nitrate as an electron acceptor is accomplished mathematically with a switching function, f_s (Henze et al., 1987):

$$f_s = \left[\frac{K_o}{K_o + O} \right] \qquad (9)$$

which mimics oxygen inhibition of nitrate as a terminal electron acceptor. Widdowson et al. (1988a,b) proposed an expression for the overall rate of growth in which the aerobic and anaerobic growth rates are additive. A noncompetitive, reversible enzyme inhibition function (I) influences the rate of growth for anaerobic, nitrate-based respiration:

$$\mu_n = \mu_{mn} \left[\frac{S}{K_s + S} \right] \left[\frac{N}{K_n + N} \right] I(O) \tag{10}$$

in which μ_{mn} is the maximum specific growth rate under nitrate-based respiration and the function $I(O)$ is given as:

$$I(O) = \left[1.0 + \frac{O}{K_c} \right]^{-1} \tag{11}$$

where K_c is a coefficient of inhibition $[M/L^3]$. Bouwer and Cobb (1987) proposed the use of a competitive inhibition model to simulate the transition from aerobic to nitrate-based respiration in a biofilm. Advantages of inhibition models are rapid convergence when numerical methods are employed and the flexibility to "control" the influence of oxygen inhibition, i.e., the range of oxygen concentration values through which the transition to denitrification occurs.

The effect of mineral nutrient limitation on microbial activity is addressed with modified Monod kinetics in Benefield and Molz (1984) in which aerobic heterotrophic growth is limited by nitrogen and phosphorous as growth nutrients:

$$\mu = \mu_m \left[\frac{S}{K_s + S} \right] \left[\frac{O}{K_o + O} \right] \left[\frac{A}{K_a + A} \right] \left[\frac{P}{K_p + P} \right] \tag{12}$$

in which A and P are biomass phase ammonia nitrogen and phosphorous concentration and K_A and K_P ammonia nitrogen and phosphorous half-saturation constants. Assuming no growth in the absence of any one of these four constituents may not be entirely valid in the subsurface (e.g., natural bioresto-ration) and an appropriate equation for nutrient limited growth is not apparent. Other researchers (e.g., Panikov, 1979) take the approach of Roels (1983) in which the minimum Monod function is employed (e.g., Equation (5)) when growth is limited by at least two nonsubstitutable nutrients.

Autotrophic biomass growth and decay rate equations in the literature are patterned after equations reflecting Herbert's theory for heterotrophic microor-ganisms. For example, the growth rate for nitrifiers (μ_a) in Henze et al. (1987) is expressed using Equation (4) in which ammonia serves as the substrate (S) and maximum specific growth rate and half-saturation constants for autotrophic biomass replace (μ_m and K_S). A model for the growth rate of ammonia and nitrite oxidizers involving pH-based inhibition functions and temperature-dependent maximum growth rate constants is given in Leggett and Iskandar

(1981). Gee et al. (1990) applied the Haldane Equation (6) to simulate ammonia and nitrite oxidation with inhibition as a two-step process.

Factors controlling transient behavior of specific growth rates and biomass decay rates are physiological adaptations to changes in nutrient and substrate availability. Baltzis and Fredrickson (1988) present data indicating microbial growth parameters change significantly during utilization of a given substrate adding to the further complexity of models. These changes are not instantaneous and impact both microbial growth and decay. Often the responsible mechanisms are not well understood. Daigger and Grady (1983) found that transfer function models were not adequate to predict the transient growth response of *Pseudomonas putida*. Baek et al. (1989) proposed decay rate constants dependent on nutrient availability in unsaturated soils to account for a large initial death rate followed by a slow decay during periods of high stress conditions. This approach has not been employed in groundwater microbiology models in which microbial decay rates for oligotrophic aquifer environments are often orders of magnitude less than typical literature values.

Equation development for microbial growth in soil matrices also reflects a variety of conceptual models relating to microscale habitation and structure in the subsurface (e.g., microcolonies, biofilms, etc.). Recently, van Loosdrecht et al. (1990) identified four stages of microbial colonization as (i) free-living cells, subject to advective-dispersive transport in pore fluid, (ii) reversibly adhered cells, (iii) attached cells, and (iv) colonized cells either as microcolonies or biofilms. Few microbial growth models are developed to simulate the complete process of structure development including both attached and free-living cells. Corapcioglu and Haridas (1984) applied theory developed for particle transport toward the development of a mathematical model for microbial transport coupled to single substrate-limited growth. Models that address the mechanisms controlling attachment/detachment of suspended microorganisms with microbial growth dynamics suffer from a reliance on many empirically derived parameters and can prove difficult to apply to real problems. Typically, growth models in the literature are developed under a preassumed community structure and new or modified approaches are needed.

Biofilm models vary in complexity from steady-state cell density models (McCarty et al., 1984; Bouwer and Cobb, 1987) to multispecies models (Wanner and Gujer, 1986) and transient cell density models (Benefield and Molz, 1985). Steady-state approach to biofilm modeling assumes equal rates of biomass growth and decay yielding a constant biofilm thickness with time (McCarty et al., 1984). Biofilm thickness (L_f) is assumed to be a function of substrate flux (J) into the biofilm across a diffusion boundary layer (Rittman and McCarty, 1980; Bouwer and Cobb, 1987):

$$L_f = J \, Y \, / \, b \, X_f \tag{13}$$

where Y is a biomass yield coefficient [-], and X_f is the biomass cell density
$[M/L^3]$. Data on substrate utilization from laboratory porous media column
experiments suggest the steady-state assumption is valid after a relatively short
period of time (Taylor and Jaffé, 1990). The transient growth case is developed
in Benefield and Molz (1985) for a fixed-film system. Cell density and biofilm
thickness are time-dependent variables. Substrate and oxygen concentrations are
variable with depth within the biofilm. A model of similar characteristics is
presented in Wanner and Gujer (1986) for the multiple microbial species case.
Models of this nature have application to the simulation of permeability and
porosity reduction because of biofouling. Recently, Taylor et al. (1990)
proposed a biofilm model using spherical geometry to account for changes in
physical properties of a porous medium. However, the circumstances under
which the biofilm approach is physically meaningful in a low-nutrient,
subsurface environment remains a source of debate (Baveye and Valocchi, 1989;
Widdowson, 1991), particularly when the assumptions of steady-state and
continuum are employed. Unlike biofilm models, the microcolony concept
assumes discrete, non-continuous growth communities. The microcolony concept
is developed in Molz et al. (1986) in which the viable biomass is conceptualized
as average bacterial colonies attached to aquifer sediments with an assumed
geometry and dimension. Updated versions of the model have no preassumed
colony geometry and a parameter for the diffusional surface area per unit
biomass (Widdowson et al., 1988a) eliminates many input requirements.

Macroscale variables assigned to account for microbial population size depend
to some extent on the conceptual model of microbial distribution at the
microscale. Microbial growth models following theories of Herbert or Pirt
typically express population size in terms of a biomass concentration as opposed
to a cell count per unit volume. Biomass concentration may be expressed in units
of biological solids per unit volume of fluid (Rozich et al., 1983), biomass per
unit mass of soil (Baek et al., 1989), and biomass per unit bulk volume of
porous media (Widdowson et al., 1988a,b). Models developed with macroscale
variables for biomass concentration typically ignore time-dependent changes in
geometries at the microscale. Colony growth models that simulate changes in
size and density have been proposed (Watson and Gardner, 1986) but are
difficult to verify.

B. Degradation of Organic Carbon

Most reported rate estimates for microbially mediated mineralization of organic
compounds in groundwater derive from controlled microcosm assays involving
radiolabeled substances. Rate estimates involving aquifer sediments or
groundwater recovered aseptically from the subsurface have been referred to as
in situ measurements. However, since these measurements can involve
substantive changes in conditions, the microbial community, and even the pool
size of available organic carbon, they are not in a true sense *in situ* measure-

ments, but rather conditional estimates of metabolic potential. Mineralization or biotransformation assays involving microcosms of aquifer material and indigenous microbial populations can provide valuable information on the relative biodegradability of various organic compounds under specified conditions. However, these conditional estimates likely overestimate the rates at which organic compounds will become mineralized or biodegraded in the aquifer. It was recently demonstrated that the rate of carbon dioxide production from ^{14}C acetate- and glucose-amended aquifer sediments derived from deep anaerobic aquifers in the Atlantic Coastal Plain of South Carolina were several orders of magnitude higher than would be predicted from geochemical modeling of changes observed in groundwater chemistry along the aquifer flow path (Chapelle and Lovley, 1990a). However, geochemical modeling accounts only for the net effects of microbial metabolism and does not account for recycling of carbon by cryptic growth, autotrophic incorporation of carbon dioxide deriving from heterotrophic respiration, or bacteria-protozoa interactions. It is likely that more accurate estimates of organic carbon mineralization in the aquifer will necessitate field experiments in which labeled organic compounds at low concentrations (that do not substantively affect the pool size of that compound) are coinjected into the aquifer along with a conservative tracer and degradation monitored as the compounds are transported downgradient.

Although little data is available on the degradation rates of naturally occurring organic carbon *in situ*, several groundwater injection experiments have been conducted which focus on the potential of indigenous bacteria to transform selected xenobiotic compounds in shallow aquifers. In the absence of bioremediation, the residence time for some xenobiotic organic contaminants in the aquifer can be as long as decades or centuries (Wilson et al., 1986). Other contaminant organic compounds appear to be degraded in the aquifer at measurable rates. Disappearance of bromoform, dichlorobenzene, and hexachloroethane from a sandy aquifer during a two-year natural gradient injection test involving a shallow, sandy, freshwater aquifer could be approximated by first-order rate constants of 0.003, 0.004, and 0.02 day^{-1}, respectively (Roberts et al., 1986). However, the fraction of the loss in organic mass due to biodegradation could not be accurately determined. Often, there are significant lag times before biotransformations take place (Wilson, 1985). In an earlier injection experiment, a lag time of 50 days was required before substantive degradation of chloroform and related compounds occurred in an anoxic zone of a coastal aquifer (Roberts et al., 1982). Due to environmental concern over the persistence of organic contaminants in groundwater, there is much interest in chemically stimulating specific bacterial populations in organically contaminated aquifers to enhance biodegradation. The biotransformation rates of chloroethenes in a shallow aquifer in Mountain View, California were apparently enhanced several fold by alternate, pulse injections of oxygen and methane under forced gradient conditions (Semprini et al., 1990). The rationale for adding oxygen and methane to the aquifer was to stimulate methanogenic bacteria, some of which are known to degrade chloroethenes. Similarly, biodegradation of benzene, toluene, ethylbenzene, and xylene (BTEX) was significantly enhanced within portions of an anaerobic plume of jet fuel in a sandy aquifer in Traverse City, Michigan (Hutchins et al., 1991) by addition of nitrate. In the latter experiment, addition of nitrate stimulated the activities of denitrifying bacteria, which were much better equipped metabolically to effect breakdown of the BTEX.

Numerous factors affect the rate of mineralization/degradation of organic carbon in the aquifer, including the type and quantity of organic carbon and terminal electron acceptor, exposure history, geohydrology, and physiochemical conditions. The fate of dissolved organic compounds may also depend upon the nature of and interactions among the various indigenous populations. This is depicted schematically in Figure 1. Where there is excess dissolved organic carbon (DOC), reduction of terminal electron acceptors in the groundwater would be expected to occur in the following sequence based on thermodynamic principals (Stumm, 1967): O_2, nitrate, Mn(IV), Fe(III), sulfate, bicarbonate, and N_2 (Champ et al., 1979), depending upon relative acceptor availability. For aquifers contaminated with organic compounds, the different electron accepting processes can be separated both spatially and temporally along the direction of flow (Baedecker et al., 1989). The progression of more reducing conditions along the direction of flow may be observed in deep, pristine aquifers (Lovley, 1991) and results from the preferential utilization of terminal electron acceptors according to the above sequence. However, this progression of microbially mediated processes along the path of groundwater flow can be reversed for some well-developed contaminant plumes in shallow aquifers, particularly those involving a point source of organic compounds in high concentrations. In such systems, reducing conditions develop near the source of contamination, due to the biodegradation of organic matter, and the oxidation potential increase with distance downgradient. An increase in oxidation potential with distance downgradient coincident with a progressive zonation of CH_4, NH_4^+, Fe^{2+}, Mn^{2+}, HCO_3^-, and NO_3^- along the flow direction was observed for at least one contaminant plume originating from a Delaware landfill (Baedecker and Back, 1979). For some contaminated and most pristine aquifers, the number and activity of microbially mediated electron-accepting processes is limited by a dearth of organic substrate. For the 4-km long plume of sewage-contaminated groundwater at the USGS study site in Cape Cod, dentrification activity is clearly controlled by availability of organic carbon (Smith and Duff, 1988).

A number of factors have hampered the study of specific electron-accepting processes in the aquifer, including the slow rate at which organic carbon is often degraded in groundwater and representative sampling. In organically contaminated aquifers, a given electron-accepting process can be restricted to very narrow zones vertically, since there is little vertical mixing of groundwater. Levels of dissolved oxygen and nitrate in water table aquifers have been observed to change significantly over distances of less than a meter (Smith et al., 1991a; Smith et al., 1991c; Ronan et al., 1987). The existence of very steep vertical gradients in chemical constituents, bacterial populations, bacterial activity, and electron acceptors in contaminant plumes can necessitate the use of networks of

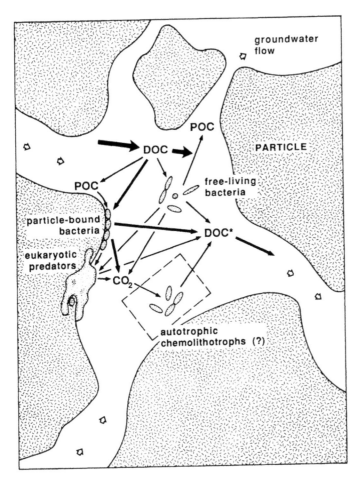

Figure 1. Schematic representation of the partitioning and interactions of dissolved organic compounds with microbial populations in an aerobic aquifer. POC is particulate organic carbon. DOC* is dissolved organic carbon that has been microbially altered.

closely spaced vertical sampling of groundwater (Smith et al., 1991a), which can be problematic and costly to install.

The slow rates of organic carbon degradation have led, in part, to the use of stable isotope fractionation, in which inference about the microbially mediated electron-accepting process is made based upon the stable isotope fractionation of the reduced product(s) of the electron acceptor of interest. Such studies have been particularly useful in aquifers where the reduced product of a given electron acceptor may also result from abiotic (geochemical) processes and where mineralization of organic carbon may take place over geological time.

McMahon et al. (1990) provide evidence from ^{13}C determinations that link bacterial CO_2 production and dissolved inorganic carbon pools in anoxic coastal plain aquifers of South Carolina. Furthermore, it was demonstrated that much of the CO_2 dissolved in the groundwater was derived from microbially mediated breakdown of organic carbon. Assessment of stable isotope fraction has also been successfully used in determining microbial sources of methane in oil and lignite-containing aquifers (Belyaev and Ivanov, 1983; Grossman et al., 1989) and in the *in situ* measurement of methane oxidation potential (Smith et al, 1991b).

A number of recent studies have utilized nitrogen isotope fraction, which has long been used to investigate microbially-mediated changes in inorganic nitrogen in surface environments, to study denitrification in contaminated and uncontaminated groundwater (e.g., Bengtsson and Annadotter, 1989; Wilson et al., 1990; Bottcher et al., 1990; Mariotti et al., 1988). Stable N isotope fraction measurements can be combined with other geochemical assays to yield useful insight about the significance of denitrification with respect to degradation of organic carbon within the aquifer. Mariotti (1986) lists three geochemical methods for identification of denitrification in groundwater: (1) identification of a redox gradient which generates the appropriate sequence of oxidation-reduction reactions; (2) determination of dinitrogen gas in oversaturation; and (3) the use of ^{15}N natural isotope abundance tracing. In the Smith et al. (1991b) study, a combination approach using field mass balance, stable istope analysis, and laboratory incubations led to the determinations that denitrification was much more significant than dissimilatory reduction of nitrate to ammonium and was the major terminal electron-accepting process in the upgradient portion of the 4 km-long plume of sewage-contaminated groundwater in Cape Cod, MA.

There is growing evidence to suggest that core samples recovered from a variety of aquifers harbor substantial numbers of bacteria in all major metabolic categories, even though only one electron-accepting processes may be occurring at a given site. Jones (1989) reported that viable methanogens and sulfate reducing bacteria were evident all along a 300 m vertical transect through aquifer sediments in South Carolina, even though the sediments were not predominantly anaerobic. Ward (1985) demonstrated rapid mineralization of NTA in the same shallow aquifer sediments under either aerobic or denitrifying conditions, and investigations at the DOE groundwater site in South Carolina suggest that metabolically active aerobes and anaerobes coexist in the same deep subsurface environments (Madsen and Bollag, 1989). Although these results are not surprising, the microbiological potential for effecting subsurface degradation of a number of organic contaminants under a variety of redox conditions is clearly indicated.

Although the distribution of predominant electron-accepting processes in aquifers may be ascertained from redox conditions, *in situ* Eh measurements in groundwater are problematic and subject to considerable error. Hydrogen concentrations may be useful as an indicator of specific bacterially mediated processes to document the predominant, electron-accepting process in groundwater systems. Chapelle and Lovley (1990b) report that hydrogen concentrations in iron-reducing, sulfate-reducing, and methanogenic zones in coastal plain aquifers in South Carolina were distinctly different and were within ranges observed for corresponding zones in surface-water sediments (Lovley and Goodwin, 1988).

Several recent studies have involved heterotrophic competition among major terminal electron-accepting processes in the aquifer as a function of terminal electron acceptor availability. Chapelle and Lovley (in press) report on the competitive exclusion of sulfate reduction by iron-reducing bacteria in high iron groundwater in South Carolina. Beeman and Suflita (1987) observed that the location of sulfate-reducing and methanogenic zones in anoxic, contaminated groundwater underneath a landfill in Oklahoma appeared to be determined by availability of sulfate; methanogenesis did not occur where there were high levels of sulfate. Within the oxygen-depleted plume of nitrate- and organics-contaminated groundwater at the USGS site on Cape Cod, sulfate reduction does not appear to be a significant process, even though there is abundant sulfate (R.L. Smith, personal communication). Clearly, more research is needed regarding the *in situ* biochemical potential of aquifer microbial populations under a variety of chemical conditions.

In spite of numerous knowledge gaps concerning subsurface microbial metabolism, mathematical description of organic carbon biodegradation kinetics has been developed to describe various microbial influences occuring under an array of environmental conditions. However, application of degradation models to subsurface environments often necessitates untested assumptions, particularly when organic compounds are present at trace concentrations. Important considerations in the selection of degradation kinetics are (i) biodegradability of the compound as a growth-limiting substrate, (ii) multiple growth substrates, (ii) presence of inhibitory and catalytic factors, (iv) cometabolism potential, and (v) electron acceptor and mineral nutrient availability. Difficulties arise because of a lack of data for model verification and a lack of knowledge concerning the basic mechanisms of organic carbon removal. It is not surprising that the variety of mathematical models developed (but rarely verified) in the literature led researchers to conclude that no one equation will adequately describe organic carbon biodegradation (Widdowson et al., 1988a,b; Grady, 1990).

Kinetics for growth substrate biodegradation have been addressed in the above section on microbial growth models. Mathematical expressions for microbial growth dynamics are directly applicable to cases in which the organic carbon compound of interest also serves as a carbon/electron donor source. Rates of organic carbon utilization (r_s) are generally written as functions of the specific growth rate (μ) and yield coefficient (Y) for biomass production (Grady, 1990):

$$r_s = \frac{\mu}{Y} \qquad (14)$$

in which the utilization rate is expressed in units of mass per biomass per unit time. The overall rate of removal from the aqueous phase thus depends on the biomass per unit bulk volume of porous media actively degrading and utilizing the compound as a growth substrate. Expressions incorporating terminal electron acceptor availability are also discussed in the previous section. Generally, equations for the rate of electron acceptor utilization follow Equation (14) in form. For the case of inhibitory substrates (e.g., chlorinated benzenes), a maximum concentration above which biodegradation ceases is given in Gantzer (1989), based on the Haldane equation (see the "Growth" section). Grady (1990) summarized appropriate kinetic models (either Monod or Haldine) for biodegradation of 20 xenobiotic compounds (mainly chlorinated phenols) based on single-substrate experiments. Equations describing the effect of toxicants on biodegradation are discussed for methanogenic (Parkin and Speece, 1982) and for aerobic (Benefield and Reed, 1985) systems.

Modeling the simultaneous biodegradation of several growth-inducing organic substrates presents a difficult challenge. A common approach involves measures of degradable organic carbon, such as chemical oxygen demand (COD). This approach fails to address inhibitory effects and the preferential use of substrates where multiple terminal electron acceptors are involved (e.g., one substrate biodegrades under aerobic conditions but does not serve as a growth substrate for anaerobes or facultative microbes). Widdowson et al. (1988b) proposed a model for multiple substrate biodegradation in which the specific growth rate reflects the sum total of each of several contributing energy and carbon donors. This and other approaches require knowledge of the preferential biodegradation of compounds and careful analysis of batch data. Multiple substrates can also result in the inhibition of microbial breakdown of another compound. Grady (1990) discussed evidence suggesting the model proposed in Han and Levenspiel (1988) is appropriate to predict biodegradation rates under similar conditions of inhibition.

Of particular interest are models for cometabolism or secondary utilization in which compounds are removed simultaneously during the utilization of a primary growth substrate. Scow et al. (1986) found that Monod-type functions may not be adequate for describing mineralization of organics at low concentrations. Biodegradation kinetics for compounds not supporting growth are found in a variety of forms (Alexander, 1985; Schmidt et al., 1985). However, these models do not simulate the removal of the growth substrate and may not be reliable as tools for prediction. A model for the simulation of both primary and secondary utilization was proposed by Widdowson et al. (1988b) and applied to the biodegradation of aromatic hydrocarbon compounds in saturated porous media.

Models for organic carbon biodegradation are often selected on a best-fit basis to laboratory batch data. Godsy et al. (1990) proposed single Monod kinetics for simulating creosote-derived phenolic compound biodegradation by methanogens. The initial lag period required for enzyme adaptation was simulated with a relatively low initial population value. Srinivasan and Mercer (1988) employed similar kinetics to simulate TCE degradation in batch experiments, but neglected endproduct generation. Models applied to experimental data with multiple electron donors and acceptors are recent additions to the literature. Strand et al. (1989) employed dual kinetics to model the degradation of chlorinated solvents coupled to methanotropic respiration. In a field study of the biostimulation of a methanotropic population, Semprini et al. (1991) employed a form of the Haldane equation coupled to electron acceptor availability to simultaneously model inhibition effects and secondary transformation of chlorinated solvents.

Models for coupling microbial activities to transport of organic compounds in porous media have been proposed for one, two, and three spatial dimensions for a variety of applications. Brief discussions of the literature are found in reviews by Lee et al. (1988) and Abriola (1987). Borden and Bedient (1986) modeled two-dimensional organic contaminant transport by assuming instantaneous degradation reaction between the contaminant, a rate-limiting electron acceptor (oxygen), and a microbial population that remains constant in size over time. Rifai and Bedient (1990) present simulations showing that instantaneous reaction assumptions are appropriate when groundwater velocities are relatively slow. Celia et al. (1989) presented a one-dimensional model for biodegradation limited by two substrates, and Widdowson et al. (1988a,b) presented transport models for facultative-based biodegradation of single and multiple organic compounds, respectively. Multi-dimensional models in two (Widdowson et al., 1987) and three (MacQuarrie and Sudicky, 1989) dimensions have important application to transport in stratified and heterogeneous aquifers. Molz and Widdowson (1988) and MacQuarrie and Sudicky (1989) demonstrated the importance of aquifer heterogeneities in enhancing mixing of hydrocarbons and electron acceptors and the relative roles of longitudinal and transverse dispersion on this process. Recently proposed models by Istok and Woods (1990) and Hossain and Corapcioglu (1990) account for coupled biodegradation of chlorinated phenols, BTX, and halogenated solvents, respectively. Semprini and McCarty (1991) present a model for the coupled transport of methane and dissolved oxygen in biostimulation applications.

Often ignored in coupled process modeling is the role of interphase transport on biodegradation. Utilization rates can be coupled to an interphase mass transport function in the macroscopic solute transport equation. Expressions of mass transport are developed to relate the pore fluid concentrations (Darcy-scale) to microbial phase (micro-scale) concentrations (Molz et al., 1986; Widdowson et al., 1988a). Other models are developed in which biodegradation rates are based on pore fluid concentrations (Borden and Bedient, 1986; Celia et al., 1989; MacQuarrie and Sudicky, 1989). Analysis in Baveye and Valocchi (1989) addressed this question, but no comparison of models was offered. These considerations are equally important in the unsaturated regions of the vadose zone.

IV. Movement

There is an increasing interest in delineating transport behavior of microorganisms through groundwater, in part, because of continued widespread contamination of shallow, drinking-water aquifers by microbial pathogens and chemical wastes. Several recent reviews discuss bacterial and viral transport through subsurface environments (e.g., Yates and Yates, 1988; Harvey, 1991; Bitton and Harvey, 1992; Gerba and Bitton, 1984). A major public health concern involving domestic use of untreated groundwater is the migration of disease-causing bacteria and viruses from contamination sources upgradient, particularly from domestic and municipal waste disposal (septic tanks, waste lagoons, landfills, and on-land application of domestic effluents). A number of reports indicate that bacteria can move through a variety of subsurface materials over substantial distances. Although it has been known for some time that bacteria can readily move through aquifers dominated by preferred-flow path structure (e.g., limestone (Kingston, 1943) and fractured bedrock (Allen and Morrison, 1973)), pathogenic and enteric "indicator" bacteria have been reported to move over long distances (400 to 1000 m, horizontally) through aquifers consisting of pebbles, gravels, sand and gravel, stony silt loam, and even fine sand (Dappert, 1932; Merrell, 1967; Sinton, 1980; Anan'ev and Demin, 1971; Kudryavtseva, 1972; Martin and Noonon, 1977). Because the source term(s) in these studies is generally ill-defined, such results are difficult to interpret.

A number of controlled *in situ* transport experiments involving a variety of microorganisms have been conducted in several geohydrologically different aquifers. Typically, these experiments involve a coinjection of a conservative tracer that accounts for advection and dispersion. By comparison of breakthrough curves (concentration histories) of the test organism to that of the conservative tracer, the loss/immobilization rate and the relative retardation and dispersion of the microbe can be determined under *in situ* conditions. Table 1 lists several *in situ* transport experiments in which microbes are coinjected with bromide into an aquifer and their relative transport behavior monitored as they move downgradient by advection. These experiments have typically been conducted on a limited scale (less than 50 m horizontal travel distance), since the rates of loss, immobilization, and dispersion limit the distance that a labeled microbial population can be followed. Forced-gradient tests have most often been employed where it was advantageous to control the flow field, particularly where the hydrology was not well understood. Forced-gradient tests have been most useful in the cases of fractured-rock aquifers, where flow occurs largely along preferred-flow-paths (Champ and Schroeter, 1988; Harvey et al., unpublished data). Natural-gradient experiments, which have largely been restricted to aquifers consisting of well-sorted sands (Harvey et al., 1989; Harvey and Garabedian, 1991; Havemeister et al., 1985; Bales et al., 1989), have the advantage of being easier to model (Harvey and Garabedian, 1991). This is because the flow field is more uniform and the degree of forced-dispersion is minimal.

Results of the small-scale field tracer tests in which bacterial transport has been compared to that of a dissolved, conservative tracer suggest that the dissimilarity between the respective breakthrough curves increases with increasing degrees of secondary (preferred-flow path) structure in the aquifer. The peak in bacterial abundance significantly preceded that of the conservative tracer in fractured crystalline rock (Champ and Schroeter, 1988) and in fractured, layered basalt (Harvey et al., unpublished), but can be coincident with that of a conservative tracer in well-sorted sandy aquifer sediments (Harvey and Garabedian, 1991). Differences between bromide and bacterial transport behavior in the latter system appeared to be due largely to sorptive interactions with grain surfaces, whereas differences observed in investigations involving transport in fractured rock appear to be due to a complicated pore structure. These experiments give more quantitative information about microbial transport behavior in the saturated zone, but a number of controlling factors are still poorly understood.

The possibility of using genetically engineered microorganisms for aquifer-restoration is becoming more attractive, due to recent advances in biotechnology, particularly in the area of molecular genetics. Also, the breakdown of many mobile and persistent organic pollutants in groundwater may be facilitated by cotransport of acclimating indigenous bacteria (Harvey and Barber, 1992). In many cases, the feasibility of aquifer bioreclamation schemes involving in situ treatment will depend upon the ability of the engineered or waste-adapted bacteria to reach the contaminant-affected area in the aquifer. However, predictive modeling of bacterial migration through contaminated aquifers is difficult. Also, there is a scarcity of field data from controlled, in situ experiments. Currently available models describing the fate and transport of microorganisms in the environment, with particular emphasis on the subsurface, are delineated and discussed in a recent book, edited by Hurst et al. (1991), and will not be discussed here.

Clearly, improvements in mathematical models that involve transport of bacteria in groundwater will require a better understanding of the various mechanisms and factors involved in bacterial migration through porous media, including geological and hydrological characteristics of the aquifer and biological factors including bacterial growth, predation by protozoa, parasitism by bacteria-specific viruses (bacteriophages) and predatory bacteria (Bdellovibrio), survival under nutrient-limiting conditions, changes in cell size and hydrophobicity in response to alterations in nutrient conditions, spore formation in the case of some Gram-positive species, reversible and irreversible attachment to solid surfaces, detachment from surfaces, and straining. A number of factors controlling transport of bacteria in groundwater are interrelated through other

R.W. Harvey and M.A. Widdowson

Table 1. Differences in apparent transport velocities of microspheres, microbes, and bromide in small-scale groundwater tracer tests

Reference	Microbe	Aquifer	Type of Test	Distance (meters)	[1]Retardation Factor
Champ and Schroeter, 1988	Escherichia coli	Fractured crystalline rock	Forced-Gradient	12.7	0.1
Harvey, Voss, and Souza (unpublished)	Indigenous isolate (Bacillus sp.)	Layered basalt	Forced-Gradient	27	0.6
Wood and Ehrlich, 1978	Saccharomyces cerevisia (yeast)	Sand/Gravel (with clay and carbonate)	Forced-Gradient	1.5	0.7
Harvey, 1988	Indigenous bacterial population	Well-sorted sand/gravel	Forced-Gradient	1.7 3.2	1.0 [2]0.6
Harvey and Garabedian, 1991	Indigenous bacterial population	Well-sorted sand/gravel	Natural-gradient	6.7	1.0
	Microspheres	Well-sorted sand/gravel	Natural-gradient	6.7	
	Plain latex				1.0
	Polyacrolein				1.3
	Carboxylated				1.4

[1]Calculated as the ratio of transport velocity (at peak abundances) of bromide to that of the microorganisms.
[2]Reasons for the retardation factors substantially less than one for the sampler at 3.2 m is not clear.
(Modified from Harvey, 1991.)

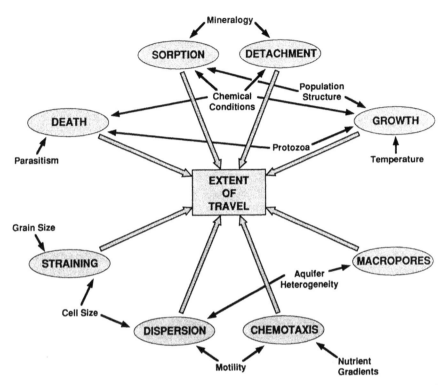

Figure 2. Schematic representation of the interrelationships of parameters and factors involved in the transport of bacteria through sandy aquifers. (From Harvey, 1991.)

processes. A schematic representation of the nature of these interrelationships is depicted in Figure 2.

Information about abiotic processes affecting microbial transport through aquifers has been gained through the use of microbe-sized microspheres in field transport experiments. In the Cape Cod experiments (Harvey et al., 1989), there was no clear relationship between the degree of immobilization and retardation for microspheres of similar size, but of different composition, even though increased contact with aquifer sediments occurring as a result of increased retardation should allow for greater opportunities to adsorb. Also, for a given type of microsphere, immobilization in the bacteria size range was inversely related to size, in accordance with colloid filtration theory. These experiments give more quantitative information about microbial transport behavior in the saturated zone, but a number of controlling factors are still poorly understood. In particular, the effects of nutrient and geohydrologic conditions upon bacterial transport and the manner in which this transport affects the fate of organic contaminants in contaminated aquifers have not been well studied.

Laboratory transport experiments involving columns packed with subsurface material have allowed much greater control than is possible under field conditions. Although microbial transport behavior in aquifer sediments that have been repacked into columns can differ substantially from that observed during *in situ* tracer tests (Harvey, 1988), column experiments have allowed a clearer transport delineation of some of the controls of microbial transport. These experiments include the effects upon microbial transport of size-dependent exclusion from small porosity (Bales et al., 1989), sorption, taxis and growth (Jenneman et al., 1985; Bosma et al., 1988; Reynolds et al., 1989), organic matter, ionic strength (Gannon et al., 1991b), preferred flow paths and cell surface properties (Gannon et al., 1991a).

V. Conclusions

In summary, there have been many recent advances in our understanding of the structure, activities, and movement of microorganisms that inhabit the terrestrial subsurface. Much of the information has only been made available within the last ten years. However, there is much that is still unknown about these organisms, their *in situ* activity, their ecology, their metabolic potential, and their ability to migrate through porous media. Of particular interest is the role of this community in the fate of subsurface contaminants. Although much work has already been done on the biodegradability of a variety of organic contaminants under a number of chemical conditions, much is unknown about *in situ* biotransformations of these compounds in aquifers. The indirect role of protozoa in the fate of subsurface contaminants has yet to be examined. The application of analytical biochemical assays to whole microbial populations in aseptically collected subsurface samples promises to yield important information about groundwater microbial communities. *In situ* tracer experiments have resulted in important new information about microbial transport behavior in aquifers, but many of the biological and environmental controls are still poorly understood. Finally, significant progress has been achieved in modeling the growth and movement of subsurface microorganisms and modeling degradation of specific organic contaminants. Substantial improvements in these models will likely result from a better understanding of the processes involved.

References

Aamand, J., C. Jorgensen, E. Arvin, and B.K. Jensen. 1989. Microbial adaptation to degradation of hydrocarbons in polluted and unpolluted groundwater. *J. Contam. Hydrol.* 4:299-312.

Abriola, L.M. 1987. Modeling contaminant transport in the subsurface: an interdisciplinary challenge. *Reviews of Geophysics.* 25:141-147.

Aelion, C.M., C.M. Swindoll, and F.K. Pfaender. 1987. Adaptation to and biodegradation of xenobiotic compounds by microbial communities from a pristine aquifer. *Appl. Environ. Microbiol.* 53:2212-2217.

Aelion, C.M., D.C. Dobbins, and F.K. Pfaender. 1989. Adaptation of aquifer microbial communities to the biodegradation of xenobiotic compounds: Influence of substrate concentration and preexposure. *Environ. Toxicol. Chem.* 8:75-86.

Alexander, M. 1985. Biodegradation of organic chemicals. *Environ. Sci. Technol.* 19:106-111.

Allen, M.J. and S.M. Morrison. 1973. Bacterial movement through fractured bedrock. *Ground Water* 11:6-10.

Anan'ev, N.I. and N.D. Demin. 1971. On the spread of pollutants in subsurface waters. *Hyg. Sanit.* 36:292.

Azam, R., J.R. Beers, L. Cambell, A.F. Carlucci, O. Holm-Hansen, R.M.H. Reid, and D.M. Karl. 1979. Occurrence and metabolic activity of organisms under the Ross Ice Shelf, Antarctica at Station J9. *Science.* 203:451-453.

Baedecker, M.J. and W. Back. 1979. Modern marine sediments as a natural analog to the chemically stressed environment of a landfill. *J. Hydrol.* 43:393-414.

Baedecker, M.J., D.I. Siegel, P. Bennett, and I.M. Cozzarelli. 1989. The fate and effects of crude oil in a shallow aquifer. I. The distribution of chemical species and geochemical facies, p. 13-20. In: G.E. Mallard and S.E. Ragone (ed.), U.S. Geological Survey Water Resources Division Report 88-4220. U.S. Geological Survey, Reston, VA.

Baek, N.H., L.S. Clesceri and N.L. Clesceri. 1989. Modeling of enhanced biodegration in unsaturated soil zone. *J. Environ. Engr.* 115:150-172.

Bales, R.C., C.P. Gerba, G.H. Grondin and S.L. Jensen. 1989. Bacteriophage transport in sandy soil and fractured tuff. *Appl. Environ. Microbiol.* 55:2061-2067.

Balkwill, D.L. 1989. Numbers, diversity, and morphological characteristics of aerobic, chemoheterotrophic bacteria in deep subsurface sediments from a site in South Carolina. *Geomicrobiol. J.* 7:33-52.

Balkwill, D.L., J.K. Fredrickson, and J.M. Thomas. 1989. Vertical and horizontal variations in the physiological diversity of the aerobic chemo-heterotrophic bacterial microflora in deep Southeast Coastal Plain sediments from a site in South Carolina. *Appl. Environ. Microbiol.* 55:1058-1065.

Balkwill, D.L. and W.C. Ghiorse. 1985. Characterization of subsurface bacteria associated with two shallow aquifers in Oklahoma. *Appl. Environ. Microbiol.* 50:580-588.

Balkwill, D.L., F.R. Leach, J.T. Wilson, J.F. McNabb, and D.C. White. 1988. Equivalence of microbial biomass measurements based on membrane lipid and cell wall components, adenosine triphosphate, and direct counts in subsurface aquifer sediments. *Microb. Ecol.* 16:73-84.

Baltzis, B.C. and A.G. Fredrickson. 1988. Limitation of growth rate by two complementary nutrients: some elementary but neglected considerations. *Biotechnol. Bioeng.* 31:75-86.

Baveye, P. and A. Valocchi. 1989. An evaluation of mathematical models of the transport of biologically reacting solutes in saturated soils and aquifers. *Water Resour. Res.* 25:1413-1421.

Beeman, R.E. and J.M. Suflita. 1987. Microbial ecology of a shallow unconfined ground water aquifer polluted by municipal landfill leachate. *Microb. Ecol.* 14:39-54.

Beloin, R.M., J.L. Sinclair, and W.C. Ghiorse. 1988. Distribution and activity of microorganisms in subsurface sediments of a pristine study site in Oklahoma. *Microb. Ecol.* 16:85-95.

Belyaev, S.S. and M.V. Ivanov. 1983. Bacterial methanogenesis in underground waters. *Environ. Biogeochem. Ecol. Bull.* (Stockholm) 35:273-280.

Benefield, L. and F. Molz. 1984. A model for the activated sludge process which considers wastewater characteristics, floc behavior, and microbial population. *Biotechnol. Bioeng.* 26:352-361.

Benefield, L. and F. Molz. 1985. Mathematical simulation of a biofilm process. *Biotechnol. Bioeng.* 27:921-931.

Benefield, L. and R.B. Reed. 1985. An activated sludge model which considers toxicant concentration: simulation and sensitivity analysis. *Appl. Math. Modeling.* 9:454-465.

Bengtsson, G. 1989. Growth and metabolic flexibility in groundwater bacteria. *Microb. Ecol.* 18:235-248.

Bengtsson, G. and H. Annadotter. 1989. Nitrate reduction in a groundwater microcosm determined by N gas chromatography-mass spectrometry. *Appl. Environ. Microbiol.* 55:2861-2870.

Bitton, G. and R.W. Harvey. 1992. Transport of pathogens through soil. p. 103-104. In: R. Mitchell (ed.), *New concepts in environmental microbiology.* Wiley Interscience, New York.

Bonde, G.J. 1987. Heterotrophic bacteria in ground water. *Stygologia.* 3:185-199.

Bone, T.L. and D.L. Balkwill. 1988. Morphological and cultural comparison of microorganisms in surface soil and subsurface sediments at a pristine study site in Oklahoma. *Microb. Ecol.* 16:49-64.

Borden, R.C. and P.B. Bedient. 1986. Transport of dissolved hydrocarbons influenced by oxygen-limited biodegradation. 1. Theoretical development. *Water Resour. Res.* 22:1973-1982.

Bosma, T.N.P., J.L. Schnoor, G.Schraa, and A.J.B. Zehnder. 1988. Simulation model for biotransformation of xenobiotics and chemotaxis in soil columns. *J. Contam. Hydrol.* 2:225-236.

Bottcher, J., O. Strebel, S. Voerkelius, and H.-L. Schmidt. 1990. Using isotope fractionation of nitrate-nitrogen and nitrate-oxygen for evaluation of microbial dentrification in a sandy aquifer. *J. Hydrol.* 114:413-424.

Bouwer, E.J. and G.D. Bouwer. 1987. Modeling of biological processes in the subsurface. *Water Sci. Tech.* 19:769-779.

Celia, M.A., J.S. Kindred, and I. Herrara. 1989. Contaminant transport and biodegradation: 1. A numerical model for reactive transport in porous media. *Water Resour. Res.* 26:223-239.

Champ, D.R., J. Gulens, and R.E. Jackson. 1979. Oxidation-reduction sequences in ground water flow systems. *Can. J. Earth Sci.* 16:12-23.

Champ, D.R. and J. Schroeter. 1988. Bacterial transport in fractured rock - a field scale tracer test at the Chalk River Nuclear Laboratories. p. 14:1-14:7. In: Proceedings of the international conference on water and wastewater microbiology, Newport Beach, CA.

Chapelle, F.H. and D.R. Lovley. 1990a. Rates of microbial metabolism in deep coastal plain aquifers. *Appl. Environ. Microbiol.* 56:1865-1874.

Chapelle, F.H. and D.R. Lovley. 1990b. Hydrogen concentrations in groundwater as an indicator of bacterial processes in deep aquifer systems. p. 2-123-2-126. In: C.B. Fliermans and T.C. Hazen (eds.) *Proceedings of the first international symposium on microbiology of the deep subsurface.* WSRC Information Services, Aiken, SC.

Chapelle, F.H. and D.R. Lovley. Competitive exclusion of sulfate reduction by Fe(III)-reducing bacteria: a mechanism for producing discrete zones of high-iron ground water. *Ground Water.* (in press.)

Chapman, S.J. and T.R.G. Gray. 1986. Importance of cryptic growth, yield factors and maintenance energy in models of microbial growth in soil. *Soil. Biol. Biochem.* 18:1-4.

Corapcioglu, M.Y. and A. Haridas. 1984. Transport and fate of microorganisms in porous media: a theoretical investigation. *J. Hydrol.* 72:149-169.

Daigger, G.T. and C.P.L. Grady. 1983. An evaluation of transfer function models for the transient growth response of microbial cultures. *Water Res.* 17:1661-1667.

Dappert, A.F. 1932. Tracing the travel and changes in composition of underground pollution. *Water Works Sewerage.* 79:265.

Fallon, R.D., S.Y. Newell, and C.S. Hopkinson. 1983. Bacterial production in marine sediments: will cell-specific measures agree with whole-system metabolism? *Mar. Ecol. Prog. Ser.* 11:119-127.

Fredrickson, J.K., D.L. Balkwill, J.M. Zachara, S.-M.W. Li, F.J. Brockman, and M.A. Simmons. 1991. Physiological diversity and distributions of heterotrophic bacteria in deep cretaceous sediments of the Atlantic Coastal Plain. *Appl. Environ. Microbiol.* 57:402-411.

Gannon, J.T., V.B. Manilal, and M. Alexander. 1991a. Relationship between cell surface properties and transport of bacteria through soil. *Appl. Environ. Microbiol.* 57:190-193.

Gannon, J., Y. Tan, P. Baveye, and M. Alexander. 1991b. Effect of sodium chloride on the transport of bacteria in saturated aquifer material. *Appl. Environ. Microbiol.* 57:2497-2501.

Gantzer, C.J. 1989. Inhibitory substrate utilization by steady-state biofilms. *J. Environ. Eng.* 115:302-319.

Gaudy, A.F. and E.T. Gaudy. 1980. *Microbiology for environmental scientists and engineers.* McGraw-Hill Book Co., New York.

Gee, C.S., M.T. Suidan, and J.T. Pfeffer. 1990. Modeling of nitrification under substrate inhibiting conditions. *J. Environ. Engr.* 116:18-31.

Gerba, C.P. and G. Bitton. 1984. Microbial pollutants: Their survival and transport pattern to groundwater. p. 65-88, In: G. Bitton and C.P. Gerba, (eds.) *Groundwater pollution microbiology,* Wiley, N.Y.

Ghiorse, W.C. and D.L. Balkwill. 1983. Enumeration and morphological characterization of bacteria indigenous to subsurface environments. *Dev. Ind. Microbiol.* 24:213-224.

Ghiorse, W.C. and J.T. Wilson. 1988. Microbial ecology of the terrestrial subsurface. *Adv. Appl. Microbiol.* 33:107-173.

Ghiorse, W.C. and F.J. Wobber. 1989. Introductory comments. *Geomicrobiol. J.* 7:1-2.

Godsy, E.M., L.M. Law, C.D. Fraley, and D. Grbic-Galic. 1990. Kinetics of phenolic compound degradation by aquifer derived methanogenic microcosms. EOS, *Transactions of the Amer. Geophys. Union.* 71:1319.

Godsy, E.M., D.F. Goerlitz, and D. Grbic-Galic. 1992. Methanogenic biodegradation of creosote contaminants in natural and simulated ground water ecosystems. *Ground Water* 30:232-242.

Goulder, R. 1977. Attached and free bacteria in an estuary with abundant suspended solids. *J. Appl. Bacteriol.* 43:399-405.

Grady, C.P.L., W. Gujer, M. Henze, G. Marais, and T. Matsuo. 1986. A model for single-sludge wastewater treatment systems. *Water Sci. Tech.* 18:47-61.

Grady, C.P.L., G. Aichinger, S.F. Cooper, and M. Naziruddin. 1989. Biodegradation kinetics for selected toxic/hazardous organic compounds. p. 141-153. In: *Hazardous waste treatment: biosystems for pollution control.* Air & Waste Management Assoc., Pittsburgh, PA.

Grady, C.P.L. 1990. Biodegradation of toxic organics: status and potential. *J. Environ. Eng.* 116:805-828.

Grossman, E.L., B.K. Coffman, S.J. Fritz, and H. Wada. 1989. Bacterial production of methane and its influence on ground-water chemistry in east-central Texas aquifers. *Geology* 17:495-499.

Han, K. and O. Levenspiel. 1988. Extended Monod kinetics for substrate, product and cell inhibition. *Biotechnol. Bioeng.* 32:430-437.

Harris, N.P. and G.S. Hansford. 1976. A study of substrate removal in a microbial film reactor. *Water Res.* 10:935-943.

Harvey, R.W., Smith, R.L., and George, L.H. 1984. Effect of organic contamination upon microbial distributions and heterotrophic uptake in a Cape Cod, Mass., aquifer. *Appl. Environ. Microbiol.* 48:1197-1202.

Harvey, R.W. and L.H. George. 1987. Growth determinations for unattached bacteria in a contaminated aquifer. *Appl. Environ. Microbiol.* 53:2992-2996.

Harvey, R.W. 1988. Transport of bacteria in a contaminated aquifer. In: G.E. Mallard and S.E. Ragone (eds.), *U.S. Geological Survey Water Resources Investigations Report* 88-4220.

Harvey, R.W., L.H. George, R.L. Smith, and D.R. LeBlanc. 1989. Transport of microspheres and indigenous bacteria through a sandy aquifer: results of natural and forced-gradient tracer experiments. *Environ. Sci. Technol.* 23:51-56.

Harvey, R.W. 1990. Evaluation of particulate and solute tracers for investigation of bacterial transport behavior in groundwater. p. 7-159-7-165. In: C.B. Fliermans and T.C. Hazen (eds.) *Proceedings of the first international symposium on microbiology of the deep subsurface.* WSRC Information Services, Aiken, SC.

Harvey, R.W. 1991. Parameters involved in modeling movement of bacteria in groundwater. p. 75-82. In: C.J. Hurst (ed.), *Modeling the Environmental Fate of Microorganisms*, Am. Soc. Microbiol., Washington.

Harvey, R.W. and S.P. Garabedian. 1991. Use of colloid filtration theory in modeling movement of bacteria through a contaminated sandy aquifer. *Environ. Sci. Technol.* 25:178-185.

Harvey, R.W. and L.B. Barber. 1992. Associations of free-living bacteria and dissolved organic compounds in a plume of contaminated groundwater. *J. Contam. Hydrol.* 9:91-103.

Havemeister, G., R. Riemer, and J. Schroeter. 1985. Felversuche zur persistenz und zum tranportverhalten von bakterien. Versuchsfeld Segeberger Forst. *Umweltbundesamt Materialien* 2/85:49-56.

Hazen, T.C., L. Jimenez, C.B. Fliermans, and G.L. De Vitoria. 1990. Comparison of bacteria from deep subsurface sediment and adjacent groundwater. p. 2-141-2-158. In: C.B. Fliermans and T.C. Hazen (eds.) *Proceedings of the first international symposium on microbiology of the deep subsurface.* WSRC Information Services, Aiken, SC.

Henze, M., C. Grady, W. Gujer, G. Marais, and T. Matsuo. 1987. A general model for single-sludge wastewater treatment systems. *Water Res.* 21:505-517.

Herbert, D. 1958. Some principles of continuous culture. p. 381-396. In: G. Tunevall (ed.) *Recent progress in microbiology*, Blackwell Scientific, Oxford, England.

Hirsch, P. and E. Rades-Rohkohl. 1988. Some special problems in the determination of viable counts of groundwater microorganisms. *Microb. Ecol.* 16:99-113.

Hissett, R. and T.R.G. Gray. 1976. Microsites and time changes in soil microbe ecology. p. 23-39. In: J.M. Anderson and A. MacFadyen (eds.) The role of terrestrial and aquatic organisms in decomposition processes. Blackwell, Oxford.

Hossain, M.A. and M.Y. Corapcioglu. 1990. Bioremediation of halogenated hydrocarbons controlled by a primary substrate and microbial activity. EOS, *Transactions of the Am. Geophys. Union.* 71:1326.

Hurst, C.J. (ed.). 1991. Modeling the Environmental Fate of Microorganisms. *Am. Soc. Microbiol.*, Washington.

Hutchins, S.R., W.C. Downs, J.T. Wilson, G.B. Smith, D.A. Kovacs, D.D. Fine, R.H. Douglass, and D.J. Hendrix. 1991. Effect of nitrate addition on biorestoration of fuel-contaminated aquifer: field demonstration. *Ground Water,* 29(4), 571-580, 1991.

Istok, J.D. and S.L. Woods. 1990. Development and verification of a numerical model for fate and transport of chlorinated phenols in groundwater. EOS, *Transactions of the Am. Geophys. Union.* 71:1324.

Jannasch, H.W. and P.H. Pritchard. 1972. The role of inert particulate matter in the activity of aquatic microorganisms. *Mem. Ist. Ital. Idrobiol. Suppl.* 29:289-308.

Jenkinson, D.S. and J.N. Ladd. 1981. Microbial biomass in soil: measurement and turnover. *Soil Biochem.* 5:415-471.

Jenneman, G.E., M.J. McInerney, and R.M. Knapp. 1985. Microbial penetration through nutrient-saturated Berea sandstone. *Appl. Environ. Microbiol.* 50:383-391.

Jimenez, L. 1990. Molecular analysis of deep-subsurface bacteria. *Appl. Environ. Microbiol.* 56:2108-2113.

Jones, R.E. 1989. Anaerobic metabolic processes in the deep terrestrial subsurface. *Geomicrobiol. J.* 7:117-130.

Keswick, B.H. 1984. Sources of groundwater pollution. p. 59-64. In: G. Bitton and C.P. Gerba (ed.) *Groundwater pollution microbiology.* John Wiley & Sons, New York.

Kingston, S.P. 1943. Contamination of water supplies in limestone formation. *J. Am. Water Works Assoc.* 35:1450.

Kinner, N.E., A.L. Bunn, R.W. Harvey, A. Warren, and L.D. Meeker. 1991. Preliminary evaluation of the relations among protozoa, bacteria and chemical properties in seage-contaminated ground water near Otis Air Base, Massachusetts. In: G.E. Mallard and D.A. Aronson (eds.) *U.S. Geological Survey Toxic Substances Hydrology Program--Proceedings of the technical meeeting,* Monterey, CA, March 11-15, 1991, Water Resour. Invest. Rpt. 91-4034.

Kinner, N.E., A.L. Bunn, L.D. Meeker, and R.W. Harvey. 1990. Enumeration and variability in the distribution of protozoa in an organically-contaminated subsurface environment. EOS, *Transactions of the Am. Geophys. Union.* 71:1319.

Kissel, J.C., P.L. McCarty, and R.L. Street. 1984. Numerical simulation of mixed-culture biofilm. *J. Environ. En.* 110:393-411.

Kjelleberg, S. 1984. Effects of interfaces on survival mechanisms of copiotrophic bacteria in low-nutrient habitats. In: M.J. Klug and C.A. Reddy (eds.) *Current Perspectives in Microbial Ecology.* Am. Soc. Micriobiol., Washington.

Kolbel-Boelke, J., E.M. Anders, and A. Nehrkorn. 1988. Microbial communities in the saturated groundwater environment. II: Diversity of bacterial communities in a pleistocene sand aquifer and their in vitro activities. *Microb. Ecol.* 16:31-48.

Kolbel-Boelke, J. 1990. Composition of aerobic, heterotrophic bacterial communities in a sandy-gravelly aquifer in the Lower Rhine Region in Germay. p. 3-15-3-28. In: C.B. Fliermans and T.C. Hazen (eds.) *Proceedings of the first international symposium on microbiology of the deep subsurface.* WSRC Information Services, Aiken, SC.

Krzanowski, K.M., C.A. Sinn, and D.L. Balkwill. 1990. Attached and unattached bacterial populations in deep aquifer sediments from a site in South Carolina. p. 5-25-5-30. In: C.B. Fliermans and T.C. Hazen (eds.) *Proceedings of the first international symposium on microbiology of the deep subsurface.* WSRC Information Services, Aiken, SC.

Kudryavtseva, B.M. 1972. An experimental approach to the establishment of zones of hygienic protection of underground water sources on the basis of sanitary-bacteriological indices. *J. Hyg. Epidemiol. Microbiol. Immunol.* 16:503.

Kurath, G. and R.Y. Morita. 1983. Starvation-survival physiological studies of a marine *Pseudomonas* sp. *Appl. Environ. Microbiol.* 45:1206-1211.

Kuznetsov, S.I., M.W. Ivanov, and N.N. Lyalikova. 1963. *Introduction to Geological Microbiology.* McGraw-Hill, New York.

Leach, L.E. 1990. An aseptic procedure for soil sampling in heaving sands using special hollow-stem auger coring. p. 2-3-2-18. In: C.B. Fliermans and T.C. Hazen (eds.) *Proceedings of the first international symposium on microbiology of the deep subsurface.* WSRC Information Services, Aiken, SC.

Lee, M.D., J.M. Thomas, R.C. Borden, P.B. Bedient, C.H. Ward, and J.T. Wilson. 1988. Biorestoration of aquifers contaminated with organic compounds. *CRC Crit. Rev. Environ. Contr.* 18:29-89.

Leggett, D.C. and I.K. Iskandar. 1981. Evaluation of a nitrification model. p. 313-3358. In: I.K. Iskander (ed.) *Modeling wastewater renovation: Land treatment,* J. Wiley & Sons, NY.

Liebezeit, G., M. Bolter, I.F. Brown, and R. Dawson. 1980. Dissolved free amino acids and carbohydrates at pycnocline boundaries in the Sargasso Sea and related microbial activity. *Oceanol. Acta.* 3:357-362.

Lovley, D.R. and S. Goodwin. 1988. Hydrogen concentrations as an indicator of the predominant terminal electron-accepting reactions in aquatic sediments. *Geochim. Cosmochim. Acta.* 52:2993-3003.

Lovley, D.R. 1991. Dissimilatory Fe(III) and Mn(IV) reduction. *Microbiol. Rev.* 55:259-287.

Lynch, J.M. and L.M. Panting. 1980. Cultivation and the soil biomass. *Soil Biol. Biochem.* 12:29-33.

MacQuarrie, K.T.B. and Sudicky, E.A. 1989. Simulation of biodegradable organic contaminants in groundwater: 2. plume behavior in uniform and random flow fields. *Water Resour. Res.* 26:223-239.

Madsen, E.L. and J.M. Bollag. 1989. Aerobic and anaerobic microbial activity in deep subsurface sediments from the Savannah River plant. *Geomicrobiol. J.* 7:93-101.

Mariotti, A., A. Landreau, and B. Simon. 1988. N isotope biogeochemistry and natural denitrification process in groundwater: application to the chalk aquifer of northern France. *Geochim. Cosmochim. Acta.* 52:1869-1878.

Mariotti, A. 1986. La denitrificaiton dans les eaux soterraines, principes et methodes de son identification: une revue. *J. Hydrol.* 88:1-23.

Martin, G.N. and M.J. Noonon. 1977. *Effects of domestic wastewater disposal by land irrigation on groundwater quality of central Canterbury plains.* Water Soil Tech. Publ. No.7, Water and Soil Division, Ministry of Work and Development, Wellington, N.Z.

Marxsen, J. 1988. Investigations into the number of respiring bacteria in groundwater from sandy and gravelly deposits. *Microb. Ecol.* 16:65-72.

Marxsen, J. 1981. Bacterial biomass and bacterial uptake of glucose in polluted and unpolluted groundwater of sandy and gravelly deposits. *Verh. Int. Ver. Limnol.* 21:1371-1375.

Mason, C.A. and G. Hamer. 1987. Cryptic growth in Klebsiella pneumoniae. *Appl. Microbiol. Biotechnol.* 25:577-584.

McCarty, P.L., B.E. Rittman, and E.J. Bouwer. 1984. Microbial processes affecting chemical transformations in groundwater. p. 89-115. In: G. Bitton and C.P. Gerba (eds.) *Groundwater pollution microbiology.* J. Wiley & Sons, NY.

McMahon, P.B., D.F. Williams, and J.T. Morris. 1990. Production and carbon isotopic composition of bacterial CO_2 in deep coastal plain sediments of South Carolina. *Ground Water.* 5:693-702

Molz, F.J., M.A. Widdowson, and L.D. Benefield. 1986. Simulation of microbial growth dynamics coupled to nutrient and oxygen transport in porous media. *Water Resour. Res.* 22:1207-1216.

Molz, F.J. and M.A. Widdowson. 1988. Internal inconsistencies in dispersion-dominated model that incorporate chemical and microbial kinetics. *Water Resour. Res.* 24:615-619.

Novitsky, J.A. 1983a. Heterotrophic activity throughout a vertical profile of seawater and sediment in Halifax Harbor, Canada. *Appl. Environ. Microbiol.* 45:1753-1760.

Novitsky, J.A. 1983b. Microbial activity at the sediment-water interface in Halifax Harbor, Canada. *Appl. Environ. Microbiol.* 45:1761-1766.

Panikov, N. 1979. Steady state growth kinetics of Chlorella vulgaris under double substrate (uera and phosphate) limitation. *J. Chem. Tech. Biotechnol.* 29:442-450.

Parkin, G.F. and R.E. Speece. 1982. Modeling toxicity in methane fermentation systems. *J. Environ. Engr.* 108:515-531.

Pedersen, K. and S. Ekendahl. 1990. Distribution and activity of bacteria in deep granitic groundwaters of southeastern Sweden. *Microb. Ecol.* 20:37-52.

Phelps, J.T., D. Ringelberg, D. Hedric, J. Davis, C.B. Fliermans, and D.C. White. 1988. Microbial biomass and activities associated with subsurface environments contaminated with chlorinated hydrocarbons. *Geomicrobiol. J.* 6:157-170.

Phelps, T.J., C.B. Fliermans, T.R. Garland, S.M. Pfiffner, and D.C. White. 1989a. Methods for recovery of deep terrestrial subsurface sediments for microbiolgy studies. *J. Microbiol. Methods.* 9:267-279.

Phelps, T.J., E.G. Raione, D.C. White, and C.B. Fliermans. 1989b. Microbial activities in deep subsurface environments. *Geomicrobiol. J.* 7:79-91.

Pirt, S.J. 1975. *Principles of microbe and cell cultivation.* John Wiley & Sons, NY.

Postgate, J.R. and J.R. Hunter. 1962. The survival of starved bacteria. *J. Gen. Microbiol.* 29:233-263.

Reeves, J.Y., R.H. Reeves, and D.L. Balkwill. 1990. Restriction endonuclease analysis of deep subsurface bacterial isolates. p. 2-115. In: C.B. Fliermans and T.C. Hazen (eds.) *Proceedings of the first international symposium on microbiology of the deep subsurface.* WSRC Information Services, Aiken, SC.

Reynolds, P.J., P. Sharma, G.E. Jenneman, and J.J. McInerney. 1989. Mechanisms of microbial movement in subsurface materials. *Appl. Environ. Microbiol.* 55:2280-2286.

Rifai, H.S. and P.B. Bedient. 1990. Comparison of biodegradation kinetics with an instantaneous reaction model for groundwater. *Water Resour. Res.* 26:637-645.

Rittman, B.E. and P.L. McCarty. 1980. Model of steady-state-biofilm kinetics. *Biotechnol. Bioeng.* 22:2343-2357.

Roberts, P.V., M.N. Goltz, and D.M. Mackay. 1986. A natural gradient experiment on solute transport in a sand aquifer 3. Retardation estimates and mass balances for organic solutes. *Water Resour. Res.* 22(13), 2047-2058.

Roberts, P.V., J.E. Schreiner, and G.D. Hopkins. 1982. Field study of organic water quality changes during ground water recharge in the Palo Alto Baylands. *Water Res.* 16:1025-1035.

Roels, J.A. 1983. *Energetics and kinetics in biotechnology.* Elsevier Science, N.Y.

Ronen, D., M. Magaritz, H. Gvirtzman, and W. Garner. 1987. Microscale chemical heterogeneity in groundwater. *J. Hydrol.* 92:173-178.

Rozich, A.F., A.F. Gaudy, Jr., and P.C. D'Adamo. 1983. Prediction model for treatment of phenolic wastes by activated sludge. *Water Res.* 17:1453-1466.

Rozich, A.F., A.F. Gaudy, Jr., and P.C. D'Adamo. 1985. Selection of growth rate model for activated sludges treating phenol. *Water Res.* 19:481-490.

Ryan, F.J. 1955. Spontaneous mutation in non-dividing bacteria. *Genetics* 40:726-738.

Saltzman, H.A. 1980. Untersuchungen uber die Veranderungen der Mikroflora beim Durchgang von Brackwasser durch die Kuhlanlagen von Kraftwerken, Ph.D. thesis, University of Kiel, F.G.R.

Schmidt, S.K., S. Simkins, and M. Alexander. 1985. Models for the kinetics of biodegradation of organic compounds not supporting growth. *Appl. Environ. Microbiol.* 50:323-331.

Scow, K.M., S. Simkins, and M. Alexander. 1986. Kinetics of mineralization of organic compounds at low concentrations in soil. *Appl. Environ. Microbiol.* 51:323-331.

Semprini, L., G.D. Hopkins, P.V. Roberts, D. Grbic-Galic, and P.L. McCarty. 1991. A field evaluation of in-situ biodegradation of chlorinated ethenes: part 3, studies of competitive inhibition. *Ground Water* 29:239-250.

Semprini, L. and P.L. McCarty. 1991. Comparison between model simulations and field results for in-situ biorestoration of chlorinated aliphatics: part 1, biostimulation of methanotrophic bacteria. *Ground Water* 29:365-374.

Semprini, L., P.V. Roberts, G.D. Hopkins, and P.L. McCarty. 1990. A field evaluation of in-situ biodegradation of chlorinated ethanes: part 2, results of biostimulation and biotransformation experiments. *Ground Water* 28:715-727.

Sinclair, J.T. and W.C. Ghiorse. 1989. Distribution of aerobic bacteria, protozoa, algae, and fungi in deep subsurface sediments. *Geomicrobiol. J.* 7:15-31.

Sinclair, J.T. and W.C. Ghiorse. 1987. Distribution of protozoa in subsurface sediments of a pristine groundwater study site in Oklahoma. *Appl. Environ. Microbiol.* 53:1157-1163.

Sinclair, J.L., S.J. Randtke, J.E. Denne, L.R. Hathaway, and W.C. Ghiorse. 1990. Survey of microbial populations in buried-valley aquifer sediments from northeastern Kansas. *Ground Water* 28:369-377.

Sinton, L.W. 1980. *Investigations into the use of the bacterial species Bacillus stearothermophilus and Escherichia coli (H S positive) as tracers of ground water movement.* Water and Soil Tech. Publ. No. 17, Water Soil Div., Ministry of Works and Development, Wellington, N.Z.

Smith, J.L. 1989. Sensitivity analysis of critical parameters in microbial maintenance-energy models. *Biol. Fertil. Soils* 8:7-12.

Smith, R.L. and J.H. Duff. 1988. Denitrification in a sand and gravel aquifer. *Appl. Environ. Microbiol.* 54:1071-1078

Smith, R.L. and R.W. Harvey. 1990. Development of sampling techniques to measure in-situ rates of microbial processes in a contaminated sand and gravel aquifer. p. 2-19-2-34. In: C.B. Fliermans and T.C. Hazen (eds.) *Proceedings of the first international symposium on microbiology of the deep subsurface.* WSRC Information Services, Aiken, SC.

Smith, R.L., R.W. Harvey, and D.R. LeBlanc. 1991a. Importance of closely spaced vertical sampling in delineating chemical and microbiological gradients in groundwater studies. *J. Contam. Hydrol.* 7:285-300.

Smith, R.L., B.L. Howes, and S.P. Garabedian. 1991b. In-situ measurement of methane oxidation in groundwater by using natural-gradient tracer tests. *Appl. Environ. Microbiol.* 57:1997-2004.

Smith, R.L., B.L. Howes, and J.H. Duff. 1991c. Denitrification in nitrate-contaminated groundwater: occurrence in steep vertical geochemical gradients. *Geochim. Cosmochim. Acta.* 55:1815-1825.

Srinivasan, P. and J.W. Mercer. 1988. Simulation of biodegradation and sorption processes in ground water. *Ground Water* 26:475-487.

Stetzenbach, L.D., L.M. Kelley, and N.A. Sinclair. 1986. Isolation, identification, and growth of well-water bacteria. *Ground Water* 24:6-10.

Stim, K.P., G.R. Drake, S.E. Padgett and D.L. Balkwill. 1990. 16S ribosomal RNA sequencing analysis of phylogenetic relatedness among aerobic chemoheterotrophic bacteria in deep aquifer sediments from a site in South Carolina. p. 2-117. In: C.B. Fliermans and T.C. Hazen (eds.), *Proceedings of the first international symposium on microbiology of the deep subsurface.* WSRC Information Services, Aiken, SC.

Strand, S., S. Woods, K. Williamson, and K. Ryan. 1989. Degradation of chlorinated low-molecular weight solvents by an enrichment culture of methanotrophs. p. 217-232. In: *Hazardous waste treatment: Biosystems for pollution control.* Air & Waste Management Assoc., Pittsburgh, PA.

Stumm, W. 1967. Redox potential as an environmental parameter; conceptual significance and operational limitation. *Adv. Water Poll. Res.* 1:283-307.

Taylor, S.W., P.C.D. Milly, and P.R. Jaffé. 1990. Biofilm growth and the related changes in the physical properties of a porous medium, 2. Permeability. *Water Resour. Res.* 26:2161-2169.

Taylor, S.W. and P.R. Jaffé. 1990. Biofilm growth and the related changes in the physical properties of a porous medium. 3. Dispersivity and model verification. *Water Resour. Res.* 26:2171-2180.

Thorn, P.M. and R.M. Ventullo. 1988. Measurement of bacterial growth rates in subsurface sediments using the incorporation of tritiated thymidine into DNA. *Microb. Ecol.* 16:3-16.

van Loosdrecht, M.C.M., J. Lyklema, W. Norde, and A.J.B. Zehnder. 1990. Influence of interfaces on microbial activity. *Microb. Rev.* 54:75-87.

Ventullo, R.M. and R.J. Larson. 1985. Metabolic diversity and activity of heterotrophic bacteria in ground water. *Environ. Toxicol. Chem.* 4:759-771.

Wanner, O. and W. Gujer. 1986. A multispecies biofilm model. *Biotechnol. Bioeng.* 28:314-328.

Ward, T.E. 1985. Characterizing the aerobic and anaerobic microbial activities in surface and subsurface soils. *Environ. Toxicol. Chem.* 4:727-737.

Watson, J.E. and W.R. Garner. 1986. A mechanistic model of bacterial colony growth response to substrate supply. p. 24. In: Summary of papers presented at the Amer. Geophys. Union Chapman conference on microbial processes in the transport, fate and in-situ treatment of subsurface contaminants.

Webster, J.J., G.J. Hampton, J.T. Wilson, W.C. Ghiorse, and F.R. Leach. 1985. Determination of microbial cell numbers in subsurface samples. *Ground Water* 23:17-25.

White, D.C., D.B. Ringelberg, J.B. Guckert, and T.J. Phelps. 1990. Biochemical markers for in-situ microbial community structure. p. 4-45-4-56. In: C.B. Fliermans and T.C. Hazen (eds.) *Proceedings of the first international symposium on microbiology of the deep subsurface.* WSRC Information Services, Aiken, SC.

Widdowson, M.A., F.J. Molz, and L.D. Benefield. 1987. Development and application of a model for simulating microbial growth dynamics coupled to nutrient and oxygen transport in porous media. p. 28-51. In: *Proceedings of the solving ground water problems with models conference and exposition.* National Water Well Assoc., Dublin, OH.

Widdowson, M.A., F.J. Molz, and L.D. Benefield. 1988a. A numerical transport model for oxygen- and nitrate-based respiration linked to substrate and nutrient availability in porous media. *Water Resour. Res.* 24:1553-1565.

Widdowson. M.A., F.J. Molz, and L.D. Benefield. 1988b. Modeling of multiple organic contaminant transport and biotransformations under aerobic and anaerobic (denitrifying) conditions in the subsurface. p. 397-416. In: *Proceedings of the petroleum hydrocarbons and organic chemicals in ground water conference and exposition.* National Water Well Assoc., Dublin, OH.

Widdowson, M.A. 1991. Comment on "An evaluation of mathematical models of the transport of biologically reacting solutes in saturated soils and aquifers" by P. Baveye and A. Valocchi. *Water Resour. Res.* 27:1375-1378.

Wilson, B. 1985. Behavior of trichloroethylene, 1,1 dichloroethylene in anoxic subsurface environments. Masters thesis, University of Oklahoma, 1985.

Wilson, J.T., J.F. McNabb, D.L. Balwill, and W.C. Ghiorse. 1983. Enumeration and characterization of bacteria indigenous to a shallow water-table aquifer. *Ground Water* 21:134-142.

Wilson, J.T., L.E. Leach, M. Henison, and J.N. Jones. 1986. In-situ biorestoration as a ground water remediation technique. *Ground Water Monit. Rev.* 6:56-64.

Wilson, G.B., J.N. Andrews, and A.H. Bath. 1990. Dissolved gas evidence for denitrification in the Lincolnshire limestone groundwaters, Eastern England. *J. Hydrol.* 113:51-60.

Wood, W.W. and G.G. Ehrlich. 1978. Use of baker's yeast to trace microbial movement in ground water. *Ground Water* 16:398-403.

Yates, M.V. and S.R. Yates. 1988. Modeling microbial fate in the subsurface environment. *CRC Crit. Rev. Environ. Contr.* 17:307-344.

Yoon, H., G. Klinzing, and H.W. Blanch. 1977. Competition for mixed substrates by microbial population. *Biotechnol. Bioeng.* 19:1193-1210.

Zapico, M.M., S. Vales, and J.A. Cherry. 1987. A wireline piston core barrel for sampling cohesionless sand and gravel below the water table. *Ground Water Mon. Rev.* 7:74-87.

Effects of Microorganisms on Phyllosilicate Properties and Behavior

J.W. Stucki, H. Gan, and H.T. Wilkinson

I. Introduction 227
II. Effects of Phyllosilicates on Microbial Activity 231
III. Phyllosilicate Dissolution by Microorganisms 234
    A. Micas 235
    B. Other 2:1, 1:1, and 2:2 Phyllosilicates 237
IV. Phase Transformations and Neoformation 238
V. Change in Phyllosilicate Microstructure by Organic
    Exudates 239
VI. Changes in Mineral Redox Status 240
VII. Summary 246
References 247

I. Introduction

Microbial activity in soils and sediments influences many natural processes that are of great significance to agriculture, the environment, and industry, and it has been the subject of literally thousands of scientific investigations. Certain aspects of the interactions between microorganisms and the mineral fraction of soils are well-established phenomena, and much has been written regarding such associations (Huang and Schnitzer, 1986), particularly as they involve the Fe oxide minerals (Münch and Ottow, 1977, 1982; Ottow, 1982; Berthelin, 1983; Robert and Berthelin, 1986; Fischer, 1988; Ferris et al., 1989) and the effects of phyllosilicates on microbial behavior. A review of the interactions between phyllosilicate clay minerals and microorganisms includes an extremely wide scope of information, including the vast literature in the respective disciplines of clay-organic chemistry, soil organic matter, mineral weathering, clay colloid chemistry, and microbial biochemistry. One quickly discovers that boundaries separating these disciplines from the processes of the mineral-microbe interface are in fact imaginary; indeed, each is an integral part of such processes. Because

ISBN 0-87371-889-5

227

of the great influence of phyllosilicates on the physical-chemical properties of soils and sediments, and because of the ubiquitous presence of microorganisms in such environments, knowledge of the interactions between these two components is vital to understanding the properties and behavior of natural systems.

The primary purpose of the present review is to describe the types of interactions that occur between phyllosilicates and microorganisms that alter the properties and behavior of the phyllosilicates. A secondary purpose is to give an overview, but not a comprehensive review, of ways in which phyllosilicates alter microbial behavior. Reference will be made to literature in most of the above-mentioned disciplines, but no attempt will be made to review that literature comprehensively.

Phyllosilicate clay minerals are among the most important structural materials of the earth's surface (Ross and Hendricks, 1945). These hydrous layer-silicate minerals have a significant role in many aspects of life, ranging from their dominion of the properties of soils and sediments to their ubiquitous commercial use (Newman, 1984; Odom, 1984; Jepson, 1984, 1988). The chemical compositions, structures, and behavior of phyllosilicates have been described and reviewed by many, including the classic textbooks by van Olphen (1966), Grim (1968), and Brindley and Brown (1980). The report of the nomenclature committee of the International Association for the Study of Clays (AIPEA), which classifies the phyllosilicates according to their layer types, charge, and interlayer properties, is duplicated in Table 1 (Bailey, 1980). This report provides a useful introduction to the phyllosilicates and a frame of reference for the various mineral names that will be used in this chapter.

Microorganisms are distinctly different from plants and animals as they consist either of a single cell or are multicellular, but lack development of organs and tissues typical of plants and animals (Boyd, 1984). They are classified generally according to the complexity of their cell structure (procaryotes and eucaryotes) and their metabolic mechanisms (photolithotrophs, photoorganotrophs, chemolithotrophs, and chemoorganotrophs). Procaryotes are the class of microorganisms that lacks a proper nucleus and has no membrane-bound organelles such as mitochondria. The procaryotic nucleus is never separated from the cytoplasm by a nuclear membrane. Examples include bacteria and blue-green algae. Eucaryotes have cell structures with a discrete nucleus enclosed within a nuclear membrane and membrane-bound organelles are present inside the cells. This structure is similar to the basic cell units of plants and animals. Fungi, protozoa, and algae are examples of eucaryotic microorganisms. Lichens are a symbiotic association between algae and fungi. All of these classes of organisms in natural environments are sometimes referred to collectively as microbiota.

Classification according to metabolic mechanisms is based on three primary factors: (1) the source of energy used as a driving force to create donor electrons (photons, e.g., sunlight, or chemical redox potentials); (2) the carbon source best used by the organism, whether CO_2 from the atmosphere or organic

Table 1. Classification scheme for phyllosilicate clay minerals as recommended by the nomenclature committee of AIPEA

Type	Group ($^3\chi$ = charge per formula unit)	Subgroup	Species[1]
1:1	Kaolinite-serpentine $\chi \sim 0$	Kaolinite Serpentine	Kaolinite, dickite, halloysite Chrysotile, lizardite, amesite
2:1	Pyrophyllite-talc $\chi \sim 0$	Pyrophyllite Talc	Pyrophyllite Talc
	Smectite $\chi \sim 0.2\text{-}0.6$	Dioctahedral smectite Trioctahedral smectite	Montmorillonite, beidellite Saponite, hectorite, sauconite
	Vermiculite $\chi \sim 0.6\text{-}0.9$	Dioctahedral vermiculite Trioctahedral vermiculite	Dioctahedral vermiculite Trioctahedral vermiculite
	Mica[2] $\chi \sim 1$	Dioctahedral mica Trioctahedral mica	Muscovite, paragonite Phlogopite, biotite, lepidolite
	Brittle mica $\chi \sim 2$	Dioctahedral brittle mica Trioctahedral brittle mica	Margarite Clintonite, anandite
	Chlorite χ variable	Dioctahedral chlorite Di,trioctahedral chlorite Trioctahedral chlorite	Donbassite Cookeite, sudoite Clinochlore, chamosite, nimite

[1]Only a few examples are given.
[2]The status of illite (or hydromica), sericite, etc., must be left open at present, because it is not clear whether or at what level they would enter the table: many materials so designated may be interstratified.
[3]χ refers to the $O_{10}(OH)_2$ structural unit.

(From Bailey, 1980.)

Table 2. Nutritional classification of microorganisms according to the carbon and energy sources

	Electron Donors and Carbon Sources	
	Inorganic	Organic
Electromagnetic energy (photons)	H_2O (e^-donors) CO_2 Photolithotrophs	$(CH_2O)_n$ Photoorganotrophs
Chemical energy	NH_4^+, S^{2-}, S°, Fe^{2+} CO_2 Chemolithotrophs	$(CH_2O)_n$ Chemoorganotrophs

(From Berthelin, 1988.)

compounds; and (3) the electron acceptor for respiration -- O_2 for aerobic respiration, and other chemical species, such as Fe(III), for anaerobic respiration. The first two of these factors relate to nutritional categories, and are summarized in Table 2. In addition to these requirements, all microorganisms require H_2O and various nutrient elements, such as N, S, P, K, and other micronutrients, to build cell structures and membranes.

The processes of microbial activity were illustrated schematically by Berthelin (1988), and are reproduced here in Figure 1. Phyllosilicates can enter the process either as an electron donor or acceptor (through oxidation or reduction of Fe in their crystal structures), as the source of inorganic nutrients or a carrier of organic substrates, or as a reactant with intermediate or end metabolic products exuded from the cell. When organisms die, the organic residue also may react with the phyllosilicates.

Studies of the interactions between phyllosilicates and microorganisms reveal that the effects occur in both directions, i.e., both the organism and the mineral may be affected. The largest body of literature in this regard deals with the effects of clay minerals on microbial activity and related biological processes, and has been reviewed extensively (Stotzky, 1972, 1980, 1985, 1986; Silver et al., 1986; Robert and Berthelin, 1986; Berthelin, 1988; Ferris et al., 1989). Less literature is available regarding the effects on mineral properties, but evidence shows that the types of effects on minerals include dissolution and chelation of constituent metal ions, neoformation, and changes in the oxidation state of Fe in the crystal structure of the phyllosilicate. Each of these topics will be considered now in more detail.

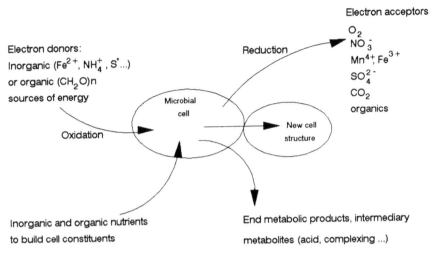

Figure 1. Diagram of general microbial activity. (From Berthelin, 1988.)

II. Effects of Phyllosilicates on Microbial Activity

The medium in which microorganisms exist has a large and even controlling influence on their activity. The presence of phyllosilicates in soils or culture media alters the ability of soil organisms to decompose organic substrates (Kunc and Stotzky, 1974; Stotzky, 1973; Novakova, 1968, 1969) and generally stimulates bacterial growth and/or respiration (Stotzky, 1966a,b; Stotzky and Rem, 1966, 1967; Filip, 1968a,b), although the specificities of these influences have yet to be clearly delineated and some inconsistencies exist.

Among the phyllosilicates, montmorillonite generally has a greater impact on microbial behavior than either kaolinite or vermiculite (Stotzky, 1967; Kunc and Stotzky, 1974). A significant reason for this influence is that microbial activity is a water-borne process and smectites retain substantial amounts of water at their surfaces, even under moisture stress, thus providing a medium for essential biochemical processes.

A good example of the influence of different clays on microbial behavior was described by Stotzky et al. (1961), who observed that the presence of a montmorillonite-like mineral in tropical soils inhibited the spread of Fusarium wilt disease on banana plantations, giving the soils a long life for banana production. The disease spread more rapidly through plantations where the soils were deficient in this mineral and more rich with kaolinite. Similar correlations were found for the spread of the human fungal pathogen *Histoplasma capsulatum* (Ajello, 1967; Stotzky and Post, 1967). Further investigations (Stotzky and Rem, 1966; Stotzky, 1966a,b) revealed that the favorable influence of montmorillonite was probably the result of its greater pH buffering capacity as compared to kaolinite. As illustrated in Figure 2, the respiration rate of

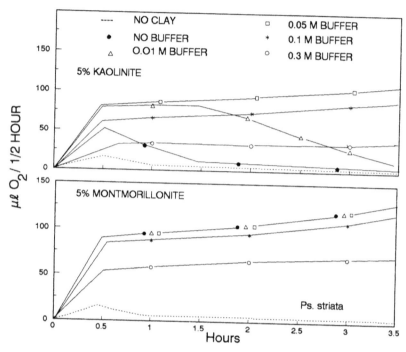

Figure 2. Effect of buffer concentration on respiration of *Pseudomonas striata* in presence of kaolinite and montmorillonite. (From Stotzky and Rem, 1966.)

Pseudomonas striata was sustained at a relatively high rate in the presence of montmorillonite, regardless of the amount of pH buffer added. In the presence of kaolinite, bacterial respiration depended on the amount of buffer added and was low in the absence of the buffer. Further, in the kaolinite system the pH dropped steadily as a result of bacterial activity. Chlorite, talc, vermiculite, and pyrophyllite behaved much like kaolinite (Stotzky, 1966a). The addition of pH buffer to the kaolinitic system inhibited Fusarium wilt, but too much buffer reduced bacterial respiration due to the hypertonic osmotic pressure created by the higher salt concentration. The addition of similar quantities of buffer to the montmorillonitic system showed only a small increase in respiration, indicating that the benefit of pH buffering had already been achieved by the clay itself, without the need for chemical amendment (Figure 2). In the presence of montmorillonite, the bacteria also had a higher tolerance for the increased salt concentrations imposed by the addition of buffer, presumably because of the higher cation exchange capacity of the montmorillonite compared to kaolinite. The basis for the favorable influence of pH buffering, as explained by Stotzky and Rem (1966), is that fungi are more tolerant than bacteria of lower pH values (3.5 to 4.5) observed in kaolinitic soils; but when the pH is nearer to neutral

(5.0 to 6.5), as is the case in the montmorillonitic soils, bacteria thrive and hold a competitive advantage over the fungus, thereby inhibiting the spread of the pathogen. Montmorillonite stimulated a broad spectrum of bacteria over all growth phases, but had a pronounced effect on shortening the lag phase. In addition to the stimulation of competing bacteria, montmorillonite adversely affected the growth of fungi when the viscosity was sufficiently high to impede the diffusion of oxygen through the system (Stotzky and Rem, 1967).

Other properties of the clay minerals that correlated directly with the inhibition of Fusarium wilt disease were surface area, cation exchange capacity, and type of exchangeable cation (Stotzky, 1966a), but these were considered to be either subordinate to or coincidental with the effect of pH. Particle size had no effect (Stotzky, 1966b).

Utilization of various substrates by heterotrophic (chemoorganotrophic) microorganisms is influenced by the presence of clay minerals in the medium. Although no general correlation has been made between the type of clay mineral and the type of substrate utilized (Kunc and Stotzky, 1974), montmorillonite seems to enhance substrate decomposition and the rate of nitrification (Macura and Stotzky, 1980), whereas kaolinite enhances oxidation (Kunc and Stotzky, 1974). Novakova (1968) reported that more glucose is mineralized by soil microorganisms in the presence of bentonite than when kaolinite is present. However, in a sand amended with bentonite, peptone utilization decreased, whereas kaolinite had no effect (Novakova, 1969).

Clay minerals appear to sequester some proteins by surface complexation, some of which then become available as a carbon source for bacteria while others become inactivated. When complexed to the clay, casein, chymotrypsin, and lactoglobulin were utilized by the microorganisms, but complexation had just the opposite effect on catalase, invertase, and pepsin by preventing their utilization (Stotzky, 1973). No explanation has been offered for these observations. Other general effects of phyllosilicates on the activity of heterotrophic microorganisms have been indicated by Novakova (1968, 1969), Filip (1968a,b, 1969, 1973), and Stotzky (1972). Enzymic reactions in soils may also be influenced by clay minerals (Ambroz, 1969; Galstyan et al., 1968; Skujins, 1967; Stotzky, 1972).

The biochemical activity of soil microorganisms, including respiration, is modified by their adhesion to clay particles (Lahav, 1962; Lahav and Keynan, 1962; Filip, 1973). Replacement of surface hydroxyl groups on the clay by negatively charged groups, such as carboxyl, at the cell surface is a possible mechanism for adhesion. Organic or inorganic polymers are also of critical importance in determining the extent of adhesion between clay surfaces and microorganisms. Hermesse et al. (1988) observed that strain 8246 of the bacterium *Acetobacter aceti* adhered strongly to freshly cleaved, but untreated mica surfaces, whereas adhesion was absent in other strains. When the mica was preconditioned by reaction with a $Fe(NO_3)_3$ solution, all strains adhered well to the mineral surface. Strain 8246 apparently is encapsulated by polymers that act as binding agents to the mica surface, and such capsules are absent in the other strains. The reaction with $Fe(NO_3)_3$ causes a polymeric species of $Fe(OH)O-$ to deposit on the mica surface, making it more positive and, thus, creating a stronger electrostatic attraction for the more negatively charged microbial surfaces. These phenomena tend to immobilize the organism, and may also affect metabolic processes.

Often the size of clay particles is much smaller than that of the organisms, resulting in the adsorption of clay particles onto the cell walls of the organism. Electron micrographic views (Figure 3) of such phenomena have been reported (Schmit and Robert, 1984; Schmit et al., 1989). From an environmental viewpoint, this can be detrimental because the surfaces of both the clay minerals and the microorganisms, including cell wall and envelop fragments from dead organisms, have a high affinity for toxic metal cations. If these surfaces and sites are tied up with one another by adsorption of either the clays onto the organic components or vice versa, then the capacity of the biological and inorganic materials to immobilize hazardous cations in soils and sediments will be greatly attenuated (Walker et al., 1989; Flemming et al., 1990; and Mullen et al., 1990).

Sometimes clays are a source of nutrients to microorganisms. Stotzky and Rem (1966) reported that montmorillonite and kaolinite provided inorganic nutrients for bacterial growth in similar, but limited, amounts. Mojallali and Weed (1978) observed that the inoculation of mica-rich soils with mycorrhizae induced considerable weathering and K release from the mica, but soybeans grown on the soils still exhibited K deficiencies when no other source of K was available, indicating that the rate of K release was insufficient to supply the needs of the plants. When KCl was added, no deficiencies were observed. The amounts of K available to the plants from various micas are compared in Table 3 and follow the same trend as K release in the presence of 1 N $CaCl_2$ (Figure 4). The mycorrhizae must, therefore, establish an ion exchange mechanism to scavenge K from the mica interlayers.

III. Phyllosilicate Dissolution by Microorganisms

The dissolution of soil minerals by microbial or fungal activity has been studied extensively and several excellent reviews are available (Dacey et al., 1981; Berthelin, 1983, 1988). The dissolution of phyllosilicates by biological agents may be either direct or indirect, which refers, respectively, to reaction of the phyllosilicate either with enzymes in the organism membrane (enzymatic reaction) or with an exudate compound (non-enzymatic reaction) (see Figure 1). Indirect dissolution generally occurs by either acid attack or chelation with organic compounds that are produced either during the metabolic processes of living organisms or released into the mineral environment upon cell death (Dacey et al., 1981).

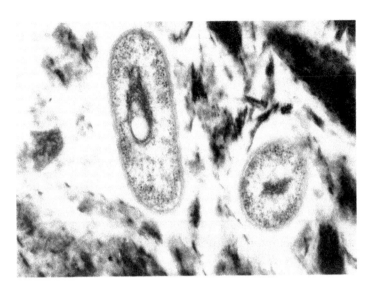

Figure 3. Electron micrograph of *Pseudomonas solanacearum* in montmorillonite extracted from Guadalupe vertisol (From Schmit et al., 1989.)

A. Micas

The dissolution of biotite and other micas by microbial and biological activity has received considerable attention, and involves both acid attack and complexation reactions (Berthelin and Belgy, 1979). Fungi are generally regarded as being more effective than bacteria in mica dissolution (Weed et al., 1969; Berthelin, 1983), but the more important property appears to be the complexing ability of the organic acid produced (Razzaghe-Karimi and Robert, 1975; Robert and Razzaghe-Karimi, 1975; Berthelin, 1983). The mechanism for degradation by microbial activity occurs either by a cation exchange process or by direct dissolution of the octahedral sheet. Berthelin and Boymond (1978) reported that polycarboxylic and hydrocarboxylic acids, which are strong complexing agents for metal cations, produced by microorganisms (including non-pigmented *Pseudomonas*, yeasts, and *Bacillus licheniformis*, *B. cereus*, *B. lenus*, *B. polymyxa*, and *B. megaterium*) in partially sterilized brown forest soils completely destroyed micas in the soil, but the micas were not destroyed by more weakly complexing acids. Robert and Razzaghe-Karimi (1975) reached a similar conclusion in studies of the breakdown of micas by organic acids, and Boyle et al. (1967) observed that dissolution results were similar with the addition of either organic acids or bacteria.

Oxalic acid is believed to be the most effective organic agent for dissolution of micas because of its combined acidic and chelating abilities (Wagner and Schwartz, 1967; Wilson and Jones, 1983; Boyle et al., 1967, 1974). It dissolves

Table 3. Potassium uptake by soybean plants from different K sources

Potassium source	Potassium uptake, mg/pot
Check (zero K added)	10.5
Muscovite	15.5
Phlogopite	15.2
Biotite B2	23.0
Biotite B3	45.4
KCl	72.4

Mycorrhizal and nonmycorrhizal treatments were averaged.
(Mojallali and Weed, 1978.)

octahedral cations, either converting biotite to vermiculite or leaving only a brittle residue of Si and Al (Berthelin and Boymond, 1978), and it removes more polyvalent cations from biotite than either citric, malonic, malic, lactic, or propionic acid (Boyle et al., 1967, 1974). However, Razzaghe-Karimi and Robert (1975) found that citric acid removed Al from phlogopite at a faster rate than did oxalic acid. *Aspergillus niger* produces both oxalic and citric acids and causes substantial dissolution of biotite, especially where the hyphal strands invade the fissures and cracks of biotite flakes (Boyle et al., 1967). The dissolution activity of *A. niger* could well be attributable to the organic acids produced. Wilson and Jones (1983) showed that lichens, which infested the surfaces and cracks of granites, dissolved Mg, Fe, K, and Al from trioctahedral micas in the granite, presumably by the same organic acid dissolution mechanism.

A similar phenomenon was observed by Frenkel (1977), who reported that microorganisms in estuarine sands caused biotite to break up and weather faster than hornblende. Microscopic examinations of particles revealed that bacterial populations in biotite were concentrated in depressed areas on mineral grains or particles, occupying holes, edges, and stepped cleavages, whereas flat surfaces were sparsely inhabited, giving evidence that biochemical attack may have occurred where the bacteria were located. The bulk of the microorganisms were observed between the layers rather than around the edges and outer surfaces, and many wedge-shaped openings were observed.

Other organic acids, although generally less effective than oxalic, can produce significant levels of dissolution and mineral alteration. Table 4 summarizes the effects of various organic acids, and reveals that citric and tartaric acids behave similarly to oxalic acid. Salicylic acid solubilizes micas by chelation, and quinic, lactic, formic, etc., acids tend to expand mica interlayers. Aspartic, alaninic, and butyric acids apparently have little or no effect on phyllosilicates. The order

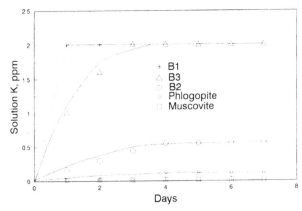

Figure 4. Release of potassium by several micas to 1 N CaCl$_2$ solution over a 7-day period. (From Mojallali and Weed, 1978.)

of activity is directly related to the complexing capacity of the organic acid (Berthelin, 1983). Microorganisms grown in a glucose medium produce 2-ketogluconic acid which, in turn, may have been responsible for the dissolution of about 10% of the K and 7 to 8% of the Al from muscovite reported by Duff et al. (1962), indicating that both the structural and interlayer cations were susceptible to attack.

The attack on minerals by microorganisms could be considered a somewhat passive process, i.e., as merely a coincidental reaction between the minerals and organic exudates. However, the uptake of inorganic nutrients supplied by the dissolution of minerals creates a strong incentive for the organism to be equipped with such compounds, and the removal of metal ions from the soil solution by the cells establishes a thermodynamic driving force for further mineral dissolution. In the previously cited study by Mojallali and Weed (1978), K was released from micas (biotite, muscovite, and phlogopite) in potted soils by the mycorrhizae associated with soybean roots in the soil. The weathering of the micas by K release was greatly accelerated, even though that rate was insufficient to supply the needs of the plants as compared to the addition of KCl. Potassium release from the mica occurred primarily around the edges and in cracks. Seven different fungi were also observed to provide a sink for K, which became the driving force for the exchange of K by Na in four different micas (Weed et al., 1969).

B. Other 2:1, 1:1, and 2:2 Phyllosilicates

Non-micaceous phyllosilicates are also susceptible to dissolution by microbial activity, as evidenced by the removal of modest amounts (2 to 21%, depending on the specific mineral) of structural Mg and Si from talc, serpentine, saponite,

Table 4. Influence of different organic acids on trioctahedral phlogopite

Potassium source	Potassium uptake, mg/pot
Check (zero K added)	10.5
Muscovite	15.5
Phlogopite	15.2
Biotite B2	23.0
Biotite B3	45.4
KCl	72.4

(From Robert and Razzaghe-Karimi, 1975.)

and vermiculite (Duff et al., 1962) by 2-ketogluconic acid exuded from microorganisms grown on glucose medium. Chlorite was completely removed and vermiculite partially removed from a granite rock by oxalic and high-molecular-weight acids (Berthelin and Boymond, 1978; Berthelin and Belgy, 1979).

In acid sulfate soils, where Fe(II), *Thiobacillus ferroxidans*, and basic Fe(III) sulfates are abundant, phyllosilicates glauconite and illite are dissolved and supply alkali cations for the formation of jarosites (Ivarson et al., 1978). Under the same conditions kaolinite resists dissolution, thus explaining why kaolinite is the principal phyllosilicate in acid sulfate soils (van Breemen, 1973). Kaolinite was also resistant to dissolution by aerobic bacteria in activated sludge (Dolobovskaya and Remizov, 1971), whereas montmorillonite was susceptible to only minor cation exchange. Oxalic acid also appears to dissolve these other phyllosilicates as it does micas, but little evidence is provided in the literature (Henderson and Duff, 1963; Berthelin et al., 1974; Berthelin and Belgy, 1979). The order of solubility of these phyllosilicates relative to organic acids is illite > montmorillonite > chlorite-vermiculite > kaolinite (Berthelin, 1983).

IV. Phase Transformations and Neoformation

Many hypotheses assert that microorganisms may have had, and continue to have, an important role in the formation and transformation of various minerals in the earth. Such proposed processes were reviewed by Ferris et al. (1989). Among the evidence supporting these hypotheses is the intimate association that exists between minerals and organic compounds in the soil (see, for example, Wang and Huang, 1986). The result of some microbial activity is either the neoformation of phyllosilicates from primary minerals or the transformation of one phyllosilicate to another, such as the transformation of biotite to vermiculite (Hernandez and Robert, 1975) by microbially produced galacturonic acid or first to vermiculite then to beidellite. In weakly complexing agents such as lactic acid, biotite transforms to hydroxy-aluminum vermiculite (Hernandez and Robert, 1975). Berthelin and Boymond (1978) observed that anaerobic and facultative anaerobic bacteria such as *Clostridium, B. cereus, B. licheniformis,* and *B. polymyxa* transformed illite-vermiculite into montmorillonite.

Weed et al. (1969) reported extensive alteration of mica to vermiculite by various strains of fungi. They noticed that the ability of the fungi to convert biotite to vermiculite, or to expand the mica, was related to the growth phase and history of the fungi. They proposed that the fungi served as a sink for K, and thus extracted K from the mica interlayers. Sodium deficiency in solution retarded the expansion process, indicating that Na was important as a replacement exchangeable cation for the extracted K.

Xianmo (1984) wrote an extensive treatise on the theory of soil formation based on biological processes. He proposed four biological steps or phases in the development of soils, based on empirical evidence. The first phase involves the attachment of lichens or other microorganisms to the faces of bare rocks, resulting in the beginning of a weathering sequence due to biological activity and influence. In the second phase, or lichen stage, hyphae from the lichens intrude into primary silicate minerals, utilizing the elements taken from the silicate and producing a secondary mineral (probably an illitic type of phyllosilicate) as a by-product of the metabolic processes of the organism. Such clay minerals are, thus, referred to as being of "biological origin". In the third phase, bryophytaes grow in the lichen relics, or remains, and the initial secondary mineral weathers to other phyllosilicates such as vermiculite, montmorillonite, and kaolinite. Such biophysical processes cause invasion of cracks, fissures, and defects in mineral crystals, which not only expands the surface area available for biological activity, but also retains moisture and provides channels for inorganic weathering processes to occur, including reactions that are pH dependent.

Boyer (1982) found that some types of termites synthesize within their mounds "artificial" illite, with lesser amounts of hydrobiotite, vermiculite, halloysite, smectite, and occasionally, sepiolite and attapulgite. These results were attributed to the combination of grinding and salivary trituration processes, where minerals already present in the soil were acted upon by the termites.

V. Change in Phyllosilicate Microstructure by Organic Exudates

The effects of microorganisms on the microstructure of phyllosilicates arise primarily from the interaction of the clay with polysaccharide exudates from the organisms, which include both chemical sorption and physical entanglement. These phenomena are not unique to phyllosilicates nor to microbial exudates, and contribute significantly to soil structure. Any source of organic compounds potentially can contribute to organo-mineral associations. A review of the large body of literature covering organo-mineral interactions, even if limited to phyllosilicate-organic reactions, is beyond the scope of this treatise, and the reader is referred to classic works such as those by Martin (1945), MacCalla (1946), Stevenson (1982), Theng (1974, 1979), and Mortland (1986). But such interactions are relevant to a discussion of clay-microbe processes because microorganisms are the origin of many natural organics. One series of experiments will be described here to illustrate the type of results that are observed.

Recent work by Chenu and co-workers (Chenu et al., 1987; Chenu, 1989; Chenu and Jaunet, 1990) showed that some fungal polysaccharides enhanced the porosity of Ca-montmorillonite and Ca-kaolinite. In Ca-montmorillonite gels, the clay was organized into ribbons or particles about 9 to 11 layers thick, having relatively uniform d-spacings of ≤ 2.0 nm, and being of large lateral extent. Each such particle or stack of layers was associated with up to 2 or 3 other, similar particles to form a quasi-crystal about 40 layers thick. The distance between particles was greater than the distance between the individual layers in each particle, forming a micropore 3.0 to 5.0 nm long. In the presence of polysaccharide, the number of particles in the quasi-crystal was about 15, or less than half that in the untreated clay, indicating that the organic molecule inhibited the face-to-face association of one clay particle to another. The lateral extent of the quasi-crystals was also decreased by the polysaccharide, which further demonstrates its inhibitory effect on clay interlayer attractive forces.

VI. Changes in Mineral Redox Status

Microorganisms are well known for their capacity to oxidize or reduce Fe in solution (Takai and Kamura, 1966), in chelates (Cox, 1980), and in Fe oxide minerals (Münch and Ottow, 1977, 1982; Ottow, 1982; Lovley and Phillips, 1986, 1987; Fischer, 1988); but only three studies have reported microbial redox reactions with phyllosilicates (Komadel et al., 1987; Stucki et al., 1987; Wu et al., 1988). Stucki et al. (1987) reported that the indigenous microbial population in ferruginous smectite, when provided a nutrient source, successfully reduced structural Fe(III) (Figure 5; Table 5). About 0.30 mmole of Fe(III)/g of clay, or about 8% of the total structural Fe, was reduced to Fe(II) *insitu* by the indigenous organism(s) in unsterilized, freeze-dried samples of ferruginous smectite SWa-1 simply by incubating in a nutrient broth (NB) suspension at room temperature for two weeks (Figure 5). Extended incubation (71 days) increased the Fe(II) content to 0.40 mmole/g or about 11% of total Fe. Exclusion of O_2 from the suspensions decreased the extent of reduction by half (compare curves C in Figures 5 and 6). Subsequently, a single, gram-negative organism (denoted P-1) was isolated from the clay, then sterilized samples of

Table 5. Iron(II) content achieved in various phyllosilicates by microbial reduction

Phyllosilicate	Total Fe	Glucose[1]		MT[2]		NB[3]	
	%	$Fe(II)/Fe_{Tot}$	m mole/g	$Fe(II)/Fe_{Tot}$	m mole/g	$Fe(II)/Fe_{Tot}$	m mole/g
API 25	3.01	n.d.[4]	n.d.	0.24	0.13	0.51	0.27
API 33A	23.46	0.41	1.72	n.d	n.d.	0.06	0.25
SWa-1	20.07	n.d.		0.18	0.65	0.09	0.32
Jelsovy Potok	2.07	n.d.		0.25	0.09	n.d	n.d.
Kadan Rokle	10.05	n.d.		0.16	0.29	n.d	n.d

[1]Incubation in glucose at room T, for 14 days inoculated with saturated extract from Chinese rice paddy soil (Wu et al., 1988).

[2]Incubation in modified Thornton's medium (see text) for 7 days at room T, inoculated with *Pseudomonas fluorescens* 13-79 (Komadel, et al., 1987).

[3]Incubation in nutrient broth for 28 days at room T, inoculated with indigenous organism P-1 from sample SWa-1 (Stucki et al., 1987).

[4]n.d. - not determined

<u>Note</u>: Because the total Fe content varies widely in these clays, the corresponding values for Fe(II) content, when expressed as percent of total Fe, differed greatly among samples, i.e., 50, 6, and 9%, respectively. Comparisons between clays must, therefore, be made on an absolute scale (e.g., mmole/g), whereas percent of total Fe can be used when comparing different treatments of the same clay.

242J.W. Stucki, H. Gan, and H.T. Wilkinson

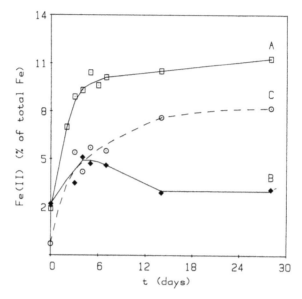

Figure 5. Reduction of structural Fe(III) to Fe(II) in unpurged sample of unsterilized (A) and sterilized (B) SWa-1 smectite incubated in nutrient broth (NB) for various time periods at room temperature. Curve C is the mathematical difference between A and B. (From Stucki et al., 1987.)

three smectites (Upton, Wyoming, montmorillonite, API 25; Garfield, Washington, nontronite, API 33a; and ferruginous smectite, SWa-1) were inoculated with an enrichment culture of that organism. After 28 days of incubation in a NB suspension, the absolute Fe(II) content was about the same in all three clays (Table 5): 0.27, 0.25, and 0.32 mmole/g in API 25, API 33a, and SWa-1, respectively. These levels were comparable with that obtained with SWa-1 by incubating the original unsterilized clay (0.30 mmole/g).

The level of reduction achieved depended on the choice of growth medium, organism, and, perhaps, on the total Fe content of the clay. In modified Thornton's medium (MT) (recipe modified from that described by Thornton (1922) by deleting $FeCl_3$ and agar), for example, two different strains of *Pseudomonas fluorescens* (WB2-79 and WBRu15-b) were less effective than in a nutrient broth (NB) suspension (Figure 7). Incubation of the same or similar organisms in glucose resulted in measurable, but lower, levels of reduction (not shown).

Even greater differences were observed when the organism was changed or a consortium of organisms was used. Wu et al. (1988) found more than 1.68 mmole Fe(II)/g clay in sample API 33a when saturated extracts from Chinese paddy soils were incubated for 14 days at room temperature with the clay

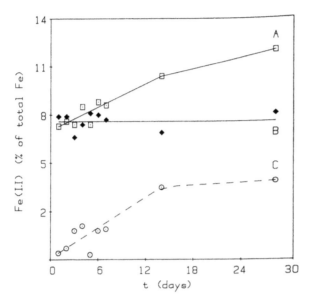

Figure 6. Reduction of sturctural Fe(III) to Fe(II) in N_2-purged sample of unsterilized (A) and sterilized (B) SWa-1 smectite incubated in nutrient broth (NB) for various time periods at room temperature. Curve C is the mathematical difference between A and B. (From Stucki et al., 1987.)

suspended in sucrose (Table 5), compared with only 0.25 mmole/g reduction (Stucki et al., 1987) when the same clay was incubated with an enrichment culture of the indigenous organism, P-1, isolated from sample SWa-1 (Table 5). A consortium of organisms probably was present in the soil extracts, indicating that the reduction potential of this single organism was less than the combined potential resulting from collaborative microbial activity in the soil extract. The highest levels of Fe reduction, thus, may require a consortium of organisms whose reducing capabilities are distributed both spatially and temporally. Measuring the activity of only one organism from the consortium could, therefore, result in either no reduction or only a fraction of the total consortium potential. Further evidence to support this hypothesis is found in the recent results of Gates et al. (W. P. Gates, J. W. Stucki, and H. T. Wilkinson, unpublished results, University of Illinois, and Agronomy Abstracts 1991:365), which revealed that the combination of four different *Pseudomonas* bacteria produced as much as 2.13 mmole Fe(II)/g clay in SWa-1, whereas the maximum reduction achieved by any of the constituent bacteria acting alone was only about 0.71 mmole Fe(II)/g clay.

The ultimate controlling factor for microbial reduction is, thus, the reducing potential of the bacteria, whether in pure culture or in concert. Hence, the

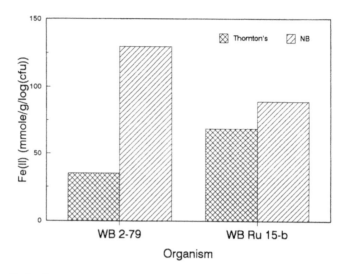

Figure 7. Reduction of structural Fe(III) to Fe(II) in unpurged samples of sterilized SWa-1 smectite incubated with *Pseudomonas fluorescens* strains WB2-79 and WBRu 15-b in either Thornton's medium or nutrient broth (NB). (From Komadel, Wilkinson, and Stucki, unpublished results.)

bacterial composition of the consortium will determine its reduction capacity. Komadel et al. (1987) found that *P. fluorescens* strain WB13-79 reduced about the same amount of Fe in SWa-1 (0.64 mmole Fe(II)/g clay) as did the bacteria from three Arkansas paddy soil extracts (0.46 to 0.71 mmole Fe/g clay), so the bacterial constituents from the Arkansas consortia were no more effective than the single organism WB13-79, and their reduction potential was much less than that found by Wu et al. (1988) in the China soil extract.

In terms of the absolute levels of Fe reduction achieved in pure cultures, four of the seven values from pure cultures reported in Table 5 (MT and NB) were about equal (0.27 to 0.32 mmole/g), even though four clays, two organisms, and two different media were represented by these values. These results may indicate that the maximum reduction level is limited in phyllosilicates, but notice from the MT results in Table 5 that the level of Fe(II) produced in four different phyllosilicates by *P. fluorescens* WB13-79, under identical growth conditions, increased with increasing total Fe content. The absolute amount of Fe reduction that can be achieved may, thus, depend on the total Fe content of the clay. More work is required before definitive conclusions can be drawn regarding the factors which optimize Fe reduction in phyllosilicates.

The oxidation and reduction of structural Fe by inorganic chemical agents greatly alters the physical-chemical properties of smectite phyllosilicates (Stucki, 1988; Scott and Amonette, 1988). An important question is whether microbial reduction invokes the same types of changes in clay properties. Chemical reduction is capable of reducing virtually all of the Fe in smectite structures

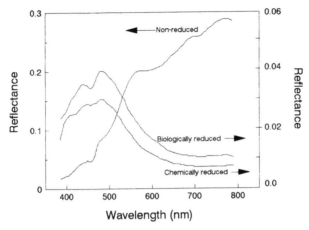

Figure 8. Reflectance spectra of nonreduced, biologically reduced, and chemically reduced Na-nontronites. (From Wu et al., 1988.)

(Komadel et al., 1990), and the levels of microbial reduction observed by Wu et al. (1989) (0.41 mmole Fe(II)/g clay) and recently by Gates et al. were significant and comparable to common levels attained chemically. Wu et al. (1988) further observed that the color of the biologically reduced sample, as measured by spectral reflectance (Figure 8), was similar to the chemically reduced sample. This suggests, according to the arguments of Lear and Stucki (1987), that Fe reduction occurs randomly throughout the octahedral sheet of the clay, and that intervalence electron transfer and magnetic exchange interactions within the microbially reduced clay layer are probable (Lear and Stucki, 1987, 1990). During chemical reduction, the interlayer attractive forces in the clay are greatly increased, causing superimposed layers to collapse (Wu et al., 1989; Lear and Stucki, 1989; Khaled and Stucki, 1991; Stucki and Tessier, 1991). A similar phenomenon appears to occur with microbial reduction, as evidenced by the collapsed layers (Figure 9) observed by Wu et al. (1988). If microbial reduction similarly affects all properties already known to change with chemical reduction, such as specific surface area (Lear and Stucki, 1990), swellability (Foster, 1953; Egashira and Ohtsubo, 1983; Stucki et al., 1984b), cation fixation (Chen et al., 1987; Stucki and Lear, 1989; Khaled and Stucki, 1991), and cation exchange capacity (Stucki and Roth, 1977; Stucki et al., 1984a), the potential for controlling or modifying, at least temporarily, the physical and chemical properties in natural systems is tremendous. The recent study by Gates et al. revealed that microbial reduction does, indeed, decrease the swelling pressure of smectite clays.

The potential significance of microbial reduction of phyllosilicates is great because virtually all phyllosilicates contain Fe in their crystal structures, although the amount ranges from only a trace in hectorite to almost all of the octahedral sites being occupied by Fe in lepidomelane. Notable exceptions,

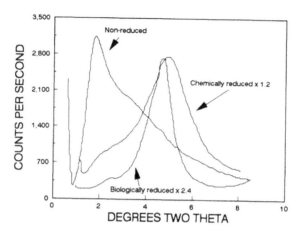

Figure 9. X-ray diffraction patterns of Na-nontronites in different states of reduction (1.2 and 2.4 represent scale expansion factors used for the respective patterns. (Wu et al., 1988.)

which contain no Fe, are talc and pyrophyllite, the pure Mg and Al end members, respectively, of the 2:1 type layer silicates. The presence of Fe in the crystal structure makes phyllosilicates susceptible to oxidation and reduction *insitu*, which changes the redox potential of the surrounding soil or sediment and induces numerous changes in clay properties and behavior (Stucki, 1988; Stucki and Lear, 1989). The redox potential found in sediments and soil profiles can have great practical impact on water quality because the fate and mobility of redox-sensitive pollutants, such as Cr, often varies with its oxidation state. A knowledge of the mechanisms by which the redox potential of the surrounding minerals are controlled could enable the *insitu* manipulation of the soil or sediment to produce the most favorable state for the percolating ion, thus enhancing the quality of effluent waters. The recent observations that microorganisms can change the oxidation state of phyllosilicate minerals strongly suggests that such methods may be feasible, but more work is needed in this area.

VII. Summary

Reviewed in this chapter are the effects of microorganisms on the properties of phyllosilicate clay minerals. Also included is an overview of the changes in microbial behavior that are induced by the presence of phyllosilicates. Dissolution of phyllosilicates occurs either by acid attack from organic acids exuded by the microorganisms, or by nutrient utilization in which elements, such as K, are taken up by the cell. As the elements are removed from the mineral solution, the thermodynamic equilibrium is shifted toward dissolution. Direct contact between the mineral and the cell wall may also occur, in which case enzymic reactions selectively remove elements from the clay structure. In a few instances, microorganisms are believed to produce new or different phyllosilicate as by-products of metabolic dissolution or utilization of other minerals. Clay fabric or microstructure may be altered by the interaction of microbially produced polysaccharides with external surfaces of clay particles.

The microbial reduction of structural Fe in phyllosilicates was recently observed, and appears to change the physical-chemical properties of smectites in ways which parallel chemical reduction. The fraction of partially collapsed layers relative to the fully expanded fraction increases with microbial reduction, indicating that a more consolidated, less-swelling structure exists. The changes in color also reveal that Fe reduction occurs randomly throughout the clay layer. While more work is required to fully characterize the effect of microbial reduction on the behavior of phyllosilicates, the potential for using microorganisms for *insitu* modification of soil and sediment properties is great.

Acknowledgment

Financial support of H. Gan by a cooperative agreement (CR816780-01-2) from the U.S. Environmental Protection Agency is gratefully acknowledged.

References

Ajello, L. 1967. Comparative ecology of respiratory mycotic disease agents. *Bacteriol. Rev.* 31:6.

Ambroz, Z. 1969. On the effect of bentonite on the enzymatic activity of the soil microflora. (In Czech). *Acta Univ. Vyroba (Praha)* 12:209.

Bailey, S.W. 1980. Summary of recommendations of AIPEA Nomenclature Committee. *Clays Clay Miner.* 28:73-78.

Berthelin, J. 1983. Microbial weathering processes. pp. 223-262. In: W.E. Krumbein (ed.) *Microbial Geochemistry.* Blackwell Scientific, Oxford.

Berthelin, J. 1988. Microbial weathering processes in natural environments. pp. 33-59. In: A. Lerman and M. Meybeck (eds.) *Physical and chemical weathering in geochemical cycles.* Kluwer Academic, Dordrecht, Netherlands.

Berthelin, J. and D. Boymond. 1978. Some aspects of the role of heterotrophic microorganisms in the degradation of minerals in waterlogged acid soils. pp. 659-673. In: W.E. Krumbein (ed.) *Environmental biogeochemistry and geomicrobiology.* Vol. 2. Ann Arbor Science, Ann Arbor, MI.

Berthelin, J. and G. Belgy. 1979. Microbial degradation of phyllosilicates during simulated podzolization. *Geoderma* 21:297-310.

Berthelin, J., A. Kogblevi, and Y. Dommergues. 1974. Microbial weathering of a brown forest soil. Influence of partial sterilization. *Soil Biol. Biochem.* 67:393-399.

Boyd, R.F. 1984. *General Microbiology.* Times Mirror/Mosby College Publishing, St. Louis. 807 pp.

Boyer, P. 1982. Quelques aspects de l'action des termites du sol sur les argiles. *Clay Miner.* 17:453-462.

Boyle, J.R., G.K. Voigt, and B.L. Sawhney. 1967. Biotite flakes: alteration by chemical and biological treatment. *Science* 155:193-195.

Boyle, J.R., G.K. Voigt, and B.L. Sawhney. 1974. Chemical weathering of biotite by organic acids. *Soil Sci.* 117:42-45.

Brindley, G.W. and G.Brown (eds.) 1980. *Crystal structures of clay minerals and their X-ray identification.* Monograph No. 5. Mineralogical Society, London.

Chen, S.A., P.F. Low, and C.B. Roth. 1987. Relation between potassium fixation and the oxidation state of octahedral iron. *Soil Sci. Soc. Am. J.* 51:82-86.

Chenu, C. 1989. Influence of a fungal polysaccharide, scleroglucan, on clay microstructures. *Soil Biol. Biochem.* 21:299-305.

Chenu, C. and A.M. Jaunet. 1990. Modifications de l'organisation texturale d'une montmorillonite calcique liées à l'adsorption d'un polysaccharide. *C. R. Acad. Sci. Paris* 310:975-980.

Chenu, C., C.H. Pons, and M. Robert. 1987. Interaction of kaolinite and montmorillonite with neutral polysaccharide. pp. 375-381. In: L.G. Schultz, H. van Olphen, and F.A. Mumpton (eds.) *Proc. Int. Clay Conf.*, Denver, 1985. The Clay Minerals Society, Bloomington, IN.

Cox, C.D. 1980. Iron reductases from *Pseudomonas aeruginosa. J. Bacteriology* 141:199-204.

Dacey, P.W., D.S. Wakerley, and N.W. Le Roux. 1981. *The biodegradation of rocks and minerals with particular reference to silicate minerals: A literature survey.* Report No. LR 380 (ME), Warren Spring Laboratory, Department of Industry, P.O. Box 20, Gunnels Wood Road, Stevenage, Hertfordshire SG1 2BX, UK.

Dolobovskaya, A.S. and V.I. Remizov. 1971. Aerobic transformations of clay minerals by heterotrophic microflora. *Doklady Akad. Nauk SSSR* 206:211-213.

Duff, R.B., D.M. Webley, and R.O. Scott. 1962. Solubilization of minerals and related materials by 2-ketogluconic acid-producing bacteria. *Soil Sci.* 95:105-114.

Egashira, K. and M. Ohtsubo. 1983. Swelling and mineralogy of smectites in paddy soils derived from marine alluvium, Japan. *Geoderma* 29:119-127.

Ferris, F.G., W. Shotyk, and W.S. Fyfe. 1989. Mineral formation and decomposition by microorganisms. p. 413-441. In: T. J. Beveridge and R. J. Doyle (eds.) *Metal Ions and Bacteria.* John Wiley & Sons, New York.

Filip, Z. 1968a. The influence of small supplements of bentonite on the development of certain groups of microorganisms in a soil culture. *Rostlinna Vyroba* 14:209-216.

Filip, Z. 1968b. Respiration activity of soil microflora in media with a different bentonite content. *Rostlinna Vyroba* 14:963-968.

Filip, Z. 1969. Characteristics of humic substances in a soil incubated with additions of bentonite. *Rostlinna Vyroba* 15:377-390.

Filip, Z. 1973. Clay minerals as a factor influencing the biochemical activity of soil microorganisms. *Folia Microbiol.* 18:56-74.

Fischer, W.R. 1988. Microbiological reactions of iron in soils. pp. 715-748. In: J.W. Stucki, B.A. Goodman, and U. Schwertmann (eds.) *Iron in Soils and Clay Minerals.* D. Reidel Publ., Dordrecht, Netherlands.

Flemming, C.A., F.G. Ferris, T.J. Beveridge, and G.W. Bailey. 1990. Remobilization of toxic heavy metals adsorbed to bacterial wall-clay composites. *Applied Environ. Microbiol.* 56:3191-3203.

Foster, M.D. 1953. Geochemical studies of clay minerals: II. relation between ionic substitution and swelling in montmorillonites. *Am. Mineral.* 38:994-1006.

Frenkel, L. 1977. Microorganism induced weathering of biotite and hornblende grains in estuarine sands. *J. Sediment. Petrol.* 47:849-854.

Galstyan, A.S., G.S. Tatevosian, and S.Z. Havoundjian. 1968. Fixation of enzymes by soil fractions of different particle sizes. *9th Int. Congr. Soil Sci. Trans. III*, pp. 281-288.

Gates, W.P., J.W. Stucki, and H.T. Wilkinson. Unpublished results, University of Illinois, and Agronomy Abstracts 1991:365.

Grim, R.E. 1968. *Clay Mineralogy.* McGraw-Hill, New York.

Henderson, M.E.K. and R.B. Duff. 1963. The release of metallic and silicate ions from minerals, rocks and soils by fungal activity. *J. Soil Sci.* 14:236-246.

Hermesse, M.P., C. Dereppe, Y. Bartholome, and P.G. Rouxhet. 1988. Immobilization of *Acetobacter aceti* by adhesion. *Can. J. Microbiol.* 34:638-644.

Hernandez, M.A.V. and M. Robert. 1975. Transformation profonde des micas sous l'action de l'acide galacturonique. Problèm des smectites des podzols. *C.R. Acad. Sci. Paris* 281:523-526.

Huang, P.M. and M. Schnitzer (eds.) 1986. *Interactions of Soils Minerals with Natural Organics and Microbes.* Soil Sci. Soc. Am. Spec. Publ. Vol. 17, Soil Science Society of America, Madison, WI.

Ivarson, K.C., G.J. Ross, and N.M. Miles. 1978. Alterations of micas and feldspars during microbial formation of basic ferric sulfates in the laboratory. *Soil Sci. Soc. Am. Proc.* 42:518-524.

Jepson, W.B. 1984. Kaolins: their properties and uses. *Phil. Trans. R. Soc. Lond.* A 311:411-432.

Jepson, W.B. 1988. Structural iron in kaolinites and associated ancillary minerals. pp. 467-536. In: J.W. Stucki, B.A. Goodman, and U. Schwertmann (eds.) *Iron in soils and clay minerals*. D. Reidel, Dordrecht.

Khaled, E.M. and J.W. Stucki. 1991. Effects of iron oxidation state on cation fixation in smectites. *Soil Sci. Soc. Am. J.* 55:550-554.

Komadel, P., J.W. Stucki, and H.T. Wilkinson. 1987. Reduction of structural iron in smectites by microorganisms. pp. 322-324. In: E. Galán, J.L. Pérez-Rodriguez, and J. Cornejo (eds.) *Proc. Sixth Meeting of the European Clay Groups, Seville, 1987.* Sociedad Española de Arcillas, Sevilla.

Komadel, P., P.R. Lear, and J.W. Stucki. 1990. Reduction and reoxidation of iron in nontronites: Rate of reaction and extent of reduction. *Clays Clay Miner.* 38:203-208.

Kunc, F. and G. Stotzky. 1974. Effect of clay minerals on heterotrophic microbial activity in soil. *Soil Sci.* 118:186-195.

Lahav, N. 1962. Adsorption of sodium bentonite particles on *Bacillus subtilis*. *Plant Soil* 17:191.

Lahav, N. and A. Keynan. 1962. The influence of bentonite and attapulgite on the respiration of *Bacillus subtilis*. *Can. J. Microbiol.* 8:565.

Lear, P.R. and J.W. Stucki. 1987. Intervalence electron transfer and magnetic exchange interactions in reduced nontronite. *Clays Clay Miner.* 35:373-378.

Lear, P.R. and J.W. Stucki. 1989. Effects of iron oxidation state on the specific surface area of nontronite. *Clays Clay Miner.* 37:547-552.

Lear, P.R. and J.W. Stucki. 1990. Magnetic ordering and site occupancy of iron in nontronite. *Clay Miner.* 25:3-13.

Lovley, D.R. and E.J.P. Phillips. 1986. Availability of ferric iron for microbial reduction in bottom sediments of the freshwater tidal Potomac River. *Appl. Environ. Microbiol.* 52:751-757.

Lovley, D.R. and E.J.P. Phillips. 1987. Competitive mechanisms for inhibition of sulfate reduction and methane production in the zone of ferric iron reduction in sediments. *Appl. Environ. Microbiol.* 53:2636-2641.

MacCalla, T.M. 1946. Influence of some microbial groups on stabilizing soil structure against failing water drops. *Soil Sci. Soc. Am. Proc.* 11:260-263.

Macura, J. and G. Stotzky. 1980. Effect of montmorillonite and kaolinite on nitrification in soil. *Folia Microbiol.* 25:90-105.

Martin, J.P. 1945. Microorganisms and soil aggregation. *Soil Sci.* 59:163-174.

Mojallali, H. and S.B. Weed. 1978. Weathering of micas by mycorrhizal soybean plants. *Soil Sci. Soc. Am. J.* 42:367-377.

Mortland, M.M. 1986. Mechanisms of adsorption of nonhumic organic species by clays. pp. 59-76. In: P.M. Huang and M.Schnitzer (eds.) *Interactions of Soils Minerals with Natural Organics and Microbes.* Soil Sci. Soc. Am. Spec. Publ. Vol. 17, Madison, WI.

Mullen, M.D., T.J. Beveridge, F.G. Ferris, D.C. Wolf, and G.W. Bailey. 1990. *Sorption of heavy metals by intact microorganisms, cell walls, and clay-wall composites.* EPA Environmental Research Brief, Environmental Research Laboratory, Athens, Georgia, EPA/600/M-90/004.

Münch, J.C. and J.C.G. Ottow. 1977. Model experiments on the mechanism of bacterial iron-reduction in water-logged soils. *Z. Pflanzenernaehr. Bodenkd.* 140:549-562.

Münch, J.C. and J.C.G. Ottow. 1982. Effect of cell contact and iron(III) oxide form on bacterial iron reduction. *Z. Pflanzenernaehr. Bodenkd.* 145:66-77.

Newman, A.C.D. 1984. The significance of clays in agriculture and soils. *Phil. Trans. R. Soc. Lond.* A 311:375-389.

Novakova, J. 1968. Einfluss der tonminerale auf die mikrobiologische aufnehmbarkeit der kohlenwasserstoffe und N-haltigen stoffe. I. Mitteilung: Ausnützung der glukose durch die komplexe bondenmikroflora in sandkulturen. *Zentr. Bakteriol. Parasitenk. II.* 122:233-241.

Novakova, J. 1969. Einfluss der tonminerale auf die mikrobiologische aufnehmbarkeit der kohlenwasserstoffe und N-haltigen stoffe. II. Ausnützung des peptons durch die komplexe mikroflora in sandkulturen. *Zentr. Bakteriol. Parasitenk. II.* 123:56-63.

Odom, I.E. 1984. Smectite clay minerals:properties and uses. *Phil. Trans. R. Soc. Lond.* A 311:391-409.

Ottow, J.C.G. 1982. Significance of redox potentials for reduction of nitrate and iron(III) oxides in soils. *Z. Pflanzenernaehr. Bodenkd.* 145:91-93.

Razzaghe-Karimi, M.H. and M. Robert. 1975. Altération des micas et géochimie de l'aluminium: rôle de la configuration de la molécule organique sur l'aptitude à la complexation. *C.R. Acad. Sci., Paris* 280:2645-2648.

Robert, M. and M.H. Razzaghe-Karimi. 1975. Mise en évidence de deux types d'évolution minéralogique des micas trioctaédriques en présence d'acides organiques hydrosolubles. *C.R. Acad. Sci., Paris* 280:2175-2178.

Robert, M. and J. Berthelin. 1986. Role of biological and biochemical factors in soil mineral weathering. p. 453-459. In: P.M. Huang and M. Schnitzer (eds.) *Interactions of Soils Minerals with Natural Organics and Microbes.* Soil Sci. Soc. Am. Spec. Publ. Vol. 17, Madison, WI.

Ross, C.S. and S.B. Hendricks. 1945. *Minerals of the montmorillonite group. Their origin and relation to soils and clays.* U.S. Geol. Surv., Profess. Paper 205b:23-79.

Schmit, J. and M. Robert. 1984. Action des argiles sur la survie d'une bactérie phytopathogène *Pseudomonas solanacearum.* E.F.S. *C.R. Acad. Sci., Paris* 299 (II, No. 11):733-738.

Schmit, J., P. Prior, H. Quiquampoix, and M. Robert. 1989. Studies on survival and localization of *Pseudomonas solanacearum* in clays extracted from vertisols. pp. 1001-1009. In: Z. Klement (ed.) *Plant Pathogenic Bacteria, Proc. 7th Int. Conf. Plant Path. Bact., Budapest, Hungary, 1989.* Akadémiai Kiadó (Publishing House of the Hungarian Academy of Sciences), Budapest.

Scott, A.D. and J.E. Amonette. 1988. Role of iron in mica weathering. pp. 537-623. In: J.W. Stucki, B.A. Goodman, and U. Schwertmann (eds.) *Iron in soils and clay minerals.* D. Reidel, Dordrecht.

Silver, M., H.L. Ehrlich, and K.C. Ivarson. 1986. Soil mineral transformation mediated by soil microbes. pp. 497-519. In: P. M. Huang and M. Schnitzer (eds.) *Interactions of Soils Minerals with Natural Organics and Microbes.* Soil Sci. Soc. Am. Spec. Publ. Vol. 17, Madison, WI.

Skujins, J.J. 1967. Enzymes in soil. pp. 371-413. In: A.D. McLaren and G. H. Peterson (eds.) *Soil Biochemistry.* Marcel Dekker, New York.

Stevenson, F.J. 1982. *Humus chemistry: genesis, composition, reactions.* Wiley-Interscience, New York.

Stotzky, G. 1965. Microbial respiration. pp. 1550-1570. In: C.A. Black et al. (eds.) *Methods of Soil Analysis, Part II. Chemical and microbiological properties.* Monograph 9. American Society of Agronomy, Madison, WI.

Stotzky, G. 1966a. Influence of clay minerals on microorganisms. II.Effect of various clay species, homoionic clays, and other particles on bacteria. *Can. J. Microbiol.* 12:831-848.

Stotzky, G. 1966b. Influence of clay minerals on microorganisms. III. Effect of particle size, cation exchange capacity, and surface area on bacteria. *Can. J. Microbiol.* 12:1235-1246.

Stotzky, G. 1967. Clay minerals and microbial ecology. *Trans. N. Y Acad. Sci.* 30:11-21.

Stotzky, G. 1972. Activity, ecology, and population dynamics of microorganisms in soil. *CRC Critical Reviews in Microbiology,* November 1972:59-137.

Stotzky, G. 1973. Techniques to study interactions between microorganisms and clay mineral *in vivo* and *in vitro. Bull. Ecol. Res. Comm. (Stockholm)* 17:17-28.

Stotzky, G. 1980. Surface interactions between clay minerals and microbes, viruses and soluble organics, and the probable importance of these interactions to the ecology of microbes in soil. pp. 231-247. In: R.C.W. Berkeley, J.M. Lynch, J. Melling, P.R. Rutter, and B. Vincent (eds.) *Microbial adhesion to surfaces.* Ellis Horwood, Ltd., Chichester, UK.

Stotzky, G. 1985. Mechanism of adhesion to clays, with reference to soil systems. A review. pp. 195-253. In: D.C. Savage and M. Fletcher (eds.) *Bacterial Adhesion: Mechanisms and Physiological Significance.* Plenum Press, New York.

Stotzky, G. 1986. Influence of soil mineral colloids on metabolic processes, growth, adhesion, and ecology of microbes and viruses. pp. 305-428. In: P.M. Huang and M. Schnitzer (eds.) *Interactions of Soils Minerals with Natural Organics and Microbes.* Soil Sci. Soc. Am. Spec. Publ. Vol. 17, Madison, WI.

Stotzky, G. and L.T. Rem. 1966. Influence of clay minerals on microorganisms. I.Montmorillonite and kaolinite on bacteria. *Can. J. Microbiol.* 12: 547-563.

Stotzky, G. and L.T. Rem. 1967. Influence of clay minerals on microorganisms. IV. Montmorillonite and kaolinite on fungi. *Can. J. Microbiol.* 13:1535-1550.

Stotzky, G. and A.H. Post. 1967. Soil mineralogy as possible factor in geographic distribution of *Histoplasma capsulatum. Can. J. Microbiol.* 13:1-7.

Stotzky, G., J.E. Dawson, R.T. Martin, and G.H.H. ter Kuile. 1961. Soil mineralogy as a factor in the spread of Fusarium wilt banana. *Science* 133:1483-1485.

Stucki, J.W. 1988. Structural iron in smectites. pp. 625-675. In: J.W. Stucki, B.A. Goodman, and U. Schwertmann (eds.) *Iron in soils and clay minerals.* D. Reidel, Dordrecht.

Stucki, J.W. and C.B. Roth. 1977. Oxidation-reduction mechanism for structural iron in nontronite. *Soil Sci. Soc. Am. J.* 51:808-811.

Stucki, J.W. and P.R. Lear. 1989. Variable oxidation states of iron in the crystal structure of smectite clay minerals. pp. 330-358. In: L.M. Coyne, D. Blake, and S. McKeever (eds.) *Structures and Active Sites of Minerals.* American Chemical Society, Washington, D.C.

Stucki, J.W. and D. Tessier. 1991. Effects of iron oxidation state on the texture and structural order of Na-nontronite. *Clays Clay Miner.* 39:137-143.

Stucki, J.W., D.C. Golden, and C.B. Roth. 1984a. The effect of reduction and reoxidation on the surface charge and dissolution of dioctahedral smectites. *Clays Clay Miner.* 32:350-356.

Stucki, J.W., P.F. Low, C.B. Roth, and D.C. Golden. 1984b. Effects of oxidation state of octahedral iron on clay swelling. *Clays Clay Miner.* 32:357-362.

Stucki, J.W., P. Komadel, and H.T. Wilkinson. 1987. The microbial reduction of iron(III) in clay minerals. *Soil Sci. Soc. Am. J.* 51:1663-1665.

Takai, Y. and T. Kamura. 1966. The mechanism of reduction of waterlogged paddy soil. *Folia Microbiol. (Prague)* 11:304-313.

Theng, B.K.G. 1974. *The chemistry of clay-organic reactions.* John Wiley & Sons, New York.

Theng, B.K.G. 1979. *Formation and properties of clay-polymer complexes.* Elsevier, New York.

Thornton, H.C. 1922. *Ann. Appl. Biol.* 9:241-274.

van Breemen, N. 1973. Soil forming processes in acid sulfate soils. pp. 66-131. In: H. Dost (ed.) *Acid Sulfate Soils*. ILRI, Publ. 28 Vol. 1.Inst. for Land Reclamation and Improvement, Wageningen, The Netherlands.

van Olphen, H. 1966. *An introduction to clay colloid chemistry*. Third Printing. Interscience, New York.

Wagner, M. and W. Schwartz. 1967. Geomicrobiologische Untersuchungen VIII.Über das Verhalten von Bakterien auf der Oberflüche von Gesteinen und Mineralieu und ihre Rolle bei der Vitterung. *Z. Allg. Mikrobiol.* 7:33-52.

Walker, S.G., C.A. Flemming, F.G. Ferris, T.J. Beveridge, and G.W. Bailey. 1989. Physicochemical interactions of *Escherichia coli* cell envelopes and *Bacillus subtilis* cell walls with two clays and ability of the composite to immobilize heavy metals from solution. *Applied Environ. Microbiol.* 55:2976-2984.

Wang, M.C. and P.M. Huang. 1986. Humic macromolecule interlayering in nontronite through interactions with phenol monomers. *Nature* 323:529-531.

Weed, S.B., C.B. Davey, and M.G. Cook. 1969. Weathering of mica by fungi. *Soil Sci. Soc. Am. Proc.* 33:702-706.

Wilson, M.J. and D. Jones. 1981. Lichen weathering of minerals: implications for pedogenesis. pp. 5-12. In: R.C.L. Wilson (ed.) *Residural Deposits: Surface Related Weathering Processes and Materials*. Blackwell Scientific Publications, London.

Wu, J., C.B. Roth, and P.F. Low. 1988. Biological reduction of structural iron in sodium-nontronite. *Soil Sci. Soc. Am. J.* 52:295-296.

Wu, J., P.F. Low, and C.B. Roth. 1989. Effects of octahedral-iron reduction and swelling pressure on interlayer distances in Na-nontronite. *Clays Clay Miner.* 37:211-218.

Xianmo, Z. 1984. On the process of primary soil formation. *Scientia Sinica (Series B)* 27:1035-1045.

The Mechanics of Soil-Root Interactions

D.R.P. Hettiaratchi

I.   Introduction 255
II. Simplifications 257
III. Critical State Model 258
     A. Moisture History 258
     B. Elements of the Model 259
     C. Unsaturated Soils 261
     D. State Path Tracing 261
IV. Root Channel Formation 262
     A. Pore Space Considerations 262
     B. Soil-Root Contact Stresses 264
     C. Cellular Architecture of the Root Apex 268
     D. Forces on Root Cap 269
     E. Penetrometer Testing 271
V.   Role of Cell Biomechanics 273
     A. Root Cell Growth 273
     B. Mathematical Model 275
     C. Biomechanics of the Root Apex 277
     D. Commentary on Root Growth Model 281
VI. Conclusion 282
     A. Summary 282
     B. Discussion 282
References 286

I. Introduction

Our understanding of the fundamental aspects of the growth and proliferation of plant roots within their soil environment is far from complete. The reason for this is not difficult to comprehend when one considers the vast complexity of the various interacting processes involved in the study of root growth. The other chapters in this volume have addressed several specialized aspects of this problem and each one deals with a set of interactions of great intricacy. This

ISBN 0-87371-889-5
© 1992 by Lewis Publishers

review concentrates on the mechanical interactions of the growing regions of plant roots with their surroundings.

The living tissue which makes up the root has to be accommodated within the voids present in a given soil volume. There is no hinderance to root growth when the geometry and size of the interconnected pore spaces in the soil can contain this biological material. If this is not the case, then the biological material has to create the necessary space for itself by deforming the soil and increasing its pore space. How successful the root is in accomplishing this crucial step dictates its rate of growth, and in the event it is unable to create this space, root growth will cease altogether. The complex physical interactions directly involved in the realization of this simple requirement for root growth can be dealt with under the following three topics:

(a) Volume Change Behavior of Soil: Pore space is a soil parameter of crucial importance to root growth. This is usually estimated in non-dimensional form as void ration (e), specific volume (v), or porosity (n). Bulk density is also an indirect indicator of pore space. The way this parameter is controlled by stresses, applied either externally or internally, is therefore central to any analysis of root growth. The Cambridge critical state model (Roscoe et al., 1958) establishes this correlation.

(b) Strength Characteristics: The manner in which the soil mineral particles, which form the pore space, interact with each other determines the stability and strength of the soil matrix. The physical nature of the bonds between the mineral particles (or crumbs) determines the volume change behavior of the soil and hence its critical state parameters. The moisture and stress history of the soil are largely responsible for the nature of these bonds. Allowance must be made for the influence of these microstructural features of the soil skeleton on the overall strength and volume change behavior of the soil.

(c) Biomechanical Considerations: When pore space is limiting an extending root attempts to alter soil pore space by exerting internal forces on the soil matrix. These forces are generated by the growing cells within the root apex. An understanding of the way these forces are developed and their influence on the growth pattern of the root is the third essential component in the study of the physical interaction between roots and soil.

It will be recognized that the two factors (a) and (b) are associated with the physics and mechanics of the soil environment and (c) deals with plant physiological matters. The study of root growth thus requires a multi-disciplinary approach, and a comprehensive perception of the problem cannot be developed without taking into account all three of these interacting processes in parallel.

II. Simplifications

It is necessary to introduce fairly severe simplifications to the real interactive processes encountered in the study of root growth. This is a requirement which is dictated by our present rudimentary understanding of most of the basic factors involved. Additionally, it would be impossible to make any sense of the overall effects on root growth of a large number of permutations and combinations of a bewildering number of complex interactions taking place simultaneously.

The soil environment is assumed to be a homogeneous continuum and the presence of fissures, cracks, and channels made by earthworms or those left by decaying roots are disregarded. Ideally the soil is considered as an unstructured continuum composed of pore spaces and solid material particles. However, within this context it is still feasible to deal with aggregated soils which exhibit a bimodal pore structure consisting of macropores formed by a system of microporous aggregates. In this event, it is necessary to simplify the real situation further by considering the micropore volume as being unavailable for occupation by the main root axes.

These simplifications pave the way for the formulation of mathematical models which describe the mechanical behavior of the soil environment. These models can then be applied to gain some understanding of factors (a) and (b) in Section I. Essentially these models must be able to quantify how the pore space in the soil environment alters with applied stresses. The following two aspects of this behavior have to be considered:

(a) Macroscopic influences on pore space (loosening and compaction) resulting from both engineering operations (tillage and tractor running gear interactions) and natural causes (frost action). The critical-state model can be used in this analysis to assess the availability of the space required for root growth within a seed bed.

(b) The microscopic factors which govern pore space and the forces arising from the deformations induced by the passage of roots through the soil. These considerations can be modeled on the elasto-plastic expansion of cavities within a soil medium.

It is well established that root growth is sensitive to several factors other than mechanical confinement such as the availability of water and nutrients to the growing root, soil temperature and aeration, the influence of pathogens and mycorrhiza, and adequate photosynthate transport from the shoot. The present analysis will concentrate on root growth conditions in which none of these factors are limiting.

Evidently there must be cross interactions between some or all of these factors. For example, changes in pore space would have an influence on the movement of water toward the growing root. Yet again, pore space change could limit the availability of oxygen at the root apex and both of these effects could restrict cell growth. At the proposed level of simplification it is imperative that such cross interactions be ignored. This makes it possible to concentrate and isolate the effects of the main physical interactions outlined in I(a), I(b) and I(c). Once these interactions are fully understood, it would be possible to advance the investigations to include more complex conditions that will bring the analysis closer to the real situation.

III. Critical State Model

A. Moisture History

The Cambridge critical state model (Roscoe et al., 1958) was developed to trace the volume change behavior of saturated soils when subjected to known stresses. Descriptions of the model have been given elsewhere (Atkinson and Bransby, 1978; Hettiaratchi and O'Callaghan, 1980, 1985; Hettiaratchi, 1987, 1989) and the reader is referred to these for a detailed statement. In the present context the primary application of this model is the prediction of the status of the void ratio of a soil when it is subjected to mechanical loading and can thus be used to assess the space available for accommodating a growing plant root within the soil medium.

The original concepts of the critical state model were established for saturated soils. For use in root growth studies it is necessary to reevaluate these concepts for partly saturated soils. This exercise raises a fundamental ambiguity in the definition of partial saturation (see Hettiaratchi, 1987). If, for example, we wish to prepare a soil specimen at some specified moisture content, then this can be approached in an infinite number of ways. However, two repeatable laboratory techniques for achieving this objective can be devised. A known weight of dry mineral matter can be thoroughly mixed with a calculated amount of distilled water. This material is then used to make up a sample having a specified void ratio. Alternatively, a soil sample at the same void ratio can be brought to near saturation and then allowed to either air dry or drain on a tension table until the moisture content is identical to the earlier sample.

It is evident that the two samples, although at identical moisture content and comparable void ratio, need not necessarily have the same physical properties. The essential physical differences between these two samples is that the bonds between the mineral particles of the first specimen (the "remolded" sample) have been destroyed during sample preparation whereas in the second sample (the "cemented" sample) strong inter-particulate bonds have been allowed to form. Under real field conditions there is a possibility that an infinite number of combinations of these two microstructural states could exist simultaneously in a seed bed (see Hettiaratchi and O'Callaghan, 1985, for an analysis of these states). This simple illustration highlights how both stress and moisture history govern the mechanical behavior of the soil in a seed bed.

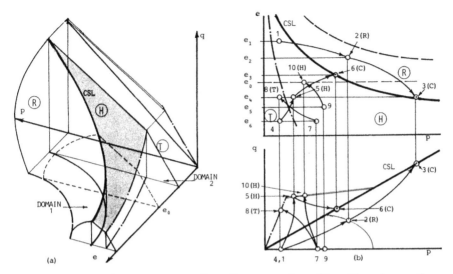

Figure 1. (a) Critical tate space. R = Roscoe surface; H = Hvorslev surface; T = Tension surface; Unimpeded root growth in Domain 1. (b) Projections of state paths in e-p and q-p planes. Letters in brackets signify that a point lies on a state boundary.

In extending the critical state concepts, developed for saturated soils, to cover partly saturated conditions it has been necessary to limit the investigations to the two simple stress and moisture history states represented by "remolded" or "cemented" conditions discussed above (Liang, 1985; Hatibu, 1987; Hettiaratchi, 1987). These investigations have identified systematic modifications that can be applied to the original saturated soil models to accommodate partial saturation in the remolded and cemented states. A summary of these modifications is given in Section III.C.

B. Elements of the Model

A brief outline of the relevant ideas of the critical state concept will be dealt with here insofar as they impinge on the present discussion. Referring to Figure 1a, the complete stress and pore volume state of a soil specimen can be depicted on a three-dimensional co-ordinate system. In this representation the combination of both the stress p (an equal all around stress component) and q (the shear stress component) taken together completely represent all possible states of stresses acting on a soil element. A formal definition of these two stress components is contained in Equations (1) and (2) in Section IV.

If these two stress components are assigned two axes of the three-dimensional system, then the remaining axis, labeled e, can be used to provide information on the pore volume of the soil. As plant roots occupy the voids in the soil, it is appropriate to quantify this parameter in terms of the void ratio (e) in preference to the more usual specific volume (v). Thus, points within p-q-e-space (referred to as "state space") completely determine the relationship existing between the pore volume of a soil that is in equilibrium with a given stress state. Paths traced out in state space by such combinations of p,q, and e are referred to as "state paths".

The central feature of the Cambridge model is the definition of a line in state space, referred to as the critical state line (CSL). The position of this line is unique to a given soil and represents the locus of end points reached by state paths traced out by soil undergoing shear failure (see Figure 1). This is, indeed, a very powerful concept in that it literally fixes the heading of the state path of soil element as it is loaded from some initial point in state space to ultimate failure on the CSL. The model also prescribes several rules to be observed while tracing out such state paths and these can be summarized as follows:

(a) Not all combinations of p,q, and e are possible. State paths cannot cross any of the three state boundary surfaces currently identified as the Roscoe surface (R), Hvorslev surface (H) and the Tension surface (T) (see Figure 1a).

(b) State paths lying wholly within these surfaces are associated with elastic or recoverable deformations. Some deformations induced by growing roots could generate such state paths [see Section III.D(d)].

(c) State paths associated with surfaces R,H, and T will induce contractile, dilatant, and fracture behavior, respectively.

(d) State paths traversing these surfaces induce permanent plastic deformations. These deformations are responsible for changing the pore space available for root extension.

(e) Once a state path arrives on the CSL, soil will continue to deform without any further change in p, q, or e.

Thus, if the state boundary surfaces can be delineated for any given soil at a specified moisture content and microstructural state, then the mechanical behavior of that soil is completely defined and the manner in which its pore space alters can be predicted. The practical application of the model to real situations is, however, fraught with two major difficulties. In the first instance the definition of state space boundaries for a range of moisture contents and stress and moisture histories requires time consuming and tedious laboratory experimentation. Second, the precise definition of the stress components p and q imposed on the soil by engineering operations and growing roots are difficult to quantify, even at a greatly simplified level. Discouraging as these problems may seem, it is nonetheless possible to use the model quite effectively to gain an insight into the diverse ways the space available for accommodating growing roots is regulated.

C. Unsaturated Soils

In general, the basic critical state concepts for saturated soils discussed in the previous subsection also hold true for the unsaturated state. Changes in moisture content and microstructural state (whether remolded or cemented) regulate the size and disposition of the state boundary surfaces.

(a) For both cemented and remolded conditions the volume of state space occupied by partly saturated soils is larger than that taken up by saturated soils. The volume of state space associated with elastic deformation is thus larger, and for a given value of e the magnitude of both p and q are larger.

(b) The dominant state surface in the remolded condition is the compaction inducing Roscoe surface (labeled R in Figure 1a). This primacy of the Roscoe surface persists over a wide range of moisture contents.

(c) The corresponding dominant state surface for cemented and dry remoulded soils is the soil loosening or cracking Hvorslev and Tension surfaces (labeled H and T, respectively, in Figure 1a).

(d) As a general rule, the clay content of a soil has a strong influence on the relative proportions of its state boundaries. The state boundary surfaces of soils with low clay content are not very sensitive to moisture content changes (see Hettiaratchi, 1987).

D. State Path Tracing

Space precludes a detailed examination of the many applications of the critical state model for understanding soil behavior. The procedure for implementing the basic rules of the model will be illustrated here by reference to a few very simple loading conditions. The pictorial view of critical state space shown in Figure 1a can also be represented in orthographic projection by the plan and elevation given in Figure 1b. In these views, a letter in brackets against a point in state space signify that it lies on one of the state boundary surfaces. Other points lie within state space.

(a) Soil compaction: Path 1-2(R)-3(C) demonstrates elastic compression ($e_2 - e_1$) from 1 to 2(R) followed by permanent plastic compaction on the

Roscoe surface from 2(R) to 3(C). As a result of this loading, the space available for root growth has been reduced from e_1 to e_4.

(b) Soil loosening: Path 4-5(H)-6(C) shows initial elastic increases in pore volume from e_6 to e_4 followed by rapid dilation on the Hvorslev surface from 5(H) to 6(C). Note that this dilation is accompanied by an unstable decrement in q from 5(H) to 6(C).

(c) Tensile cracking: This takes place at the end of the state path 7-8(T) and is associated with a decrement in p accompanied by high q.

(d) Root action: If it is assumed that e_0 is the minimum void ratio required to accommodate the root axes of a particular crop, then the soil deformation induced by the root can be traced along 9-10(H). The pore space increment is from e_5 to e_0 and is elastic. Plastic deformation will then take place, if necessary, on the Hvorslev surface with the path heading for a point on the critical state line.

(e) Mechanical confinement: In Figure 1 the $e = e_0$ plane divides state space into two domains (labeled 1 and 2). A root growing in a seed bed whose soil state is located within the state space domain furthest from the origin of coordinates (domain 1) will not be subjected to any mechanical impedance to its growth. All other soil states (in domain 2) will cause mechanical confinement to some degree or other. The severity of the confinement increases the further the state point is from the origin and the e_0 plane.

IV. Root Channel Formation

A. Pore Space Considerations

The mathematical model discussed in the previous section provides the basis for predicting the manner in which soil pore space (e) changes with applied boundary stresses (p and q). It is particularly well suited to forecasting the relative balance between the space available for accommodating root growth (the macropores) and the mineral particles or aggregates which make up the seed bed when the soil is subjected to known external loading. These loads are applied to large areas of the soil during engineering operations and comprise a macroscopic external loading system. The model can thus be used to gauge the capacity of a prepared seed bed to support unimpeded growth of plant roots.

The value of the root extension cut-off void ration, e_0, discussed in Section III.D(d) and (e) and set out in Figure 1, depends largely on the dimensions of the fully grown apical regions of the roots of the crop. When dealing with root crops a further allowance has to be made for the increase in the bulk volume of the storage root. In either event, soil states predicted by the critical state model to lie above the $e = e_0$ line will not impede root growth.

The next point of interest is the nature of the soil-root interactions when the state point of the seed bed lies within the state space with $e < e_0$. In this instance the root must overcome the restriction to spatial requirements by increasing the local void ratio back to e_0. In attempting to do, so the root must exert stresses on the soil [see state path 9-10(H) in Figure 1b]. Unlike the stresses generated on external soil boundaries during engineering operations, these stresses act exclusively on internal boundaries on a microscopic scale and are localized at the sites of root extension.

In the context of the critical state model, the only intrinsic difference between stresses applied either internally or externally is the sign of the principal stresses associated with each stress system. Essentially, a compressive internal stress (+ve) acting outwards is equivalent to a tensile stress (-ve) applied externally and vice versa. For example, a negative internal pore-water suction in a soil has a similar effect, an equal all round compressive stress acting on the external boundaries of the soil element (the intergranular contact stresses are increased in each case).

The volume change effects of these two loading systems, as predicted by the critical state model, are dramatically different. This contrasting effect is associated with the influence these stresses have on the magnitude of the stress components p and q:

$$p = (\sigma_1 + \sigma_2 + \sigma_3)/3 \tag{1}$$

$$q = [(\sigma_1 - \sigma_2)^2 + (\sigma_2 - \sigma_3)^2 + (\sigma_1 - \sigma_3)^2]^{\frac{1}{2}}/\sqrt{2} \tag{2}$$

It is evident that the sign of the principal stresses governs the magnitude of both p and q. As a simple illustration consider the case when $\sigma_1 = 10$, $\sigma_2 = 8$ and $\sigma_3 = \pm 4$ (arbitrary units). When $\sigma_3 = +4$, $p = 7.3$, $q = 5.3$, and when $\sigma_3 = -4$ these values become $p = 4.7$, $q = 13.1$. Thus, as the minor principal stress (σ_3) changes from compression (+4) to tension (-4) the value of p decreases and that of q increases.

It can be seen from the critical state model shown in Figure 1b that for the compressive loading associated with high values of p acting on a soil with high void ratio, the state path 1-2(R)-3(C) leads to compaction. In contrast, the alteration of one principal stress to tension reduces the magnitude of p enabling state paths such as 7-8(T) and 9-10(H) to be generated. When such state paths originate in regions of low void ratio (compact soil), the decreasing nature of p places the tension and Hvorslev surfaces within easy reach. Recall that deformations taking place on these state surfaces induce soil dilation. Thus, the internal stresses exerted by roots on the soil (characterized by decreasing p in association with high q) provide an optimal combination for soil loosening.

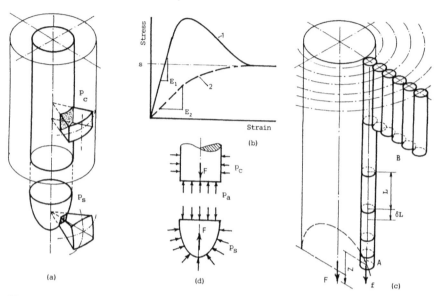

Figure 2. (a) Simplified geometry of a root apex showing spherical and cylindrical deformation boundaries; (b) typical stress-strain characteristics of soil: 1 - dense/cemented, 2 - loose/remolded; (c) simplified model of cells in the root cortex; (d) stresses acting on elongating root cells.

As pointed out in Section III.C, the state boundaries of a cemented soil enclose a larger volume of state space than that commanded by a remolded soil. Consequently, higher values of q are required by the state paths to reach the state boundaries to initiate permanent deformations in the soil. The internal expansion mode associated with growing roots has been shown to generate high q and the characteristics of this form of stressing are, thus, well adapted to deal with the dilation of stronger cemented soils. However, if these shearing stresses cannot be developed by the expanding root, then growth will cease. The factors which determine this limit will be discussed in Section V.

B. Soil-Root Contact Stresses

The volume change behavior of soils presented so far has featured the two stress components p and q. The physical interpretation of the component p as an equal all round stress or pressure is easy to visualize, but associating q with shear stresses is less obvious. Furthermore, these two stress components, although they fit admirably well into the critical state model, are somewhat obscure functions of the actual contact stresses acting on the surface of the root. These contact stresses constitute the vital boundary conditions required to extend the analysis from soil related phenomena, discussed so far, to the behavior of living tissue within the root.

In order to facilitate the development of a simple model to evaluate these contact stresses, it is necessary to approximate the kinematics of the soil deformation at the root apex in terms of simple, mathematically quantifiable forms. With a view to this, the physical configuration of a typical root tip may be approximated by two regular curvilinear solids of revolution. As depicted in Figure 2a, the root cap can be matched by a cone, parabaloid, or semi-ellipsoid. As discussed in Section IV.C, the precise shape chosen for the physical model of the root cap is not very critical to the analysis. The root shank extending behind the cap approximates to a right circular cylinder (see Figure 2a).

The thrusting of the root cap into the soil represents a familiar problem in soil mechanics referred to as "punch indentation". A readily recognizable example of this form of soil penetration is the advance of a pile tip driven into the ground. Bishop et al. (1945) first proposed that the pressure required to deform a deep hole in a metal is a function of the pressure necessary to expand a cavity of similar volume within the metal. Gibson (1950) proposed the extension of this idea to the punch indentation problem in soil mechanics. There have been many subsequent applications of this concept in analyzing the performance of indenters in soil, and Farrell and Greacen (1966) first drew the parallel with root growth.

The two simple deformation modes utilized in the application of cavity expansion theory to estimate the stresses acting on the root surfaces are

(a) The expansion of a spherical cavity from zero to finite radius in an infinite elastic-plastic medium. The root cap deforms the soil in this spherical mode and the stress (p_s) acting on the surface of the root cap is estimated on this basis (see lower part of Figure 2a).

(b) The enlargement of a right circular semi-infinite cylinder from finite radius in an infinite elastic-plastic medium. Diametral growth of the root shank follows this mode of soil deformation and root shank contact stress (p_c) is calculated on this basis (see upper part of Figure 2a).

The theory of cylindrical deformation of soil was presented by Gibson and Anderson (1961). Ladanyi published an analysis of both modes of deformation for a clay medium (Ladanyi, 1963) and a brittle medium (Ladanyi, 1967). Vesic (1972) presented a very useful procedure for calculating these stresses in terms of the Mohr-Coulomb parameters of a soil. In his procedure the contact pressure (or stress) p* acting at the surface of the cavity is given by the equation:

$$p^* = cF_1 + q_0F_2 \tag{3}$$

where c is the cohesion of the soil and q_0 is the ambient geostatic stress level in the soil at the rooting depth. The non-dimensional coefficients F_1 and F_2 in Equation (3) are functions of the soil rigidity index, $I = G/(c + q_0 \tan \phi)$, which is a function of the angle of internal friction ϕ of the soil, its elastic shear modulus G, and the mode of deformation of the soil (spherical or cylindrical). Values of these coefficients for both cylindrical and spherical enlargement modes were given by Vesic (1972), for I and ϕ in the intervals $5 < I < 500$ and $0 < \phi < 45°$.

Vesic's analysis can be used to make an order of magnitude estimate of the contact pressures acting on the surface of plant roots when they deform the soil. Before attempting this calculation it is necessary to examine the practical significance of the soil rigidity index I which, by definition, is the ratio of the soil shear deformation modulus G to the soil shear strength s at the depth of the root apex. The latter is simply the Mohr-Coulomb shear stress at failure for the ambient horizontal geostatic pressure q_0 at the rooting depth. The shear modulus G is a parameter connected with the elastic behavior of the soil and is more difficult to measure. The shear modulus is related to Young's modulus E by the expression $G = E/[2(1+v)]$ so that for a fixed value of Poisson's ratio v the shear modulus G is directly proportional to the other elastic part parameter E. This parameter is the slope of the initial elastic part of the well-known, direct stress-strain curve and this is, therefore, a measure of the magnitude G.

An idea of the nature of the rigidity index I can be obtained by studying the direct stress-strain curves of soils. The stress-strain curves of cemented soils generally exhibit the work-softening characteristic shown by the curve (1) in Figure 2b where the residual strength s lies below the peak value. On the other hand, a remolded soil exhibits the work-hardening characteristic typified by the curve labeled (2) on this diagram. For similar shear strengths s of these two types of soils, the ratio of $G/s = I$ would be higher for the cemented soil than for the remolded one $(E_1 > E_2)$. Thus, in the present context the rigidity index for cemented soils would, in general, be higher than that for remolded soils.

Table 1 sets out the steps for using Equation (3) to evaluate the contact stresses acting on the surface of the root cap (spherical expansion pressure $p^* = p_s$) and the root shank (cylindrical expansion pressure $p^* = p_c$). In this calculation the ambient geostatic stress q_0 in a shallow seed bed is assumed to be of the order of 2 kPa (based on a soil unit weight of 20 kN/m^3, coefficient of earth pressure at rest of 0.5, and a root apex depth of 200 mm). The calculations are based on typical values of c and ϕ for a silty loam (soil type 1) and a heavy clay soil (soil type 2). As discussed in the previous paragraph, $I = 50$ is used for both soil types in a remolded state and a higher value of $I = 100$ is assigned to the cemented state. It should be noted that these values and the chosen value of the geostatic stress q_0 are all order of magnitude estimates only.

The calculated contact stresses in both spherical and cylindrical modes are the minimum values required to cause permanent plastic deformation in a zone around the expanding cavity. The approximate upper limit of the ratio of the radius of this zone to the root radius is about 7 for the spherical expansion and 15 for a cylindrical cavity (see Vesic, 1972). The cavity contains the living root; hence, these are also the pressures which act on the surfaces of the growing

Table 1. Steps for using Equation (3) to evaluate contact stresses

Soil	I	C	$\phi°$	Spherical			Cylindrical			P_s/P_c
				F_1	F_2	P_s kPa	vF_1	F_2	P_c kPa	
1R	50	2	30	15.71	10.07	51.3	8.31	5.80	28.2	1.83
2R	50	20	5	7.96	1.70	162.6	5.58	1.49	114.6	1.42
1C	100	10	35	24.46	18.13	280.9	11.52	9.07	133.3	2.11
2C	100	35	15	14.16	4.80	505.2	8.42	3.27	302.6	1.67

root. The following two distinct trends in the magnitudes of these stresses are worthy of note:

(a) The contact stress p_s in spherical expansion mode is always greater than the companion cylindrical mode stress p_c. The ratios of p_s/p_c given in Table 1 show that in the frictional soil this ratio approaches a value of 2 while in the cohesive soil this is closer to 1.5. In either case the spherical expansion mode generates higher contact stresses than the cylindrical mode and this trend is generally true for all soil types.

(b) The stresses estimated in both spherical and cylindrical enlargement modes in cemented soils are much greater than those generated in the remolded state. This effect is not entirely unexpected in that plastic deformation of the cemented soil would entail fracturing of interparticulate bonds, while in the remolded state deformations can be accommodated by frictional sliding between the solid particles or crumbs.

The cavity expansion analysis outlined above predicts the stresses acting on the interface between the soil environment and the plant root. These stresses have been assessed purely on soil mechanics and kinematic criteria. However, we must not lose sight of the most important fact that it is the root apex that is responsible for the expansion of the cavities in the soil. The interaction of the root cap with the soil initiates plastic failure in a near spherical zone around the root tip while the growth of the root shank is responsible for bringing a cylindrical collar of soil around it to failure. These deformations must, in turn, be linked directly to the growth of certain groups of cells in the root apex. It is therefore imperative that a precise correlation be established between the cavity expansion stresses and the stresses carried or initiated by the cells in the root apex.

C. Cellular Architecture of the Root Apex

The cellular arrangement within the apical regions of roots is well documented (see, for example, Easu, 1965; Fahn, 1990). The basic anatomical features can be incorporated into a simple mechanical model of a growing root tip. The cells, particularly those in the cortical zones, are arranged in a number of concentric cylindrical files that terminate distally in the quiescent center of the apical meristem. A schematic arrangement of these cells is shown in Figure 2c. The cells in these files are not unlike right circular cylinders with their longitudinal axes aligned parallel to the central axis of the root. The aspect ratio (ratio of length to diameter) of the cells recently formed by cell division at the meristem are near unity (cell A in Figure 2c). The cells get older and their aspect ratio increases as we proceed proximally along the files. One such file is shown in Figure 2c. The aspect ratio reaches a maximum value of about 4 to 5 at the end of the elongation zone of the root (cell B in Figure 2c). The mature lignified cells making up the root immediately above this region do not extend any further.

An examination of this arrangement leads to the inescapable conclusion that the normal extension growth of the root is geared to the elongation of the cells in this region of the root apex. To draw an engineering analogy, each elongating cell is a sub-miniature hydraulic jack which is extended irreversibly by the influx of water. The series arrangement of these jacks in each file provides a stroke amplification system. This is necessary because the individual axial elongation or stroke δL ($= \lambda L$ described in Section V.A) of each cell is small. By this means, microscopic length changes can be converted to a macroscopic movement of the root cap ($z = \Sigma \delta L$). The parallel arrangement of the files is a force amplifier. The small force f, generated by individual cells in each file of cells, is intensified by the number of rows in the root apex to generate the large force F required at the root cap ($F = \Sigma f$). Continuous extension is maintained by a supply of fresh retracted jacks (daughter cells) added to the files by cell division at the surface of the quiescent center in the apical meristem.

The cells in the root cap are not actively engaged in the root elongation process. The root cap doubles as a protective appendage and the soil "cutting tool" which guides the root tip through the hostile soil environment. Cells abraded off the cap are renewed from the distal surface of the quiescent center of the meristem. The force required to thrust the root cap through the soil is developed almost exclusively by the longitudinal elongation of the cells in the root cortex. The biomechanics of this process is dealt with in Section V.

Under normal conditions the pressure acting on the shank of the root apex would be the geostatic stress (q_0 in Section IV.B) mobilized in the soil after the root cap has formed a space for the root shank. It is easy to recognize that the contact stress transmitted to the cylindrical surfaces of the elongating cells would increase from q_0 to p_c (the cylindrical cavity expansion pressure) as a result of any cylindrical enlargement of the root shank. The significance of this change will be discussed in Section V.

The thrusting of the root cap into the soil generates stresses on its surface commensurate with the enlargement of a spherical cavity. If this spherical cavity expansion pressure p_s is assumed to be distributed uniformly over the entire surface of the root cap, then by hydrostatic analogy the force F (Figure 2d) acting along the axis of the root is given by $F = p_s \times A$, where A is the projected area of the root cap in the axial direction. Note that the precise shape of the root cap does not enter into this calculation, only its projected area in the longitudinal direction. The projected area A of the root cap is also the sectional area of the root shank. However, the average pressure p_a on the root cross-section is F/A and it therefore follows that $p_a = p_s$. It is important to recognize that this analysis neglects the frictional forces between the root cap and the soil, and the soil reaction stress p_s on its surface is assumed to be normal to it (without any associated shear). The stress $p_a = p_s$ acting on the circular ends of the idealized cylindrical cells making up the root apex has been arrived at on this basis. If frictional effects on the root cap are allowed for, then $p_a > p_s$.

D. Forces on Root Cap

The simplified analysis discussed in Section C outlined the technique for making theoretical predictions of the axial force acting on a growing root cap in terms of the Mohr-Coulomb soil parameters c and ϕ and a soil rigidity index I. The latter was shown to be dependent on the nature of the bonding between soil particles. The earliest attempt to measure the axial forces developed by plant roots can be attributed to Pfeffer (1893). Contemporary investigations along these lines were carried out by Eavis (1967), Eavis et al. (1969), Taylor and Ratliff (1969), Whiteley et al. (1981), and Misra et al. (1986). The techniques employed by these investigations were to measure the thrust exerted by seminal roots either by direct measurement of the force exerted by the root against a suitable "anvil" or by estimating the dead weight which the root reaction could lift. The models for a static analysis of the various force components in these investigations are summarized in Figure 3a through e. In the schematic arrangements the root is represented as a single acting hydraulic jack with its inlet port valved for extension-only operation. The sketch in Figure 3b shows the equivalent system for the experiments which employ a force measuring transducer and Figure 3c the corresponding dead-load arrangement. The "anvils" used are either stiff blocks of plaster (A) or soil cores (S).

Two factors have to be taken into account when interpreting the published results of the experiments devised for measuring axial root forces. First, the transducer and dead weight methods used to measure the axial force exerted by the root are not transposable. In the transducer method, where the thrust of the root is taken up by a modern weighing machine (Misra et al., 1986) or an instrumented flexible beam (Eavis, 1967), the actual extension of the root is arrested by the very high stiffness of the force measuring instrument. Referring to the hydraulic analogy given in Figure 3b, it will be seen that as the jack is

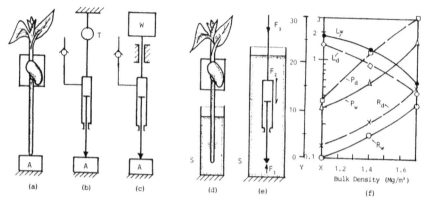

Figure 3. (a) Root apex on rigid anvil A. (b) Schematic of (a) with transducer force measuring system. (c) Schematic of (a) with dead-load system. (d) Root apex in soil core S. (e) Schematic of (d). (f) Penetrometer and root tip forces and root growth (From Eavis, 1967): pressure on transvere section of root (R) and penetrometer (P); refer to scale X, with pressure in MPa. Root length (L); refer to scale Y, with root length in mm. Subscripts: w, wet; d, dry.

operated (simulating root growth) the force recorded by the transduced T will keep on increasing until either the fluid pressure applied to the jack reaches its maximum value or the anvil fractures. Translating this to the case of the live root shown in Figure 3a, the measured force will be the limiting force developed by the extending cells in the root apex. The interpretation discounts the possibility of fracture of the plaster anvil or the failure of the root by buckling.

The dead weight system, on the other hand, imposes only a force constraint on growth but does not restrict extension. It is therefore an arrangement which effectively incorporates a force transducer which is infinitely soft; the transducer method is, in comparison, infinitely stiff. Consider the simple mechanical analogy given in Figure 3c. If the weight W is less than the peak force capacity of the ram then the ram, will extend and displace the weight upwards. If, on the other hand, the weight is in excess of the peak force capacity of the ram (or the extending cells), then extension (root elongation) will cease, but the weight used will not necessarily be the peak force of the ram (the cells). In order to find the peak force it would be necessary to keep increasing the weights until extension has just ceased. This technique must therefore have some means of altering the weight in very small steps (or infinitely variable) together with a sensitive transducer to monitor the movement of the weight.

The second point worth considering is the nature of the "anvil" against which the root develops its thrust. If, as shown in Figure 3a, this is a stiff block of plaster (Misra et al., 1986), then its interference with the force measuring system is minimal. Consider the case shown in Figure 3d where the root is resting on a soil core (Whiteley et al., 1981). The nature of the measured force will depend on whether the root has actually penetrated the core or not and how far the growth process has taken place within the soil core. If the root has grown any distance into the soil core, then the externally measured force will not necessarily be a measure of the thrust of the root cap.

The reason for this is illustrated in Figure 3e where $F_3 = (F_1 - F_2)$. A measure of F_1 cannot be obtained without a knowledge of the drag force F_2 and there is no convenient way of finding this out. Translating this example to the live root situation (Figure 3d), imponderable forces arise when the site of the generation of the forces (the elongation zone just behind the root apex) is located within the soil. Thus, when the anvil is a soil core the force measured externally is not necessarily the root growth force. The externally measured force is therefore a function of the embedment of the root into the core and the degree to which the root hairs and the root shank itself takes up the root cap thrust.

A useful summary of the recorded axial root growth pressure developed by several species has been presented by Misra et al (1986). These pressures vary over a range from 1.6 to 0.4 MPa. The wealth of available data on the magnitude of p_a has to be assessed in terms of the factors discussed in the previous paragraphs. Not all the details of the relevant boundary conditions are available in the published literature. An unambiguous value could be arrived at by the method illustrated in Figure 3a and b. The experiments of Whiteley et al. (1981) approach this, but the anvil used was a soil core. The root penetration was claimed to be only 4 mm so that the problem illustrated in Figure 3e may not be operative in these experiments. Their work also shows that the value of p_a ranged from 0.15 to 0.33 MPa. In any event, it is evident that roots can sustain quite large axial pressures before growth is arrested. The simple example worked out in Table 1 demonstrates that the axial pressure of 0.5 MPa predicted for an arbitrarily chosen cemented clay soil is of the right order of magnitude.

E. Penetrometer Testing

Most of the experimental work on the axial forces developed by roots has been closely associated with penetrometer testing. It would therefore seem appropriate to review these findings at this stage of the discussion on root forces. There is a strong physical similarity between a microprobe and the apex of a root. Almost all previous investigations have attempted to make direct comparisons between the living root apex and inanimate probes [Farrell and Greacen (1966); Eavis (1967); Cockroft et al. (1968); Greacen et al. (1968); Whiteley et al. (1981); Misra et al. (1986); Bengough and Mullins (1988)]. The common conclusions drawn from these investigations is that the measured root cap pressure p_a ($= p_s$) is always smaller than the comparable penetrometer pressure p_p (the cone index of a cone penetrometer).

The graph given in Figure 3f has been constructed from Eavis' extensive data [Eavis, 1967, Figs. 39.10, 39.11, and 39.13) and shows the experimental variation of the projected axial pressure acting on a 1 mm diameter probe (curves labeled P_d, P_w) and pea roots (curves labeled R_d, R_w) in a sandy loam soil. Two soil moisture levels and a range of packing bulk densities have been selected from the data to prepare the graph. The observed trends are

(a) Axial pressure increases with bulk density.

(b) The wetter soil (lower moisture tension) shows lower axial pressure.

(c) The penetrometer pressure (p_p) is consistently greater than the measured root reaction pressures (p_a). The ratio $R = p_p/p_a$ is invariably greater than unity.

(d) Root extension increases with decrease in bulk density and increase in moisture content.

The trends demonstrated by these experiments are fairly typical of many experimental investigations comparing penetrometer performance with root extension forces. Observations (a), (b), and (d) are as expected and can be explained in terms of the discussion presented in Section IV.B. The discrepancy (c) has also been noted in many other parallel investigations (see Whiteley et al., 1981, who recorded $1.6 < R < 5.3$). The fact that the ratio R is always very much greater than unity points to a fundamental problem in establishing this connection between the performance of penetrometers and roots. Refinements to move R closer to unity have been introduced. These include allowing for friction (see, for example, Bengough and Mullins, 1988), modification to shape and size of penetrometer and relieving the penetrometer, shank diameter relative to the conical tip. These modifications, however, do not appear to get rid of the problem.

These difficulties can be completely circumvented if it is recognized that the passage of an inert penetrometer is not representative of the way a plant root grows in its soil environment. The penetrometer force readings are of the same order of magnitude as the force experienced by the root cap on its own. The fact that radial swelling of the root apex takes place under mechanical confinement was recorded quite early on in root growth investigations (Abdalla et al., 1969; Goss, 1977). All the penetrometers discussed so far do not have the facility to change their geometry to imitate this mode of root growth. Variable geometry penetrometers have been developed at Newcastle (Abdalla et al., 1969; Hettiaratchi and Ferguson, 1973; Nguyen, 1977) and their performance will be discussed in the context of the variable root apex geometry growth model to be described in Section V.C and D.

The correlation between axial root growth pressure (p_a) and penetrometer pressure (p_p) is poor and the reasons for this have already been outlined. However, many workers have found good agreement between penetrometer index (pressure p_p) and root growth rate (see, for example, Scott Russell and Goss, 1974; Groenevelt et al., 1984; Greacen, 1986). This performance is not difficult to explain in that penetrometer index (or cone index) is a function of several soil-related factors which also control root growth. As discussed in Section IV.B, spherical cavity expansion pressure, which is sensed by penetrometers, is a function of soil strength (c,ϕ), its rigidity modulus I, and the microstructural state of the soil. Cylindrical cavity expansion stresses are also functions of these parameters and both of these forms of soil deformation play a key role in the variable geometry root growth model to be described in Section V. On this basis one must expect a good correlation between penetrometer index and overall root growth performance. In contrast, attempts to relate penetrometer index to just one aspect of root extension on its own, its axial extension pressure, have been singularly unsuccessful.

V. Role of Cell Biomechanics

A. Root Cell Growth

A discussion of the physical interaction of plant roots with their soil environment is incomplete without taking into consideration the reciprocity of these interactions with the biological material making up the growing root. In fact, the soil interactions are passive in nature and arise as a result of active deformations imposed by the root apex. These effects are, in turn, the result of endogenous changes taking place in the cells of the root apex. There is, therefore, a chain of interactions connecting the soil, through the root cap and epidermis right into the cells making up the root.

As discussed in Section IV.C the axial growth of the root is the cumulative result of the longitudinal expansion of the daughter cells formed by mitotic activity at the boundary of the quiescent center of the apical meristem. These cells start life as squat cylinders and extend to several times their original length before they lignify and stop extending. At this stage they become part of the mature root shank. The key to this developmental pattern and the nature of the forces exerted by the root on its surroundings lies in the fundamental mechanism which dictates this ordered axial extension of the cells in the root. Although this aspect of the problem is not, strictly speaking, a soil-related problem, it is probably the most important component which impinges on a vast number of mechanical and physiological interactions taking place between plant roots and soil. In view of this, the essentials of cellular biomechanics of the root apex will be considered next.

The microstructure of plant cells and cell walls has been extensively studied and there is a wealth of published optical and electron micrographs, which unfold the intricate and beautifully ordered nature of these basic building blocks of plants (see, for example, Roelofsen, 1965; Preston, 1974). Even at a superficial level the structural design of vacuolated parenchymatous cortical cells of plant roots bears a close physical similarity to hollow, thin-walled pressurized structures.

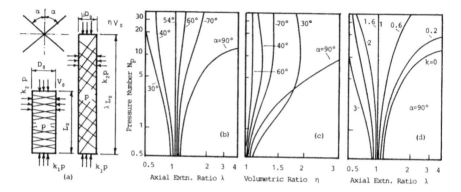

Figure 4. (a) Definitions of cell proportions, pressures, and fibril angle. Cell geometry before (left) and after (right) extension step. (b) Pressure-extension characteristic of reinforced cells. (c) Pressure-volume characteristics of reinforced cells. (d) Influence of external contact stress on pressure-extension characteristic for a reinforced cell with $\alpha = 90°$ ($k_1 = k_2 = k$).

The wall material of these cells consists of a base made up of polymeric substances and embedded in this are precisely ordered arrangements of relatively inextensible cellulosic microfibrils. To improve the authenticity of the simple physical analogy of a plant cell to a pressure vessel it is necessary to include the presence of these microfibrils in the wall structure. The simple physical model of a typical extending root cell can thus be described as a thin-walled pressurized structure with walls made of flexible polymeric material reinforced by a network of flexible, but inextensible cords.

A mathematical simulation of the behavior of the cortical cells in plant roots can be developed from this simple physical representation. In this model the microfibrils consist of a single layer of inextensible cords comprising a pair of helices making an angle α with the central axis of the cell (see Figure 4a). As large dimensional changes are involved, the mathematical models are developed as functions of extension ratios in preference to strain, the more familiar engineering parameter. Cell proportions are specified in terms of three extension ratios, λ, μ, and η, which are, respectively, axial, diametral, and volumetric changes (see Figure 4a). Note that an extension ratio of unity designates the initial state of a cell. If, for example, a cell reaches $\eta = 2$, this signifies that the internal volume of the cell has doubled and any ratio of less than unity implies a contraction.

The pressure of the fluid inside the cell vacuole is denoted by the symbol p and the contact stresses acting on the outside surfaces are as defined in Figure

4a. The cell pressure number $N_p = p/[C(\omega - 1)]$ is a dimensionless group that includes the strain energy modulus C of the wall material (Treloar, 1975) and ω, which is simply the ratio of external to internal radius of the cell and is a measure of cell wall thickness. The elastic properties of the cell wall are characterized by the strain energy modulus C and are therefore an invariant material parameter. Thus, for all practical purposes N_p may be considered as being directly proportional to the cell pressure p.

The general pressure deformation characteristics of thin-walled reinforced structures have been presented in non-dimensional form (Hettiaratchi and O'Callaghan, 1974, 1978) and numerical values have been calculated (Wu et al, 1988; Cosgrove, 1988). The strain energy modulus C of the base wall material of plant cells has not been precisely determined yet, and in view of this, a non-dimensional presentation provides a general solution to the deformation characteristics of these structures. At a future date when values of C have been evaluated, it would be a simple matter to use the general solution to calculate absolute values. The physical behavior of such structures during inflation and external loading falls within the experience of specialist engineers engaged in the design and manufacture of rubber hose and automobile tires. To the non-specialist there are some intriguing and totally unexpected ways in which these structures behave and, indeed, it is these very aspects which are relevant to the problem being considered here.

B. Mathematical Model

The basic deformation characteristics of plant cells, as deduced from the mathematical model, are summarized in Figure 4 (for details of Figs. 4b and 4c see Hettiaratchi and O'Callaghan, 1978; Figure 4d is from the author's unpublished data). The variation of the axial extension and the cell volume ratios with cell pressure (proportional to the pressure number N_p) and fibril angle are depicted in Figure 4b and c. These curves are for a cell without any external stresses ($k_1 = k_2 = 0$). The curves in Figure 4d are for the case when equal outside contact pressure is applied to the ends and the sides of the cell ($k_1 = k_2 = k$) and is for one specific fibril angle of $\alpha = 90°$ only. These solutions highlight three important points:

(a) Cell Shape: When $\alpha > 54°$ an internal pressure rise will cause a cell (free of external stresses on its surfaces) to extend axially. Conversely, for a fibril arrangement with $\alpha < 54°$ the cell contracts axially. At the so-called neutral angle when $\alpha = 54°$ the cell does not change in shape. The fibril angles of the critical cells in plant roots are oriented with $\alpha > 54°$ so that osmotic inflow of water into the cells will cause them to extend.

(b) Cell Volume: The geometry of the cell alters as discussed in (a) with an accompanying change in cell volume. As the cell internal pressure p is increased, cell volume reaches a peak value (η_m at λ_m) and thereafter the volume shrinks. The limit η_m is a function of the fibril angle α, this being infinite at $\alpha = 90°$ (and 0°) and unity at 54°.

Cell growth is sustained by the entry of water through the semi-permeable membranes into the cell along an osmotic gradient. In this context there cannot be a decrease in cell volume because such a change would require an efflux of water out of the cell. The mathematical model, therefore, predicts that at the point of peak cell volume the cell will cease to alter in shape. As already indicated, this peak volume limit is quite large for cells with fibrillar alignments approaching 90°. The fibril angles in root cells are, indeed, arranged to exploit this volume change behavior.

(c) External Stresses:
(i) Equal stresses on all surfaces.

The effect of this form of external boundary stresses on the two pressure characteristics of a cell can be reflected as an imaginary change in fibril angle. Typical pressure extension curves for the case when $\alpha = 90°$, set out in Figure 4d, show that when $k_1 = k_2 = k$ (i.e., $p_a = p_c = p$) and $k < 1$ the characteristic reverts to that appropriate for $\alpha = 54°$. For $k > 1$ the cell begins to contract with an accompanying increase in diameter. For $k = 1$ there is no volume change.
(ii) Unequal stresses on sides and ends.

A simplified approach to this case (Hettiaratchi, 1990) extends the analysis developed for modeling pneumatic tire carcass deformations. This model shows that when the displacement of the ends of the cell is prevented the cell will continue to increase in volume with radial expansion. This behavior holds true for all fibril angles greater than 54°.

There is scant experimental evidence to support the theoretical cell pressure deformation characteristics described so far. The technical difficulties in measuring these characteristics for plant cells are formidable and most of the published work relates to the behavior of giant algal cells such as *Nitella* and *Chara*. The elegant experiments carried out by Zimmerman and Steudle (1975) have been converted to pressure deformation characteristics by Cosgrove (1988). Kamiya et al. (1963) also presented the pressure volume characteristics of *Nitella* internodal cells. An indirect estimation has been made by Wu et al. (1988) for cells of sunflower, trefoil, and pepper. The wall structure of the internodal cells of *Nitella* (Probine and Barber, 1966) is almost identical to those of the much smaller vacuolated cells in the root cortex (Ledbetter and Porter, 1963; Hardham and Gunning, 1977; Hogetsu, 1986). In both cases the fibril arrangement is nearly transverse ($\alpha = 90°$) and axial extension growth is the basic function of both the giant internodal cell and the cells in the elongation zone of plant roots. The general shape of the published experimental pressure volume characteristics of the *Nitella* internodal cell agrees very well with the theoretical characteristics given in Figure 4c. This is most encouraging even though the evidence is sparse and is drawn from a giant replica of a root cell.

C. Biomechanics of the Root Apex

The mathematical models described in Section B above can now be used to explain observed growth behavior of plant roots. Three possible scenarios will be dealt with. The salient points of the basic cell growth model are discussed in the first situation, which also deals with the extension of an unimpeded cell (for a detailed account of this model see Hettiaratchi and O'Callaghan, 1974, 1978). The other two cases deal with the performance of the model when external confinement is present and when this suppresses root growth completely.

1. Unimpeded Growth

Growth of a plant root cell, characterized as simple geometric changes in its physical shape, is initiated and sustained by the osmotic influx of water from the soil into the cell vacuole. For this to take place the contents of the vacuole must be at a lower chemical potential than the water in the soil pores. The mass transfer of water into the vacuole dilutes its contents with a consequent reduction in the available potential differences. In Figure 5a curve 1 represents the available potential difference between the contents of the vacuole and the pore water in the soil. In the initial state ($\lambda = \mu = \eta = 1$) the potential difference is arbitrarily set at around 3 MPa and falls off with dilution as the cell increases in volume. Also shown in Fig. 5a as curve 2 is the pressure-extension characteristic of a typical elongating cell ($\alpha \cong 70°$). Curves 3 and 4 in Figure 5b are the companion pressure-volume and pressure-diametral change characteristics for the same cell.

Curve 3 in Figure 5b shows that the volume of the cell will increase with the inflow of water into the cell until position A is reached, at which point the cell volume has increased by about 30% ($\eta = 1.3$). During this process the cell diameter would have contracted by about 18% ($\mu = 0.82$) to point B and the axial length of the cell has increased significantly by 80% ($\lambda = 1.8$) to point C. It is worth noting that if the volume had not peaked at A, extension could have proceeded to point D as there is still an available potential gradient to drive the water into the cell. Thus, given time the pressure inside would rise from C to D without any further change in shape, and at D the influx of water will cease.

Next, consider the case where the concentration of the cell contents results in a lower potential difference at the start of the growth step. Such a situation could arise from inadequate osmoregulation coupled with a lack of aeration or some other related stress factor that could interfere with the proper functioning of the cell protoplasm. Curve 5 (Figure 5a) represents the dilution characteristic for

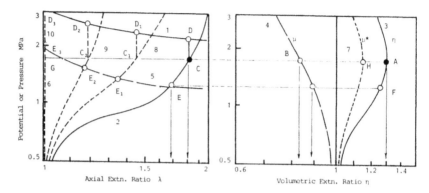

Figure 5. Growth model of plant root cells: (a) pressure-extension characteristics (2,8,9,10); chemical potential curves (1,5). (b) Variation of volumetric (η) and radial (μ) extension ratios with pressure. Note: Curve 4 is associated with curve 2 and curve 7 (μ^*) is the radial extension ratio for curves 8, 9, and 10. Cell fibril angle $\alpha \cong 70°$.

this case and starts at the lower value of around 2 MPa. In this event the cell pressure characteristic 2 intersects this curve E, at which point there is no further potential difference available to drive cell growth. Note that in this event the cell has only increased in volume to point F on curve 3, which is still short of its maximum value at A. There is no available potential difference at this point, and extension is arrested and the axial elongation of the cell is now only about 70% (at $\lambda = 1.7$). These two simple examples illustrate that each individual step in the cell extension sequence is arrested either by a volume increase limitation (point A) or a potential difference limit (point E). In either event the cell has extended an appreciable amount.

2. Growth Under Mechanical Confinement

The case discussed in Section V.C.1 ignores the presence of any external stresses on the cell surfaces and may therefore be regarded as an outline of the performance of the model for unimpeded root growth. Consider the case when axial extension of the root is arrested. This will occur when the extending cells can no longer overcome the force required to advance the root cap into the soil ahead of it. Then, according to Section V.B(c)(ii), the cell pressure extension characteristic will revert to the $\lambda = 1$ line identified as vertical line 6 in Figure 5a and the corresponding diametral characteristic will switch to curve 7 in Figure 5b. A rigorous analysis of these characteristics is still incomplete and only approximate trends are discussed here.

Following the action outlined in the previous case, cell pressure will rise on line 6 to point G and the diametral increase in the cell will take place along curve 7 to point H, where the volume limit will arrest growth. Both curves 1

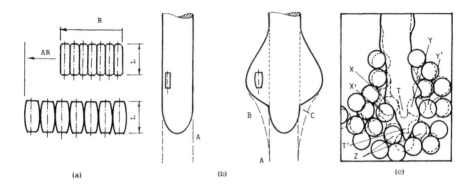

Figure 6. Variable geometry root apex growth model. (a) Cells with attenuated axial extension. (b) Root apex geometry changes and soil deformation. (c) Video time-lapse observations of root apex extension in a glass bead medium. (From Gordon et al., 1991.)

and 5 still lie above point G on line 6 and, hence, there is no potential limit to growth for this particular case. The macroscopic consequence of this revised growth activity is shown in Figure 6a where the cumulative increase in the diameter of each cell in a row results in a change in growth polarity from axial to radial ($\Delta L = 0$; $\Delta R > 0$). This results in the tip of the root swelling outward as shown in Figure 6b.

As pointed out in Section IV.B, this radial swelling will mobilize a cylindrical cavity expansion pressure on the epidermis of the root. This behavior is consistent with the analysis already presented, which demonstrated that the axial stress $p_a = p_s$ acting on the circular ends of the cells is always greater than the stress p_c developed on the root shank. Thus, the turgor pressure in the cell has reached p_s, but because $p_c < p_s$ the radial expansion can still proceed.

The consequence of this radial enlargement is that the soil in the root cap zone is loosened. A qualitative assessment of this effect can be made by reference to Figure 6b. The radial enlargement just behind the root cap will distort the imaginary cylindrical surface A in the soil to a conical shape, e.g., B. It is clear that the region C between these surfaces will be subjected to a tensile stress and, as discussed in Section IV.A, this will increase the pore space in the root cap zone [see state paths 7-8(T) and 9-10(H) in Figure 1b]. The soil in the root cap zone is thus weakened and loosened allowing the root to extend axially into a region that it could not penetrate without this expedient.

Other more rigorous theoretical and experimental evidence is available to show that interaction between the two stress fields, in the soil helps to relieve the root cap stresses (Abdalla et al, 1969; Hettiaratchi and Ferguson, 1973; Richards and Greacen, 1986; Misra et al, 1986).

As the stress relief effects of the radial expansion are short range, the root cap will soon enter a zone of soil which will again arrest axial extension. If soil strength conditions permit the cycle will repeat, and axial growth of the root will take place accompanied by an increase in root shank diameter. The swelling of root tips under mechanical impedance is well documented (see, for example, Abdalla et al, 1969; Goss, 1977). The growth model just described predicts this growth behavior.

There is a substantial amount of experimental evidence to support this variable geometry root growth model which predicts the characteristic thickening of root axes under mechanical confinement. All of these observations provide only indirect confirmation from examinations of roots grown under moderate mechanical impedance under both laboratory and field conditions. Preliminary work on a direct non-invasive observation of this growth mode has just been completed. A Newcastle triaxial rhizometer (Abdalla et al, 1969; Hettiaratchi, 1990) has been modified to enable the growth of a root to be monitored in physiological darkness by a video time-lapse camera (Gordon et al., 1991). Carefully controlled confining pressures can be applied to the growing root while factors such as temperature, aeration, and nutrient supply are maintained at optimum levels.

A tracing made from a typical photo-micrograph recorded during a preliminary run of the time-lapse video camera equipment is shown in Figure 6c. This is a composite diagram of two sequential growth stages of a *Vicia faba* root in a 0.5 mm diameter glass bead medium subjected to a controlled confining pressure of 5 kPa (Gordon et al., 1991). Chronologically, the outline shown in broken lines was recorded about an hour earlier than the one shown in solid outline. The sequence shows how the root tip T is impeded by the beads X, Y and their interacting neighbors. Radial expansion of the root apex displaces this group of beads to X", Y" and the tip resumes axial extension and is shown to be on the point of being arrested by the bead group at the bottom of the diagram. Subsequently, root growth repolarizes to the radial mode and displaces beads Z to advance its tip T" further into the medium.

3. Growth Arrested

The situation discussed in item c(i) in section V.B comes into operation when the stress p_c, generated during the radial expansion, approaches the cell pressure p. This condition is tantamount to the application of equal contact pressure to all the cell surfaces so that $k_1 = k_2 = k$ (where k_1 and k_2 are the ratios of internal pressure in the cell to the external contact stresses on the end and sides of the cell, respectively). As shown in Figure 4d, the progressive rise of k from zero to unity steadily increases the slope of the pressure volume characteristic until at $k = 1$ the cell extension behavior is identical to that of a cell with a neutral fibril angle of 54° (see Figure 4b).

This realignment of the pressure-extension characteristic shown in Figure 4d is reproduced in Figure 5a by the curves 2, 8, 9, and 10. The effect of the external confinement on cell extension behavior can be traced on this diagram by following the procedure set out in the previous case (letters with subscripts for the two $k > 0$ cases correspond to the original unsubscripted letters on curve 2.) It will be seen that as k approaches unity the cell extension is gradually reduced until at $k = 1$ extension will cease. This represents the point at which root growth is completely arrested. This situation will arise when the confinement conditions at the root apex are such that even though $p_c < p_a$ the condition $p_c > p$ will arrest growth.

D. Commentary on Root Growth Model

The variable geometry root apex growth model can be used to account for the wide discrepancy observed by many workers between penetrometer pressure and root growth pressure (see Section IV.E). As indicated earlier, the rigid penetrometer will sense p_s ($= p_a$), which is essentially the stress acting on the root cap. The growth model shows that this stress can be reduced by the radial expansion of the root apex behind the meristem. A rigid penetrometer is unable to simulate this so that its estimates of p_a must always be higher than the stress a living root can overcome.

Instrumented penetrometers with tips that can expand radially have been designed and tested (Abdalla et al., 1969; Hettiaratchi and Ferguson, 1973; Nguyen, 1977). In these experiments the axial extension tip of the penetrometer is loaded to the point where sinkage just ceases (axial extension arrested). While this load is held the radial expansion region is pressurized (radial expansion stress field set up) and the axial extension tip is observed to sink into the soil (confinement relieved by interacting stress fields). By cycling this sequence of loading it has been possible to advance the penetrometer into the soil carrying an axial load which on its own, could not induce the penetrometer to penetrate the soil.

On an overall assessment, the variable geometry root apex growth model performs adequately. However, it must not be overlooked that in its present form it is, at best, a very crude representation of the supremely intricate design of a living root apex. The model embodies the basic elements of a subtle strategy that plant roots have evolved to penetrate a hostile and unyielding environment. Plants have had a period of nearly 300 million years in which to develop this technique!

The interaction of the cells in the root with the soil is a key element in the model. However, only the kinematic aspects of their behavior as pressurized flexible structures have featured in the discussion and very little has been said about the forces developed by them. Neither has the development of the behavior of such units when unequal contact stresses are present on their surfaces been fully explored. These are obvious areas for future refinement of the theory.

VI. Conclusion

A. Summary

The simple premise that many soil-root interactions are ordained by pore space considerations is a useful starting point in the study of root growth. The critical state model can be used to ascertain the status of the pore space in a soil after it has been subjected to engineering operations. Equally, this model can be used to interpret how plant roots can adjust pore space when it is inadequate for root growth. This effort generates contact stresses against the surfaces of the root located within the pore space and these stresses can be estimated by a soil cavity expansion model. The analysis then moves across from the soil medium into the living tissue of the root.

The contact stresses on the root surfaces are derived passively as a result of active changes in the geometry of the growing root apex. These changes are the result of the growth of the individual cells in the root tissue. Simple statistical considerations require these contact stresses to be transmitted to individual root cells. These cells have been modeled as reinforced, thin-walled pressurized structures which grow in size by inflation. In the live cell, represented by this model, growth takes place by osmotic movement of water into the vacuole across the semipermeable cell membranes. The changes in geometry of the root apex and its associated growth rate are regulated by the cumulative physical behavior of these cells undergoing inflation while being subjected to the boundary contact loads mentioned earlier.

This analysis shows that the roots have evolved a sophisticated and efficient technique for penetrating strong soils. This strategy is linked to a passive change of the growth polarization of the root apex from the normal axial mode to a radial one. The stress field set up by this action interacts with the soil in contact with the root apex and increases the soil pore space in this zone. The net effect of this interaction is that the root cap can extend into the soils which it could not otherwise penetrate by means of a simple axial extension only.

B. Discussion

The review has examined an ordered chain of interactions, starting from soil-related phenomena right through to individual root cell behavior. The analysis of the performance of each link in this sequence has been made possible only because of stringent simplifications. In view of this, it is perhaps appropriate to reflect on some of the implications of this simplistic approach.

1. Soil Mechanics

The analysis of soil-related interactions was based on treating soil as an isotropic particulate continuum. A problem of scale arises when considering the interaction between the soil and the root cells. This is illustrated in Figure 6c where the proportions of the glass beads are massive in comparison with the individual root cells, and under these conditions the idea of uniformly distributed contact stresses (p_s and p_c) acting on the root surface becomes somewhat unrealistic. To some extent this effect is minimized when dealing with fine-textured soils. For a resolution of this difficulty we may follow a parallel problem encountered in the analysis of engineering structures. Implicit in many engineering stress analysis solutions is the principle of St. Venant, which provides the rationale for dealing with uniformly distributed stresses within the structure when the points of application of external loads are clearly at discrete points. There is no reason why this principle cannot also be invoked in the present small-scale structure, the root. In any event, the experiment outlined in Section V.C.2 shows that the basic action of the variable geometry root apex growth model operates effectively even in the case where pore space is bounded by coarse particles.

A brief mention must be made regarding the main practical difficulties in the application of the critical state model to determine the cut-off void ratio e_0 [Section III.D(d) and (e)]. Accurate estimates of the stress components p and q exerted on the soil during engineering operations are difficult to predict accurately. The state surfaces can be delineated satisfactorily, but not without a large expenditure of laboratory time and effort.

2. Physiological Considerations

The cell growth model discussed here has deliberately skirted the fact that each root cell is a living entity. However, this approach does not in any way detract from the substance of the model and it is perhaps opportune to briefly assess the positive contribution made by physiological factors to the performance of the soil-root interaction model.

Plant root growth is sensitive to temperature and the availability of both water and air (oxygen) to the root. The model described here can account for these esoteric effects on root growth in terms of the cell growth model. As discussed in the previous subsection, cell extension is consequent on the osmotic inflow of water abstracted from the soil matrix. Thus, water stress would most certainly limit the rate of elongation when physical interaction is not limiting and may possibly attenuate the potential force that extending cells are capable of generating. Temperature, on the other hand, influences the permeability of the plasma membranes in the cell and this controls the rate of influx of water and hence the time interval taken by the cell to reach the sequential extension steps required for growth (see Hettiaratchi and O'Callaghan, 1974, 1978). The lack of air or oxygen required for respiration interferes with the ability of the living cell to osmoregulate and hinders the osmotic mass transfer of water required to drive cell extension (compare curves 1 and 5 in Figure 5a).

It is much more difficult to infer whether there is any influence of the mechanical interaction on these physiological factors themselves. The influence of external contact stress on the pressure volume characteristic illustrated in Figure 5, for example, demonstrates one such interaction that moderates the available potential difference for water influx. Greacen and Oh (1972) have reported that plant root cells exhibit a limited ability to osmoregulate against external confining stresses, but the exact physiological mechanism involved in this behavior has not been identified.

In the cell extension model the arrangement of the wall microfibrils was represented simply as a single layer of symmetrically disposed fibrils. At the end of each extension step the fibril angle changes from some high value close to 90° to the neutral value of 54°. The living cell protoplasm deposits further microfibrils on the inside of the stretched wall at a preordained angle approaching 90°. This step is controlled by the genetic code of the cell and the mechanism involved in the orientation of the microtubules, which are the precursors of the fibrils, is still a biological mystery. However, as far as the model goes it would appear that cell walls have many layers of fibrils at different angular orientation (see, for example, the multi-net theory of Roelofsen, 1965) and this would tend to negate the simple arrangement chosen for the mathematic model. The quantification of such complex reinforcement in an inflatable structure has still not been attempted. However, a simple qualitative explanation of the action of the multi-layer reinforcement can be advanced.

According to the model each layer of microfibrils ceases extending when it reaches $\alpha = 54°$. Any newly deposited inner layer can still expand in the manner already discussed and it does this by slippage relative to the outer layer, which is required by theory to remain unchanged. In fact, an outer confining cylinder would be a strong encouragement to axial extension. Subsequent layers will repeat this effect and the consequence is rapid axial extension. However, this idea requires that the mature cell would end up with its wall structure having the bulk of its fibrils aligned at the angle of 54°. Some evidence for this has been presented by Hogetsu (1986).

It is probably important to record here that the growth models discussed highlight the crucial part played by purely physical factors. For instance, the model does not rely on a complex, as yet not clearly identified, feed-back mechanism for controlling growth polarity changes. The physical behavior of each cell is preordained entirely by the design of the wall structure and this is solely responsible for the changes in root geometry. This is not to deny that there are complex electrochemical and biochemical factors involved in the many interactions and these are indirectly woven into the model through such factors as osmoregulation, phloem transport of photosynthate, mitotic activity, cell wall synthesis, and the mechanism of water transport to and into the cells. The well understood macroscopic effects of these complex biological activities are all that is required for the model, not the less clearly understood manner in which they are achieved at a molecular level.

3. Future Developments

The models describing the various physical interactions between plant roots and soil discussed in this review are by no means fully developed. Their accuracy can be improved and their scope extended, and investigations on the following broad fronts would go a long way toward advancing the frontiers of understanding of the complex and fascinating subject of root growth.

(a) There is an urgent need to simplify the experimental procedures required for establishing critical state space of partly saturated soils. Without this quantum leap this model will simply remain a powerful research technique with little hope of it ever becoming a valuable practical tool.

(b) The design and development of special penetrometers to overcome the limitations of rigid sensing elements could provide the data required for assessing the suitability of seed beds for supporting vigorous root growth. These instruments could provide the information required for assessing minimum tillage requirements.

(c) The rudimentary kinematic cell growth models currently available need further refinement to equip them with a force prediction capability.

(d) There is a dearth of information relating to the strain energy modulus of cell walls which are assumed to behave as polymeric elastomers. Considerable experimental ingenuity would be required to evaluate these for actual root cells.

(e) Most plant breeding work appears to concentrate on above-ground attributes of crops. Perhaps the emphasis may have to be given to selecting for varieties with the most efficient soil-root interactions. Associated with this must be the development of precise techniques for identifying and measuring root functions that equip a root system to abstract the maximum benefit from its soil environment.

(f) The root growth models discussed concentrate on the growth mechanics of individual root apices. The problem of how these models could be incorporated into an overall root growth model has not been addressed. This extension requires an understanding of the mechanical and biological factors controlling root branch initiation and the hormonal signaling system involved in photosynthate partitioning between shoot and root.

References

Abdalla, A.M., D.R.P. Hettiaratchi, and A.R. Reece. 1969. The mechanics of root growth in granular media. *J. Agric. Engng. Res.* 14:236-248.

Atkinson, J.H. and P.L. Bransby. 1978. *The mechanics of soils-an introduction to critical state soil mechanics.* McGraw-Hill, London.

Bengough, A.G. and C.E. Mullins. 1988. Use of low-friction penetrometer to estimate soil resistance to root growth. *International Soil Tillage Research Organization 11th International Conference, Edinburgh,* 1:1-6.

Bishop, R.F., R. Hill, and N.F. Mott. 1945. The theory of indentation and hardness tests. *Proc. Phys. Soc.* 57:147-159.

Cockroft, B., K.P. Barley, and E.L. Greacen. 1968. The penetration of clays by fine probes and root tips. *Aust. J. Soil Res.* 7:333-348.

Cosgrove, D. 1988. In defence of the cell volumetric elastic modulus. *Plant Cell Env.* 11:67-69.

Eavis, B.W. 1967. Mechanical impedance and root growth. *Proc. Agric. Eng. Symp. Inst. Agric. Engng.* No. 4/F/39.

Eavis, B.W., L.F. Ratliff, and H.M. Taylor. 1969. Use of dead load technique to determine axial root growth pressure. *Agron. J.* 61:640-643.

Easu, K. 1965. *Plant Anatomy.* John Wiley & Sons, New York.

Fahn, A. 1990. *Plant Anatomy.* Pergamon Press, Oxford.

Farrell, D.A. and E.L. Greacen. 1966. Resistance to penetration of fine probes in compressible soil. *Aust. J. Soil Res.* 4:1-17.

Gibson, R.E. 1950. Discussion in *J. Inst. Civil Eng.* London 34:382.

Gibson, R.E. and W.F. Anderson. 1961. In-situ measurements of soil properties with the pressuremeter. *Civil Engng. Pub. Wks. Rev.* 56:615-618.

Gordon, D.C., D.R.P. Hettiaratchi, A.G. Bengough, and I.M. Young. 1991. Non destructive analysis of root growth in porous media. *Plant Cell Env.* 15:123-128.

Goss, M.J. 1977. Effects of mechanical impedance on root growth in barley (*Hordeum vulgare* L). Effects on elongation and branching of seminal roots. *J. Exp. Bot.* 28:96-111.

Greacen, E.L. 1986. Root response to soil mechanical properties. *Trans. 13th Int. Congress Soil Sci.* 5:20-47.

Greacen, E.L., R.O. Farrell, and B. Cockroft. 1968. Soil resistance to metal probes and plant roots. *Trans. 9th Int. Congress Soil Sci.* 1:769-779.

Greacen, E.L. and J.S. Oh. 1972. Physics of root growth. *Nature (New Biol.)* 235:24-25.

Groenevelt, P.H., B.D. Kay, and C.D. Grant. 1984. Physical assessment of soil with respect to rooting potential. *Goderma* 34:101-114.

Hardham, A.R. and B.E.S. Gunning. 1977. The length and deposition of cortical microtubules in plant cells fixed in glutaraldehyde-osmium tetroxide. *Planta* 134:201-203.

Hatibu, N. 1987. The mechanical behavior of brittle agricultural soils. Ph.D. thesis, University of Newcastle upon Tyne.

Hettiaratchi, D.R.P. 1987. A critical state soil mechanics model for agricultural soils. *Soil Use Mang.* 3:94-105.

Hettiaratchi, D.R.P. 1989. Critical state soil-machine mechanics. *Proc. Nordiske Jordbruksforskers Forening Seminar* no. 165, 53-68. Oslo, Norway.

Hettiaratchi, D.R.P. 1990. Soil compaction and root growth. *Phil. Trans. R. Soc. London* B329:343-355.

Hettiaratchi, D.R.P. and C.A Ferguson. 1973. Stress deformation behavior of soil in root growth mechanics. *J. Agric. Engng. Res.* 18:309-320.

Hettiaratchi, D.R.P. and J.R. O'Callaghan. 1974. A membrane model of plant cell extension. *J. Theor. Biol.* 45:459-465.

Hettiaratchi, D.R.P. and J.R. O'Callaghan. 1978. Structural mechanics of plant cells. *J. Theor Biol.* 74:235-257.

Hettiaratchi, D.R.P., and J.R. O'Callaghan. 1980. Mechanical behaviour of agricultural soils. *J. Agric. Engng. Res.* 25:239-259.

Hettiaratchi, D.R.P., and J.R. O'Callaghan. 1985. The mechanical behavior of unsaturated soils. *Proc. Int. Conf. Soil Dynamics*, Auburn, AL 2:266-281.

Hogetsu, T. 1986. Orientation of wall microfibril deposition in root cells of *Pisum sativum*. *Plant Cell Physiol.* 27:947-951.

Kamiya, N., M. Tazawa, and T. Takata. 1963. The relation of turgor pressure to cell volume in *Nitella* with special reference to mechanical properties of the cell wall. *Protoplasma* 57:502-521.

Ladanyi, B. 1963. Expansion of a cavity in a saturated clay medium. *Proc. American Soc. Civil Engng.* SM4:127-161.

Ladanyi, B. 1967. Expansion of cavities in brittle media. *Int. J. Rock Mech. Min. Soc.* 4:301-328.

Ledbetter, M.C. and K.R. Porter. 1963. A microtubule in plant cell fine structure. *J. Cell Biol.* 19:239-250.

Liang, Y. 1985. Mohr-Coulomb parameters and soil indentation tests. Ph.D. thesis, University of Newcastle upon Tyne.

Misra, R.K., A.R. Dexter, and A.M. Alston. 1986. Maximum axial and radial growth pressures of plant roots. *Plant Soil* 95:315-326.

Nguyen, P.T. 1977. Mechanics of soil deformation in relation to root growth. Ph.D. thesis, University of Newcastle upon Tyne.

Pfeffer, W. 1893. Druck and Arbeitslestung durch wachsende pflanzen. *Abhandl. Saechs. Akad. Wiss.* 33:235-474.

Preston, R.D. 1974. *The Physical Biology of Plant Cell Walls*. Chapman & Hall, London.

Probine, M.C. and N.F. Barber. 1966. The structure and plastic properties of the cell wall of *Nitella* in relation to extension growth. *Aust. J. Biol. Sci.* 19:439-457.

Roelofsen, P.A. 1965. Ultrastructure of the wall in growing cells and its relation to the direction of growth. *Adv. Bot. Res.* 2:69-149.

Roscoe, K.H., A.N. Schofield, and C.P. Wroth. 1958. On the yielding of soils. *Gèotechnique* 8:22-53.

Richards, B.G. and E.L. Greacen. 1986. Mechanical stresses in an elongating cylindrical root analogue in granular media. *Aust. J. Soil Res.* 24:393-404.

Scott Russell, R. and M.J. Goss. 1974. Physical aspects of soil fertility. The response of roots to mechanical impedance. *Neth. J. Agric. Sci.* 22:305-318.

Taylor, H.M. and L.F. Ratliff. 1969. Root growth pressures of cotton, peas and peanuts. *Agron. J.* 61:398-402.

Treloar, L.R.G. 1975. *The Physics of Rubber Elasticity.* Clarendon Press, Oxford.

Vesic, A.S. 1972. Expansion of cavities in infinite soil mass. *J. Soil Mech. Fdn. Div.*, ASCE, 99(SM3):265-289.

Whiteley, G.M., W.H. Utomo, and A.R. Dexter. 1981. A comparison of penetrometer pressure and the pressures exerted by roots. *Plant Soil* 61:351-365.

Wu, H., R.D. Spence, and P.J.H. Sharpe. 1988. Plant cell wall elasticity II: polymer elastic properties of microfibrils. *J. Theor. Biol.* 133:239-253.

Zimmermann, U. and E. Steudle. 1975. The hydraulic conductivity and volumetric elastic modulus of cells and isolated cell walls of *Nitella* and *Chara* spp: Pressure and volume effects. *Aust. J. Plant Physiol.* 2:1-12.

Index

acidity 55, 106
adsorption 45, 48, 67, 89, 94,
    97, 100-105, 104, 105, 104,
    106,108, 109, 110, 111,
    113-118,117, 118, 134-142,
    144, 145, 153, 178, 179,
    180-182, 184, 234, 248, 250
aeration 4, 132, 257, 278, 280
aerobic conditions 206
aggregate 39, 67, 73, 76, 81, 82,
    88, 120, 125
aggregation 1, 250
algae 223, 228
aluminum 43, 53, 55, 85, 86, 88,
    89, 93, 105, 109, 138, 140,
    239
amendments 37, 68, 79, 93
ammonia 127, 143, 198
ammonium 135, 204
anaerobes 204, 206
anaerobic conditions 163
anion 45, 110, 127, 136
availability 67, 95, 153, 154,
    163, 182, 184, 188, 195, 196,
    199, 202, 205, 206, 207, 225,
    250, 257, 258, 284
axial root growth pressure 271,
    273, 286

bacteria 152, 155, 156, 165, 166,
    170, 177, 182, 183, 186,
    188-190, 191-193, 195, 197,
    201, 202, 204, 205, 208-210,
    212, 214, 215, 216-225, 228,
    232, 233, 235, 236, 238, 239,
    243, 244, 248, 252, 253
bacterial activity 202, 232
bentonite 45, 233, 247, 249, 250
biodegradation 151-156, 160,
    161, 163, 164, 163-166, 171,

170, 172, 174, 175-184, 196,
    201, 202, 205-207, 214-217,
    222-224, 248
biomass 140, 151-154, 158, 162,
    165-167, 172, 175, 194-200,
    205, 206, 214, 219, 221, 222
Boltzmann equation 99
bulk density 3, 4, 6, 7, 15, 32,
    35, 53, 56, 72, 83, 113, 158,
    256, 272
bypass flow 10, 12, 23, 24, 32,
    33, 36

cadmium 104, 105, 109, 110,
    125, 135, 136
calcium 42, 44, 85-89, 91, 93,
    104, 105, 138, 142, 144
Cambridge critical state model
    256, 258
Cambridge model 260
carbon 23, 125, 155, 156, 185,
    188, 189, 192-194, 196,
    200-203, 202, 203, 204-207,
    221, 228, 230, 233
cation 42, 43, 48, 54, 66, 68,
    76, 78, 84, 86, 88-91, 94, 97,
    99, 115, 117, 127, 134-136,
    138, 142, 144, 232, 233, 235,
    238, 239, 245, 250, 252
cation exchange 66, 68, 88, 90,
    91, 97, 99, 135, 138, 142,
    144, 232, 233, 235, 238, 245,
    252
cation exchange capacity 66, 88,
    232, 233, 245, 252
cellular architecture 255, 268,
    274
chloride 104, 105, 104, 110,
    112, 117, 137, 141, 145, 163,
    217

chlorinated phenols 206, 207, 219

clay 5, 6, 10, 11, 17, 19, 20, 23-25, 34, 35, 38-44, 43-51, 53-59, 58, 59, 61, 62, 61-76, 80, 82-94, 96, 97, 100, 124, 137, 155, 184, 188, 191, 211, 227-230, 232-234, 239-254, 261, 265, 266, 271, 288

clay soil 5, 35, 49, 62, 90, 91, 266, 271

COD 180, 188, 192, 193, 202, 204-206, 212, 218

cohesion 266

colloidal properties 37, 38

column method 10, 11

community 185, 186, 189-191, 193, 199, 200, 213, 225

community structure 185, 186, 189-191, 199, 225

compaction 13, 19, 20, 25, 31, 33-36, 71, 72, 84, 257, 261, 263, 287

complexation 100, 101, 104, 106, 108-110, 136, 140, 233, 235, 251

crop 1, 6, 28-33, 36, 38, 262

crop growth 1, 28, 29, 33

crust 10, 11, 32, 34, 35, 69, 70, 79, 84, 85, 87, 88, 91-93

cultivation 12, 221, 222

decomposition 89, 179, 184, 219, 233, 248

degradation 2, 33, 75, 116, 118, 139, 143, 144, 153-157, 159-163, 165, 167, 168, 170, 176-179, 181-183, 185, 186, 191, 194, 200-202, 203-205, 207, 213, 217, 224, 235, 247

denitrification 116, 127, 128, 132, 140, 155, 181, 197, 198, 204, 221, 224, 226

depositional seals 37, 82

desorption 119, 124-126, 142,

155, 170, 178, 179, 181, 182, 184

diffusion 24, 39, 86, 106, 107, 114, 116, 119, 129, 134, 141, 142, 148, 149, 155, 159, 165-167, 172, 174, 177, 180, 182, 199, 233

dispersion 37, 38, 41-43, 42, 44, 43-45, 47-50, 55, 57, 58, 61-63, 67, 68, 69, 71-76, 79, 80, 83, 84, 86, 87, 92-94, 107, 109, 113, 114, 116, 118-120, 129, 131, 134-137, 140-143, 145, 157, 163, 167, 174, 175, 179-181, 207-209, 221

dissolution 49, 60, 61, 63, 68, 69, 73, 80, 81, 84, 89, 104-107, 109, 111, 115, 134, 135, 138, 140, 142, 145, 154, 183, 227, 230, 234, 235-238, 246, 247, 253

DOC 137, 202, 203

domain 122, 129, 132, 149-151, 155, 160, 161, 164, 167, 171, 170, 171, 173, 259, 262

double layer theory 39, 54

drainage 115, 123

drying 5, 17, 21, 79, 88

earthworms 257

electrical conductivity 56, 71, 92

electrolyte 37, 39, 42, 46, 47, 49, 50, 53-60, 64, 68, 69, 73, 77-79, 81, 82, 83-87, 89-94, 137, 145, 146

electron microscopy 42, 48, 72, 86, 91, 151, 163, 166, 172, 177, 195, 196, 197, 202-207, 221, 230, 234, 235, 245, 250, 274

electrophoretic mobility 40-42, 85

Elementary units of structure 7, 8

ELUS 7, 8

enzyme 154, 197, 207

erosion 82, 88, 90, 91, 93
ESP 40-42, 48-50, 49-51, 50,
    52, 54-62, 64, 67, 72-74, 73,
    74, 73, 74, 76, 78, 79, 81, 80,
    83, 84, 87
evaporation 85
evapotranspiration 124
exchangeable sodium percentage
    40, 56, 87, 137
expansion pressure 266, 269,
    273, 279
extraction 17

fallow 135
feldspars 54, 60, 106, 134, 249
fertility 127, 288
fertilizer 24, 75, 79, 91, 132,
    138
fibril 275-277, 279, 281, 284
field soils 107, 120, 136
fixation 43, 66, 93, 96, 245,
    248-250
flocculation 37, 42-46, 49, 56,
    58, 61, 68, 82-85, 88, 89, 91,
    92
forest soil 25, 28, 138, 145, 248
fractionation 203, 216
Freundlich 103, 104, 114-118,
    148, 162
fungi 192, 223, 228, 232, 233,
    235, 237, 239, 253, 254

geostatic stress 266, 269
gibbsite 67, 96, 138
ground water 17, 95, 127,
    135-137, 139, 141-144, 184,
    214-217, 219, 221, 222-226
groundwater 23, 88, 118, 120,
    132, 135, 136, 139, 147, 152,
    165, 177, 178, 179, 181-194,
    199-202, 204, 205, 207-211,
    213, 215-217, 218-224
growth 1, 19, 21, 28-31, 33, 34,
    36, 88, 122, 151-154, 156,

162, 165, 166, 172, 175, 181,
    182, 184-186, 189, 192-201,
    205, 206, 209, 213, 215, 216,
    218, 221-225, 231, 233, 234,
    239, 241, 242, 244, 253,
    255-259, 262-265, 268, 271,
    270, 271, 273, 276, 277, 279,
    278, 279, 278-288
gypsum 4, 5, 8, 67, 68, 80,
    88-92, 137, 138, 144

halogenated solvents 207
HC 37, 38, 45-58, 57-61, 63,
    62-64, 63-67, 69, 70, 73-78,
    82, 84
heat 109
heavy metal 104, 117
herbicide 19, 160
heterogeneity 90, 93, 95, 96,
    101, 102, 118-121, 123-125,
    127-129, 131, 133, 136, 142,
    144, 145, 158, 159, 167,
    171-173, 175, 190, 223
horizon 6, 7, 9, 13, 14, 19, 20,
    24, 25, 32, 34, 93
hydraulic conductivity 2, 5, 6, 8,
    17, 20, 22, 25, 26, 28, 32-38,
    46, 47, 50, 51, 52, 57, 68, 71,
    85-94, 115, 121, 125-128, 140,
    142, 158, 175, 288
hydroxides 43, 86, 96, 97
hydroxy-aluminum 239
hysteresis 142

illite 38, 42, 44, 63, 66, 76, 91,
    229, 238, 239
immobilization 208, 212, 249
infiltration 4, 5, 8-11, 23, 24,
    34, 35, 37, 38, 69-72, 74, 75,
    77, 80, 82, 83, 84-86, 88-91,
    93, 124, 128, 143
infiltrometer 7, 10, 11, 34
IR 37, 38, 70, 71, 73-79, 81,
    80-82, 84, 128

iron 25, 53, 55, 66, 67, 85, 86,
  88, 89, 93, 140, 205, 216,
  241, 248, 249, 250-254
irrigated 38, 56, 87, 91
irrigation 5, 35, 38, 56, 62, 63,
  70, 82, 84-87, 91, 115, 124,
  142, 221
kaolinite 38, 42, 44, 45, 54, 61,
  63, 64, 66, 67, 86, 88, 92,
  184, 229, 231, 232,
  231-234, 238-240, 248, 250,
  253
kaolinitic soil 45, 67
Ksat 6, 8-12, 20, 25
Kunsat 5, 31

land 2, 13, 16-18, 23-25, 30, 34,
  36, 60, 136, 138, 187, 208,
  220, 221, 254
land qualities 30, 36
leaching 24, 35-37, 48-50, 52,
  53, 55-58, 60, 61, 64, 84, 115,
  124, 130, 132, 134, 137, 138,
  140, 141, 143-145
lime 67, 69, 68, 69, 91, 92

macropore 4, 5, 12, 155
macropores 3-5, 9-12, 17, 23,
  33-36, 145, 158, 257, 262
macroporosity 34
magnesium 37, 42, 44, 62, 76,
  85-88, 92, 93, 142
manure 138
measurement 2, 6, 7, 10-12, 23,
  28, 32-34, 36, 40, 48, 59, 204,
  219, 224, 269
mechanical confinement 257,
  262, 273, 278, 280
Michaelis-Menten 116, 154-156,
  163-166, 180
microbial activity 48, 154, 172,
  194, 198, 219-222, 225, 227,
  230, 231, 235, 237, 238, 243,
  250

microfibril 287
micromorphology 35, 82, 83, 93
micronutrients 230
microorganisms 111, 154, 156,
  179, 185-187, 194, 198, 199,
  208, 209, 211, 213, 215, 216,
  218, 219, 227, 228, 230, 231,
  233-240, 246, 247-253
mineralization 155, 182, 183,
  192, 200-204, 206, 223
mineralogy 50, 51, 53, 54, 57,
  65, 73, 85, 88, 94, 248, 249,
  253
minimum tillage 285
model 29, 30, 33, 36, 41, 42,
  48, 54, 85, 97, 98, 100, 101,
  107-109, 110, 111, 117, 119,
  122-125, 130-132, 135,
  137-144, 148-150, 151-162,
  161-164, 163-167, 166-168,
  171-173, 176, 175, 177, 178,
  179, 180, 182-184, 195,
  198-200, 205-207, 209,
  215-217, 218-225, 251,
  255-265, 268, 273-277, 279,
  278-287
moisture retention 2, 4, 6, 16,
  17, 19-22, 25, 26, 28, 30, 32,
  33, 54, 87, 231, 239, 255,
  256, 258-261, 272
Monod equation 153, 156, 162,
  166, 175, 193, 195
montmorillonite 38-46, 49,
  54-56, 58, 66, 67, 76, 84-90,
  92, 93, 96, 137, 184, 229,
  231-233, 235, 234, 238-240,
  242, 248, 250, 251, 253
morphology 34, 35, 43, 72, 91,
  107
movement 24, 28, 34, 37, 38,
  49, 58, 62, 71, 73, 90, 92,
  119, 134, 138, 141, 144, 145,
  165, 177, 179, 182, 184-186,
  208, 213, 214, 218, 222, 223,

226, 257, 268, 270, 282
mulch 81

nitrate 36, 90, 95, 116, 119,
  127-129, 131, 130, 132, 133,
  145, 159, 166, 177, 184, 197,
  198, 202, 204, 205, 215, 216,
  219, 224, 225, 251
nitrification 135, 140, 217, 220,
  233, 250
nitrite 162, 177, 197, 198
nitrogen 35, 109, 116, 127, 137,
  138, 140, 141, 198, 204, 216
nontronite 242, 250, 253, 254
nutrient 95, 127, 151, 153, 166,
  181, 184, 189, 194, 195,
  198-200, 205, 209, 212,
  219-221, 225, 230, 240-243,
  242, 244, 246, 280

organic carbon 23, 125, 185,
  188, 193, 194, 196, 200-203,
  202-204, 205-207
organic contaminants 154, 180,
  181, 188, 189, 191, 193, 201,
  204, 213, 221
organic matter 7, 14, 15, 20, 35,
  53, 55, 86, 87, 100, 127, 148,
  178, 180, 194, 202, 213, 227
oxidation 111, 135, 162, 177,
  194, 196, 198, 202, 204, 216,
  224, 230, 233, 244, 246, 248,
  250, 253
Oxisol 141

parent material 67
penetrometer 255, 271-273, 281,
  286, 288
permeability 4, 47-50, 53, 54,
  56, 62, 65-68, 71-76, 82,
  84-86, 89-91, 94, 191, 200,
  224, 284
pesticide 109, 116-118, 124, 139,
  141, 144, 156, 162, 180, 183
pH 14, 23, 44, 46, 55, 67, 68,

86, 88, 89, 100, 102-107, 111,
  125, 134, 140, 144, 146, 157,
  180, 183, 188, 198, 223,
  231-233, 239, 287, 288
phosphate 45, 79, 104, 105, 134,
  140, 144, 180, 222
phosphorus 141
physical properties 1-3, 5-7, 9,
  15, 17, 25, 28, 29, 32-36, 38,
  45, 46, 66, 68, 71, 74, 75, 81,
  85, 87, 88, 91, 93, 96, 97,
  105, 107, 111, 115-119, 125,
  133, 152, 155, 173, 175, 176,
  193, 194, 200, 224, 228, 239,
  244, 245, 247, 256, 258, 264,
  265, 272, 273, 274, 275, 277,
  282, 284, 285, 287, 288
plant 24, 31, 33, 36, 88, 96,
  111, 137, 140, 144, 221, 250,
  252, 255, 256, 258, 259, 262,
  266, 267, 269, 272-277, 279,
  282, 284-286, 287, 288
ploughpan 13, 14, 16-20, 22, 30,
  31
Poisson's ratio 266
pollution 139, 177, 216, 217,
  219, 221, 224
pore space 7, 255-257, 260, 262,
  279, 282, 283
porosity 3, 4, 20, 25, 28, 29, 47,
  71, 72, 89, 117, 120, 125,
  200, 213, 240, 256
potassium 37, 38, 43, 65, 76, 86,
  89, 93, 236-238, 237, 248
precipitation 24, 38, 70, 97,
  104-106, 109-111, 115, 124,
  135, 138, 145
protozoa 191, 192, 201, 209,
  213, 219, 223, 228

raindrop 71, 72, 75, 79, 84, 87,
  90
rainfall 4, 9, 12, 24, 56, 68, 69,
  71, 72, 83, 85, 88, 90, 91,
  124, 143

reclamation 67, 68, 89, 91, 115, 143, 254
redox potential 224, 246
redox reactions 109, 240
reduction 28, 48-50, 52, 58, 71, 74, 84, 88, 110, 111, 156, 159, 200, 202, 204, 205, 215, 216, 221, 230, 240-243, 242-244, 243-245, 246, 245-247, 250, 251, 253, 254, 277
rigidity index 266, 269, 273
root 7, 25, 30, 96, 124, 185, 186, 255-260, 262-271, 270-274, 276, 277, 279, 278, 279, 278-288
root apex 255, 256, 258, 264-269, 271, 270, 272-274, 277, 279-281, 282, 283
root cap 255, 265-273, 278, 279, 281, 283
root cell 255, 273, 274, 277, 283
root cell growth 255, 273
root elongation 268, 270
root extension 260, 262, 263, 272, 273
root growth 255-259, 262, 263, 265, 271, 270, 271, 273, 277, 278, 280, 281, 282, 284-288
root growth model 255, 273, 280, 281, 286
root shank 265, 266, 268, 269, 271, 273, 279
rooting 23, 29-32, 34, 266, 287
runoff 71, 75, 84, 132

salinity 48, 50, 52, 56, 59, 62, 66, 68, 81, 84, 85, 89, 93, 183
sampling 7, 9, 36, 185-187, 202, 220, 224, 226
sandy soil 91, 214
saturated flow 10, 37, 46
saturation 5, 11, 17, 21, 25, 32, 34, 52, 66, 103, 121, 128,

193, 196, 197, 198, 258, 259
scale 72, 95, 101, 102, 116, 118-133, 135, 136, 142, 152, 175, 176, 181, 207-209, 211, 216, 241, 246, 263, 271, 283
scaling 124
seal 38, 68, 70-76, 79-82, 84
sealing 9, 37, 68, 70, 71, 75, 86, 93
sewage effluent 188
shear strength 82, 266
simulation 1, 2, 6, 21, 28-31, 33-36, 72, 77, 124, 134, 137, 139-141, 143, 160, 164, 163, 167, 170, 171, 178, 179, 181-183, 195, 200, 206, 215, 220, 221, 224, 274
smectite 38, 39, 45, 46, 55, 63, 66, 93, 229, 239, 240, 242-245, 251, 253
sodic soil 67, 68, 73, 89, 92, 114, 115
sodicity 48, 49, 53, 55, 58, 73, 74, 77, 84
sodium 38, 40, 44, 45, 48, 54, 56, 58-60, 78, 85-87, 89-91, 94, 115, 137, 144, 217, 239, 250, 254
sodium adsorption ratio 48, 94, 115
soil amendments 37, 79, 93
soil animals 3
soil conditioner 85
soil density 56
soil dispersion 67
soil erosion 82
soil fertility 127, 288
soil heterogeneity 95, 96, 118, 119, 127, 128, 144, 159
soil matrix 4, 10-12, 24, 47, 48, 54, 73, 96, 256, 284
soil mechanics 265, 267, 283, 286, 287
soil morphology 34, 35
soil physical properties 32, 93

soil porosity 4
soil properties 7, 37, 53, 54, 56, 86, 119, 125, 181, 287
soil strength 273, 279
soil structure 1-7, 9, 10, 12, 13, 15, 18, 19, 23, 24, 27-35, 62, 69, 70, 76, 81, 82, 86, 87, 125, 240, 250
soil taxonomy 36
soil temperature 257
soil tillage 1, 33, 36, 286
soil water 1, 21, 24, 28-31, 33-35, 51, 52, 85, 87
soil water regime 21, 31, 33, 34
soil-root contact 255, 264
sorptive solute 95, 125, 127, 136, 209
spatial variability 118, 119, 124-129, 132, 134
stability 3, 17, 34, 42, 44, 66, 67, 71, 76, 79, 81, 82, 85, 86, 89, 143, 256
statistics 103, 132
strain 233, 244, 264, 266, 274, 275, 285
stress 38, 199, 231, 256, 258-260, 263-267, 269, 275, 278, 279, 281, 282, 283, 284, 287
structure degradation 2, 33
suction crust infiltrometer 10, 11, 34
sulfate 186, 202, 204, 205, 216, 238, 250, 254
surface charge 45, 99-101, 104, 110, 253
surface soil 19, 25, 215

tactoid 39, 40, 83, 85
temperature 99, 106, 118, 135, 146, 193, 198, 240, 242, 243, 242, 257, 280, 284
texture 6, 7, 15, 24, 25, 32, 35, 54, 70, 73, 85, 253
tillage 1, 2, 7, 12, 19, 23, 24,

33, 36, 257, 285, 286
toxicity 140, 222
transformation 51, 66, 116, 147, 148, 151, 153, 154, 156-160, 162, 164, 163, 166-168, 170, 171, 179, 181, 184, 207, 238, 239, 249, 252
transport 7, 36, 95, 96, 107-111, 113, 114, 116-121, 123-125, 127, 128, 132, 133-148, 150, 152-154, 156-164, 163-167, 171, 170-172, 173-186, 188, 193, 199, 207-219, 221, 222, 225, 257, 285

unsaturated flow 5, 31, 33, 109, 135
unsaturated soil 10, 33, 34, 96, 135, 214
unsaturated zone 119-121, 123, 125, 128, 135, 136, 141, 143
urea 143

vadose 95, 142, 174, 208
vadose zone 95, 174, 208
Vanselow 97, 99, 145
vegetation 33, 95, 96, 124, 137
Vertisol 23, 25, 36, 235
volatilization 109, 143

water content 17, 42, 52, 70, 121, 128, 157, 160
water flow 34, 46, 84, 89, 119, 128, 134, 142, 143, 157, 158, 216
water quality 79, 85-88, 91, 93, 137, 222, 246
water retention 35, 41, 52
water transport 285
water uptake 43, 124
water use 35
weathering 60, 61, 91, 92, 106, 134, 135, 146, 227, 234, 237, 239, 247, 248-252, 254

X-ray diffraction 40, 246

yield 9, 10, 30, 95, 153, 191,
    194, 199, 204, 205, 213, 216
Young's modulus 266

Zinc 105, 134, 138, 140